Engineering Materials

This series provides topical information on innovative, structural and functional materials and composites with applications in optical, electrical, mechanical, civil, aeronautical, medical, bio- and nano-engineering. The individual volumes are complete, comprehensive monographs covering the structure, properties, manufacturing process and applications of these materials. This multidisciplinary series is devoted to professionals, students and all those interested in the latest developments in the Materials Science field, that look for a carefully selected collection of high quality review articles on their respective field of expertise.

Indexed at Compendex (2021) and Scopus (2022)

Deepak Punetha
Editor

Polyvinylidene Fluoride (PVDF)

Advanced Material for Multifunctional Applications

Springer

Editor
Deepak Punetha
Department of Electronics & Communication Engineering
Motilal Nehru National Institute of Technology Allahabad
Prayagraj, Uttar Pradesh, India

ISSN 1612-1317 ISSN 1868-1212 (electronic)
Engineering Materials
ISBN 978-981-95-6717-1 ISBN 978-981-95-6718-8 (eBook)
https://doi.org/10.1007/978-981-95-6718-8

This Springer imprint is published by the registered company Springer Nature Singapore Pte Ltd.
The registered company address is: 152 Beach Road, #21-01/04 Gateway East, Singapore 189721, Singapore

Foreword

Polyvinylidene fluoride (PVDF) has become one of the most significant advanced materials of the modern era, owing to its exceptional electrical, mechanical, and physicochemical characteristics. Its unique combination of piezoelectricity, ferroelectricity, flexibility, and durability has enabled transformative progress in a wide spectrum of applications ranging from energy harvesting and smart sensing to next-generation flexible and wearable electronic systems.

This book, *Polyvinylidene Fluoride (PVDF): Advanced Material for Multifunctional Applications*, offers a comprehensive and insightful overview of this versatile material. It brings together detailed discussions on PVDF's structural fundamentals, fabrication methodologies, and its integration into emerging technologies. The editors and contributors have presented a well-curated collection of recent advancements that not only highlight the current state of PVDF research but also illuminate future directions for innovation.

As interdisciplinary research continues to shape the future of science and engineering, materials like PVDF will play an increasingly central role in enabling sustainable, intelligent, and high-performance technological solutions. This book will undoubtedly serve as a valuable reference for researchers, academicians, industry professionals, and students who aspire to explore and exploit the full potential of PVDF in cutting-edge applications.

I congratulate the editors and contributing authors for their dedicated effort in producing this scholarly volume. I am confident that this book will significantly contribute to ongoing research efforts and inspire further breakthroughs in the field of advanced functional materials.

Motilal Nehru National Institute of Technology Allahabad, Prayagraj, India
IIT Madras
Chennai, India

Rama Shanker Verma

Foreword

Polyvinylidene fluoride (PVDF) has emerged as a game-changing material, exhibiting remarkable properties, such as piezoelectricity, ferroelectricity, chemical stability, and mechanical flexibility. Over the past few decades, PVDF has played a pivotal role in diverse fields, including energy harvesting, biomedical sensing, flexible electronics, and wireless communication.

This book, *Polyvinylidene Fluoride (PVDF): Advanced Material for Multifunctional Applications*, provides an in-depth exploration of the fundamental properties, fabrication techniques, and cutting-edge applications of PVDF. The authors have meticulously compiled contributions from leading researchers to present a comprehensive view of recent advancements and future prospects.

With rapid developments in nanotechnology and material science, PVDF-based materials are set to drive the next wave of technological evolution, from smart sensors to wearable electronics and beyond. This book serves as an essential resource for researchers, engineers, and students who seek to harness the full potential of PVDF for scientific and industrial applications.

I extend my heartfelt congratulations to the editors and contributors for compiling this outstanding volume, which will undoubtedly shape the future of PVDF research and its real-world implementations.

Rajiv Gandhi Proudyogiki Vishwavidyalaya
Bhopal, India
Motilal Nehru National Institute of Technology Allahabad
Prayagraj, India

Rajeev Tripathi

Preface

The last few decades have witnessed an unprecedented demand for multifunctional materials capable of powering the next generation of electronic, energy, and biomedical systems. Among these, polyvinylidene fluoride (PVDF) has emerged as a transformative polymer due to its unique combination of electroactive properties, mechanical robustness, chemical stability, and processing versatility. Its ability to simultaneously exhibit piezoelectric, pyroelectric, and ferroelectric behavior has positioned PVDF at the forefront of modern material science, enabling innovative applications across diverse fields, such as sensors, energy harvesting, photovoltaics, flexible electronics, and soft robotics.

The motivation for this book arises from the rapidly expanding role of PVDF and its derivatives in advanced technologies. With the integration of nanomaterials, hybrid composites, and scalable fabrication techniques, PVDF has transcended its traditional boundaries to become a cornerstone material for multifunctional applications. As research in this area accelerates, there is a clear need to consolidate the growing body of knowledge, critically evaluate emerging trends, and provide a road map for future innovation and industrial translation.

This book, *Polyvinylidene Fluoride (PVDF): Advanced Material for Multifunctional Applications*, brings together contributions from leading experts in academia and industry to present a comprehensive perspective on both fundamental principles and state-of-the-art applications. Spanning ten chapters, it offers a balanced blend of theoretical insights, experimental findings, and application-driven discussions.

Chapter "Fundamentals and Material Characteristics of PVDF"—Introduces PVDF's semicrystalline structure, electroactive properties, fabrication strategies, and nanocomposite improvements, emphasizing its significance in next-generation sensing and energy-harvesting technologies.

Chapter "PVDF-Based Piezoelectric and Pyroelectric Sensors"—Explores PVDF's electroactive behavior and nanocomposite enhancements, highlighting biomedical, acoustic, structural, and wearable sensor applications.

Chapter "PVDF as a Flexible Photovoltaic Material"—Highlights PVDF's role in organic and perovskite solar cells, enhancing charge transport, efficiency, stability, and flexible device fabrication.

Chapter "PVDF for Supercapacitors and Energy Storage Devices"—Examines piezoelectric energy harvesting and storage applications, emphasizing self-sustained devices for IoT, wearables, and industrial systems.

Chapter "PVDF-Based Antennas for Wireless Communication"—Explores dielectric and ferroelectric properties for flexible antennas, focusing on 5G, IoT, wearable electronics, and reconfigurable RF platforms.

Chapter "PVDF in Magnetic and Spintronic Applications"—Discusses PVDF's multifunctionality, magnetoelectric composites, and copolymer enhancements for MRAM, spintronics, sensors, and flexible electronic systems.

Chapter "PVDF-Based Actuators and Soft Robotics"—Presents electroactive properties, fabrication strategies, and biomedical applications, highlighting PVDF's potential in soft robotic and actuator design.

Chapter "PVDF in Flexible and Wearable Electronics"—Covers PVDF's structure, processing, and applications in sensors, energy harvesters, and smart textiles, emphasizing flexibility, biocompatibility, and integration challenges.

Chapter "PVDF Nanocomposites for Advanced Functional Applications"—Illustrates how nanofillers improve β-phase, piezoelectricity, and conductivity, enabling wearables, EMI shielding, PENGs, and scalable IoT platforms.

Chapter "Future Prospects and Challenges of PVDF-Based Materials"—Highlights PVDF's role in next-generation electronics and IoFT, addressing sustainability, recyclability, and scalable manufacturing, with green synthesis and circular economy strategies.

This book aims to serve as a comprehensive resource for students, researchers, and professionals in materials science, electronics, applied physics, and related fields, supporting further study and development of PVDF-based technologies. We sincerely acknowledge the contributions of all authors and the publishing team for their efforts in making this work possible.

Prayagraj, Uttar Pradesh, India Deepak Punetha

Acknowledgments

This book would not have been possible without the love, support, and encouragement of the remarkable people in my life.

First and foremost, I am deeply grateful to my parents, whose unwavering support and guidance have shaped who I am today. I also wish to thank my siblings for their constant encouragement and inspiration throughout this journey.

I am profoundly thankful to my wife, Vartika, for her patience, understanding, and unwavering support, and to my two sons, Shlok and Satvik, whose joy, curiosity, and love continually inspire me to strive for excellence.

I extend my sincere gratitude to Prof. Rajeev Tripathi for his invaluable motivation and guidance and to Prof. Asheesh K. Singh for his continuous encouragement. I am equally grateful to Prof. V.S. Tripathi and Prof. R.A. Mishra for their steadfast support and mentorship throughout my academic and professional journey.

I also wish to acknowledge all faculty members of the Department of Electronics and Communication Engineering, MNNIT Allahabad, for their cooperation, insights, and support, which have greatly enriched my work. My thanks extend to the institute, the Director, and the Dean of Research and Consultancy for providing an enabling environment for research and learning.

Finally, I am eternally grateful to God, whose blessings and guidance have made this endeavor possible. To you, the reader, thank you for taking the time to engage with this book—I hope it serves as a valuable resource and inspiration in your exploration of PVDF-based technologies.

With all my gratitude,

Deepak Punetha

Contents

Editor and Contributors

About the Editor

Deepak Punetha is a distinguished researcher and academic specializing in flexible and wearable low-cost nanocomposite-based microelectronic devices, with a strong record of translating advanced materials research into innovative electronic technologies. He earned his Ph.D. from the Department of Electrical Engineering, IIT Patna, followed by a Post-Doctoral Fellowship at IIT Bombay, focusing on next-generation flexible and wearable devices. Currently, he serves as an Assistant Professor in the Department of Electronics and Communication Engineering at MNNIT Allahabad, driving innovations in sustainable and wearable microelectronics.

His research spans nanomaterials engineering, device physics, and system-level electronics design, emphasizing flexible sensors, self-powered electronics, photovoltaic devices, triboelectric and piezoelectric technologies, and biosensors, targeting healthcare, environmental monitoring, energy harvesting, and the Internet of Flexible Things (IoFT).

Dr. Punetha has authored over 100 publications, numerous book chapters, and holds three patents. He is a Senior Member of IEEE, faculty advisor of the IEEE Sensors Council Student Chapter at MNNIT, serves on editorial boards, including Scientific Reports, and is an active reviewer and TPC member for international conferences.

Contributors

Debanjan Acharyya Department of Electronics and Instrumentation Engineering, National Institute of Technology Agartala, Agartala, India

A. Bharathi Sankar Ammaiyappan School of Electronics Engineering, Vellore Institute of Technology (VIT), Chennai, India

Praveena Balasubramaniyan School of Advanced Sciences, Vellore Institute of Technology, Chennai, India

Radha Bayya Department of Electronics and Communication Engineering, National Institute of Technology, Warangal, Warangal, India

Bharat Bhushan Bhatt Department of Metallurgical Engineering and Materials Science, Indian Institute of Technology Bombay, Mumbai, India

Dipti Gupta Department of Metallurgical Engineering and Materials Science, Indian Institute of Technology Bombay, Mumbai, India

Maimur Hossain Aiiso Yufeng Li Family Department of Chemical and Nano Engineering, University of California San Diego, San Diego, CA, USA

Manoranjan Kar Department of Physics, Indian Institute of Technology Patna, Patna, Bihar, India

Prakash Karipoth School of Design, University of Leeds, Leeds, UK

Mani Karthik Centre for Solar Energy Materials, International Advanced Research Centre for Powder Metallurgy and New Materials (ARCI), Hyderabad, India

Gulshan Kumar Department of Electrical Engineering, Indian Institute of Technology Patna, Bihta, Patna, Bihar, India

Pawan Kumar Department of Physics, Mahatma Gandhi Central University, Motihari, Bihar, India

Ravi Ranjan Kumar Department of Electronics and Communication Engineering, Motilal Nehru National Institute of Technology, Allahabad, Prayagraj, Uttar Pradesh, India

Sunil Kumar University Department of Electronic Science, Babasaheb Bhimrao Ambedkar Bihar University, Muzaffarpur, Bihar, India

Vibhor Kumar College of Science and Engineering, Troy University, Troy, AL, USA

Ankit Mayur Department of Electronics and Communication Engineering, Motilal Nehru National Institute of Technology, Allahabad, Prayagraj, India

Saurabh Kumar Pandey Department of Electrical Engineering, Indian Institute of Technology Patna, Bihta, Patna, Bihar, India

Shuvam Pawar Department of Electronics and Communication Engineering, National Institute of Technology, Warangal, Warangal, India

Deepak Punetha Department of Electronics and Communication Engineering, Motilal Nehru National Institute of Technology, Allahabad, Prayagraj, Uttar Pradesh, India

A. Rahul Narasimhan School of Electronics Engineering, Vellore Institute of Technology, Chennai, India

Harsh Ranjan Department of Electrical Engineering, Indian Institute of Technology Patna, Bihta, Patna, Bihar, India

Reena L. S. College, Babasaheb Bhimrao Ambedkar Bihar University, Muzaffarpur, Bihar, India

Sharmistha Roy Department of Electronics and Instrumentation Engineering, National Institute of Technology Agartala, Agartala, India

Department of Electronics and Communication Engineering, Dream Institute of Technology, Kolkata, India

Anand Sharma Department of Electronics and Communication Engineering, Motilal Nehru National Institute of Technology, Allahabad, Prayagraj, India

Poonam Subudhi School of Advanced Sciences, Vellore Institute of Technology, Chennai, India

Manavalan Vijayakumar Chemical Engineering and Applied Chemistry, School of Infrastructure and Sustainable Engineering, College of Engineering and Physical Sciences, Birmingham, UK

Fundamentals and Material Characteristics of PVDF

Sunil Kumar, Reena, Pawan Kumar, and Manoranjan Kar

Abstract Polyvinylidene fluoride, or PVDF, is a semicrystalline piezopolymer that has garnered significant attention due to its exceptional piezoelectric and pyroelectric properties, particularly in its β-phase. Furthermore, PVDF has significant mechanical rigidity, endurance, and chemical resistance. PVDF is an environmentally friendly polymer material and an excellent substitute for lead-based piezoceramics (PZTs) due to its lead-free composition, suitable for biomedical, structural, acoustic, and wearable sensor applications. PVDF can be fabricated into thin films or nanofibers according to specifications utilizing various facile fabrication techniques such as electrospinning, 3D printing, solution coating, and solution casting etc. Recent breakthroughs in co-polymerization and nanocomposite creation significantly boost the sensitivity of PVDF, its crystalline phase purity, and facilitate its seamless incorporation into next-generation improvements for piezo and pyroelectric device applications. This book chapter examines the physical, piezoelectric, and pyroelectric characteristics of PVDF material, its fabrication process, the adaptability of PVDF-based sensors, and the future roadmap towards the potential development route of high-performance piezoelectric and pyroelectric sensors.

Keywords Polyvinylidene-fluoride · Piezoelectric · Pyroelectric · Nanocomposite · Crystalline β-phase · Biomedical · Acoustic · Structural health monitoring

---

S. Kumar (✉)
University Department of Electronic Science, Babasaheb Bhimrao Ambedkar Bihar University, Muzaffarpur, Bihar, India

Reena
L. S. College, Babasaheb Bhimrao Ambedkar Bihar University, Muzaffarpur, Bihar, India

P. Kumar
Department of Physics, Mahatma Gandhi Central University, Motihari, Bihar, India

M. Kar
Department of Physics, Indian Institute of Technology Patna, Patna, Bihar, India

D. Punetha (ed.), *Polyvinylidene Fluoride (PVDF)*, Engineering Materials,
https://doi.org/10.1007/978-981-95-6718-8_1

1 Introduction

Polyvinylidene fluoride (PVDF) is recognized for its multifunctionality, making it a critical material in the domain of advanced functional polymers. This semi-crystalline fluoropolymer consists of alternating $-CH_2$ and $-CF_2$ units that generate strong dipole moments, contributing to its excellent dielectric, thermal, mechanical, and chemical properties [1–3]. As technological demands grow for smart, flexible, and energy-efficient devices, PVDF's structural versatility and tunable electroactive behavior have gained increased research attention in both academic and industrial settings [4, 5].

Theoretically, PVDF's ability to exhibit different crystalline phases—most notably the non-polar α-phase and the highly polar β-phase—forms the foundation of its electromechanical utility [6, 7]. The β-phase features an all-trans conformation that aligns molecular dipoles, resulting in enhanced polarization and piezoelectricity [8]. Transitioning from the α- to β-phase involves a conformational rearrangement, which can be promoted by mechanical stretching, electrical poling, or copolymerization [9, 10]. The Landau–Ginzburg–Devonshire (LGD) theory has been applied to describe phase transitions in ferroelectric polymers like PVDF, while atomistic modeling and molecular dynamics simulations provide deeper insight into how local chain orientations influence bulk behavior [11]. To realize practical electroactive properties, strategies have been developed to stabilize the β-phase. Nanostructuring and the introduction of fillers play a pivotal role to enhance the desired electrical properties [12]. When PVDF is combined with ceramic nanoparticles such as barium titanate ($BaTiO_3$), titanium dioxide (TiO_2), or tin oxide (SnO_2), the developed nanocomposites demonstrate significant enhancements in dielectric constant and ferroelectric properties [13, 14]. The interface between polymer and filler generates local electric fields that facilitate dipole alignment, while theoretical constructs like effective medium approximation (EMA) and percolation theory help predict the behavior of these heterogeneous systems [15]. Additionally, the interfacial polarization resulting from heterogeneous interfaces is often described by Maxwell–Wagner–Sillars (MWS) theory. These interfaces act as localized sites for charge accumulation, which intensifies the effective dielectric response [12, 13]. This synergy between material design and theoretical modeling enables the development of high-performance PVDF-based nanocomposites for applications requiring tunable dielectric, piezoelectric, and mechanical properties. In energy harvesting, PVDF is foundational in piezoelectric nanogenerators and hybrid triboelectric–piezoelectric systems [6, 11]. In the biomedical domain, its flexibility and biocompatibility make it suitable for wearable sensors and actuators [8]. In electronics, PVDF serves as a dielectric medium in capacitors and memory storage components [14]. The growing reliance on portable, flexible, and low-power electronics highlights the increasing importance of PVDF materials [10, 15]. Looking ahead, the trajectory of PVDF research involves multi-scale theoretical modeling, environmentally conscious synthesis approaches, and machine-learning-assisted material

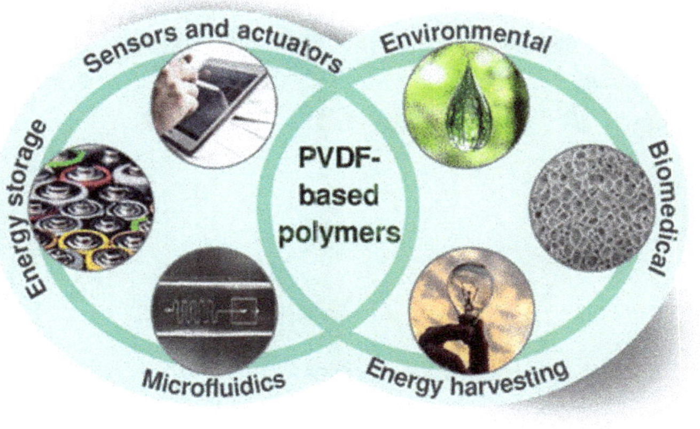

Fig. 1 Environment friendly and smart applications of PVDF and its composites [5]

optimizations as shown in Fig. 1. As a platform material, PVDF is well positioned to contribute to the next generation of soft electronics, sustainable energy systems, and smart sensor networks.

This chapter aims to provide a detailed understanding of the theoretical principles governing PVDF behavior, the mechanisms for enhancing its functional phases, and its wide-ranging applications. By bridging theory and practical engineering, it offers a foundational perspective on PVDF as a strategic material in future technologies.

2 Crystal Structure and Polymorphism

Poly(vinylidene fluoride) (PVDF) is a semicrystalline polymer derived from the monomer vinylidene fluoride (CH_2=CF_2). It belongs to the class of fluoropolymers and is characterized by its exceptional chemical resistance, thermal stability, and remarkable electroactive properties, such as piezoelectricity, pyroelectricity, and ferroelectricity [16]. These attributes make PVDF particularly attractive for applications across diverse fields including electronics, energy storage, biomedical engineering, and environmental sensing. The molecular structure of PVDF comprises alternating –CH_2– and –CF_2– groups as shown in Fig. 2. This asymmetry in the polymer backbone leads to a net dipole moment, which becomes significant in specific crystalline phases. The unique combination of mechanical flexibility and electrical responsiveness in PVDF arises from its partially crystalline nature and the orientation of dipoles in its structure. The physical properties of PVDF are not merely intrinsic to the polymer chains but are highly dependent on its crystalline phase composition and the morphology of those phases [3]. The functional performance of PVDF is largely governed by its crystalline phases, particularly in

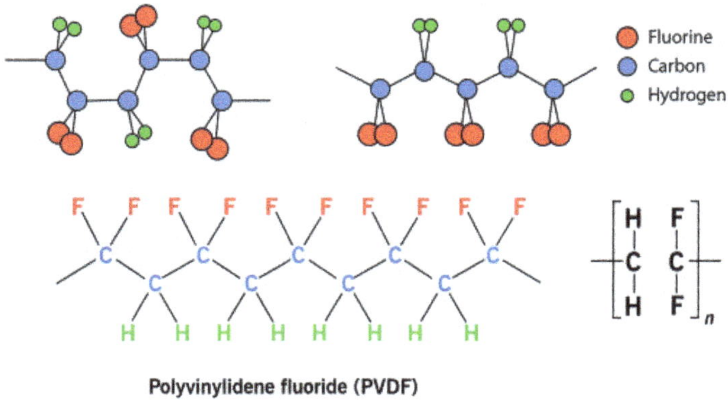

Fig. 2 Environment friendly and smart applications of PVDF and its composites [18]

applications exploiting its piezoelectric and ferroelectric properties. PVDF exhibits multiple crystalline forms, most notably the α, β, γ, δ, and ε phases [17]. Each phase has a distinct chain conformation and dipole alignment, which influences its electrical and mechanical behavior. Among them, the β-phase is the most desirable for electronic applications due to its all-trans (TTTT) zigzag planar conformation, resulting in a maximum net dipole moment per unit cell [18]. The α-phase, which crystallizes in a TGTG′ conformation, is nonpolar and thermodynamically the most stable form under standard processing conditions. The transformation from α- to β-phase, which can be induced through mechanical stretching, electrical poling, or copolymerization, is a critical step in enhancing the piezoelectric performance of PVDF polymer.

Polymorphism refers to the ability of a material to crystallize into more than one distinct crystal structure. In semicrystalline polymers like PVDF, polymorphism arises from variations in the packing arrangements and conformational sequences of polymer chains within the crystalline domains. These polymorphic forms exhibit different thermodynamic stabilities, molecular symmetries, and physical properties [19]. Unlike metals and simple inorganic crystals, polymorphism in polymers is often more complex due to the flexibility of macromolecular chains and the interplay between crystalline and amorphous regions.

In PVDF, the existence of multiple polymorphic forms enables phase engineering, which deliberate manipulation of crystalline phases to tailor material properties. Understanding the thermodynamic and kinetic factors that govern polymorphic transitions is crucial for optimizing processing techniques and enhancing the functional performance of PVDF-based materials. Moreover, theoretical modeling, spectroscopy, and diffraction techniques are widely used to investigate the structure–property relationships associated with PVDF polymorphs [20].

2.1 Molecular Structure of PVDF

Poly(vinylidene fluoride) (PVDF) is synthesized from the monomer vinylidene fluoride ($CH_2=CF_2$) through free-radical polymerization. The repeating unit in PVDF is $-(CH_2-CF_2)-$, where each monomer adds to the growing chain via a head-to-tail arrangement, although head-to-head and tail-to-tail defects may occasionally occur depending on polymerization conditions [16]. The polymer backbone consists of alternating $-CH_2-$ and $-CF_2-$ groups. The high electronegativity of fluorine atoms induces a strong dipole moment between the $-CH_2-$ and $-CF_2-$ segments, giving the polymer a polar character. The arrangement and orientation of these dipoles along the chain axis are critical in defining the electroactive properties of PVDF, particularly in its β-phase, where the dipoles align in the same direction to produce macroscopic polarization [6]. In terms of geometry, PVDF chains can adopt different conformations such as trans (T) or gauche (G) rotational isomers around the carbon–carbon bonds. These conformations TTTT (all-trans), TGTG′, and TTTGTTTG′, for instance serve as the structural basis for its crystalline polymorphs (e.g., β, α, and γ phases respectively) [7].

The interatomic bonding in PVDF consists of covalent bonds within the polymer backbone and weaker van der Waals interactions and dipole–dipole interactions between adjacent chains. The carbon and fluorine bond, with its high bond energy (~485 kJ/mol), is particularly significant in conferring thermal and chemical stability to the polymer. Additionally, the polar C–F bonds induce strong dipole interactions, which play a pivotal role in the formation of electroactive crystalline phases [18]. Conformational preferences in PVDF are governed by a balance between steric hindrance and electrostatic ©nteractions. The most energetically favorable conformation in terms of dipole alignment is the all-trans (TTTT) configuration found in the β-phase, where all dipole moments align parallel to each other, leading to the highest net polarization. In contrast, the α-phase adopts a TGTG′ conformation, which results in antiparallel dipole orientation and a nonpolar crystal [3].

2.2 Crystalline Phases of PVDF

Poly(vinylidene fluoride) (PVDF) is a semicrystalline polymer that has garnered considerable attention due to its unique combination of mechanical durability, chemical resistance, and exceptional electroactive properties such as piezoelectricity, pyroelectricity, and ferroelectricity. Crystallographically, PVDF can form at least five known polymorphic phases such as α, β, γ, δ, and ε each differing in chain conformation, symmetry, [3] and resulting properties as given in Table 1.

The α-phase is the most thermodynamically stable and is readily formed upon cooling from the melt. This phase is characterized by a TGTG′ (trans–gauche–trans–gauche) conformation, with dipoles arranged antiparallel, resulting in a net zero polarization. It crystallizes in the orthorhombic $P2_1/c$ space group and is

Table 1 Crystal structure parameters and polymorphic phases

Phase	Crystal system	Space group	Lattice parameters (Å)	Polarity	Unit cell content
α	Orthorhombic	$P2_1/c$	$a = 4.96$, $b = 9.64$, $c = 4.62$, $\beta = 90.0°$	Non-polar	TGTG′ conformation
β	Orthorhombic	*Cm2m*	$a = 8.58$, $b = 4.91$, $c = 2.56$	Highly polar	All-trans (TTTT)
γ	Monoclinic/ orthorhombic	*Cc*	$a = 4.96$, $b = 9.64$, $c = 4.62$	Weakly polar	TTTGTTTG′ conformation
δ	Orthorhombic	$P2_1cn$	Similar to α-phase	Polar	Polar version of α-phase
ε	Not fully reported	Not fully reported	Not fully reported	Polar (assumed)	Not clear

electrically inactive. In contrast, the β-phase exhibits a highly polar, all-trans (TTTT) planar zigzag conformation, where all dipoles align parallel to the crystallographic b-axis, producing a net polarization of ~131 mC/m^2. It crystallizes in the Cm_2m space group and is responsible for the highest piezoelectric, pyroelectric, and ferroelectric activity among all PVDF phases [17, 19]. The γ-phase, intermediate between α and β in structure and polarity, adopts a TTTGTTTG′ chain conformation and crystallizes in a monoclinic unit cell. The δ-phase is a polar analogue of the α-phase, formed under strong electric fields which invert the dipole directions without changing the chain conformation. Lastly, the ε-phase, structurally similar to the γ-phase, is rarely observed due to its specific formation conditions and lower thermodynamic stability [17, 20].

The most desirable electroactive β-phase over the more stable α-phase, several processing strategies are employed. Mechanical stretching is one of the most effective techniques. When PVDF films are uniaxially stretched at elevated temperatures (typically between 70–100 °C), the polymer chains become highly oriented, favoring the β-phase configuration. Electrospinning is another method that introduces a high voltage across a polymer solution, leading to the formation of aligned nanofibers rich in β-phase due to the alignment of dipoles in the high-field environment. Additionally, nanoparticle incorporation as fillers such as barium titanate, carbon nanotubes, or graphene oxide has been shown to act as a nucleating agent, significantly improving β-phase crystallization by facilitating dipole orientation through ion-dipole and hydrogen bonding interactions [21]. Solvent evaporation methods and thermal annealing further enable tuning of crystallinity and phase content by controlling chain mobility and crystallization kinetics. The α-phase is the most thermodynamically stable and non-polar. The β-phase is the most electroactive due to its all-trans chain conformation. The γ-phase is intermediate and shows some polar behaviours. The δ-phase is a polar modification of the α-phase, formed under high electric field.

The structural configuration and molecular conformation with unit cell parameters of PVDF are shown in Fig. 3 [22]. The structural and molecular as well as the unit cell dimensions for the α-phase of PVDF are approximately $a = 4.96$ Å,

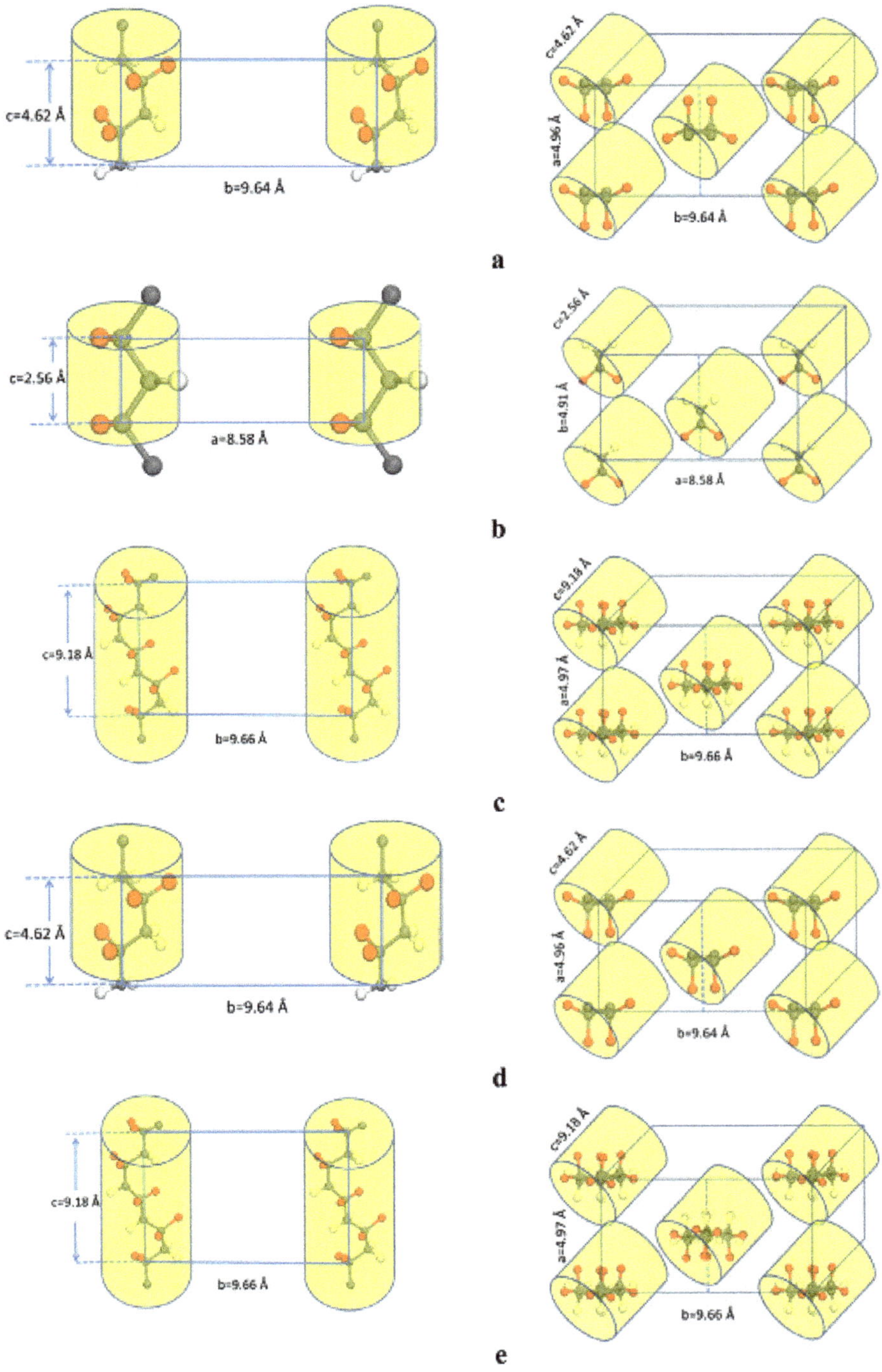

Fig. 3 Illustrates molecular conformations and unit cell structures for PVDF phases: (**a**) α, (**b**) β, (**c**) γ, (**d**) δ, and (**e**) ε. Fluorine atoms are marked in red, carbon in gray, hydrogen in white, and the cylinders represent the direction of molecular chains [22]

$b = 9.64$ Å, and $c = 4.62$ Å are shown in Fig. 3a. In comparison, the β-phase exhibits lattice parameters of $a = 8.58$ Å, $b = 4.91$ Å, and $c = 2.56$ Å, and its crystal and molecular layout is depicted in Fig. 3b, derived from the same references. The δ-phase, which shares a similar molecular conformation with the α-phase, is believed to form due to a 180° rotation and a shift of $c_0/2$ along the c-axis for alternate polymer chains in the α-phase unit cell. This rearrangement transitions the dipole orientation from antiparallel to parallel, influenced by an external electric field. The configuration for this phase is shown in Fig. 3d. The structure of the γ-phase has been extensively investigated, with two main interpretations. One theory suggests its structure closely resembles that of the β-phase, with parameters $a = 8.66$ Å, $b = 4.93$ Å, and $c = 2.58$ Å. A more widely accepted model proposes that the γ-phase maintains a and b values similar to the α-phase, while the c dimension nearly doubles, resulting in lattice parameters of $a = 4.97$ Å, $b = 9.66$ Å, and $c = 9.18$ Å, as shown in Fig. 3c. The ε-phase is regarded as a non-polar counterpart of the γ-phase and its structure is shown in Fig. 3e.

In the β-phase, the molecular chains are aligned perpendicular to the direction of the electric dipole. The consistent alignment of dipoles, which arise due to the separation between CF_2 and CH_2 groups along the chain, enhances the material's piezoelectric, pyroelectric, and ferroelectric responses. Conversely, in the **α**-phase, the anti-parallel alignment of polymer chains results in a non-polar configuration. Depending on the statistical packing of chains, the crystal lattice of the **α** and **γ** phases may exhibit either orthorhombic or monoclinic symmetry. However, recent research suggests that while the α-phase may have a monoclinic structure based on its symmetry [22].

2.3 Structure–Property Relationships

The understanding the structure–property relationships in polyvinylidene fluoride (PVDF) is crucial for optimizing its performance across diverse applications. The polymorphic nature of PVDF enables the tuning of its properties by controlling its crystalline phases. The α-phase, being non-polar and thermally stable, is primarily used in structural components and insulating materials due to its robustness and resistance to heat. In contrast, the β-phase exhibits the strongest piezoelectric and ferroelectric properties, making it ideal for use in sensors, actuators, energy harvesting systems, and non-volatile memory devices. The γ-phase possesses moderate polarity and is thus well-suited for applications in flexible electronics and electro-optic systems, where a balance between flexibility and polarization is desired [17]. The δ-phase, which is polar but forms under an applied electric field, finds use in memory devices and switchable dielectric applications, offering dynamic control over material behavior. This correlation between crystalline structure and functional performance allows PVDF to be engineered precisely for targeted technological uses as given in Table 2.

Table 2 Structure–property relationships in polyvinylidene fluoride (PVDF)

Crystalline phase	Key property	Application area
α-Phase	High thermal stability, non-polar	Structural composites, insulation
β-Phase	High piezoelectric and ferroelectric	Sensors, actuators, energy harvesters, memory devices
γ-Phase	Moderate polarity	Flexible electronics, electro-optic materials
δ-Phase	Polar (field-induced)	Memory, switchable dielectric applications

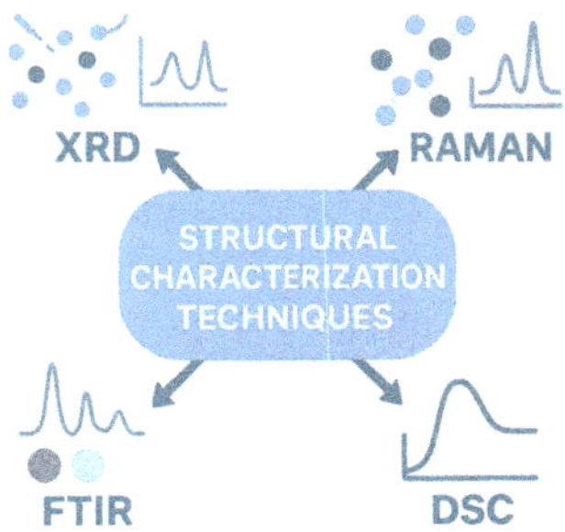

Fig. 4 Structural characterization techniques

2.4 Structural Characterization Techniques

In the study of semicrystalline polymers like poly(vinylidene fluoride) (PVDF), characterizing the internal structure is essential for understanding how processing conditions influence material properties. PVDF's polymorphic nature with at least five known crystalline phases (α, β, γ, δ, and ε) requires robust analytical tools capable of distinguishing subtle structural differences [20]. The four most commonly employed techniques for this purpose are X-ray Diffraction (XRD), Raman spectroscopy, Fourier Transform Infrared Spectroscopy (FTIR), and Differential Scanning Calorimetry (DSC) as shown in Fig. 4. Each technique provides unique and complementary insights into the crystalline structure, polymorphism, molecular conformation, and thermal behavior of PVDF. XRD provide details about the phase identification and crystallinity, lattice parameters, phase purity, orientation, degree of crystallinity. FTIR explore the molecular bonding, phase identification in mixed/ low-crystalline samples and more crystal information have been obtained though the Raman spectroscopy. More important parameters i.e. thermal parameters have been identified by employing DSC such as thermal behavior and crystallinity, melting, crystallization, glass transition temperature (T_g) and phase transformation temperatures.

2.4.1 X-Ray Diffraction (XRD) for Phase Identification

X-ray diffraction is a powerful and non-destructive technique that reveals the periodic atomic arrangement in crystalline materials. X-ray diffraction (XRD) is a widely utilized technique to examine the structural evolution and crystallographic nature of composite materials. In this study, the XRD patterns of pristine PVDF and its nanocomposites with varying concentrations of barium hexaferrite (BHF) are systematically analyzed. The composites were formulated with different weight ratios [(1 − x) PVDF − (x) BHF], where x = 0.00, 0.05, 0.15, 0.30, and 0.50. Figure 5 presents the corresponding diffraction profiles for each composition. For the BHF component, characteristic diffraction peaks were observed at 2θ values of 30.3°, 32.2°, 34.1°, 37.1°, 40.3°, 55.0°, 56.5°, and 63.0°. These reflections are consistent with the standard diffraction pattern of M-type barium hexaferrite ($BaFe_{12}O_{19}$), as indexed by the Joint Committee on Powder Diffraction Standards (JCPDS Card No. 84-0757). The diffraction features affirm the formation of a nanocrystalline, single-phase M-type hexaferrite with a hexagonal structure, crystallizing in the space group $P6_3/mmc$. The average crystallite size of BHF particles was estimated using the Williamson–Hall (W–H) method, yielding a value of approximately 48 nm, confirming their nanometric dimension.

The XRD profiles of PVDF-BHF nanocomposite films exhibit three broad peaks centered at 2θ ≈ 18.4°, 19.9°, and 26.6°, which are attributed to the α-phase of PVDF [23, 24]. A broad diffraction feature appearing around 20.4° is indicative of the β-phase. These broad peaks suggest a semi-crystalline nature of the polymer matrix and reflect the coexistence of both α and β phases. The presence of BHF nanoparticles within the PVDF matrix leads to an evident increase in the β-phase content, as evidenced by the enhancement of the β-phase diffraction intensity at 20.4°, accompanied by a relative decrease in the α-phase intensity at 18.4°. To quantify the phase transformation, the intensity ratio of the β-phase to α-phase peaks

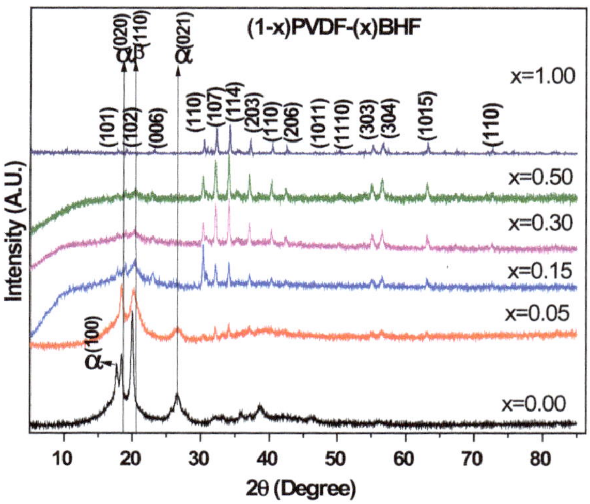

Fig. 5 XRD patterns of pure PVDF and their composites (1 − x) PVDF − (x) BHF with different weight fractions of BHF nanoparticle [16]

(I_β/I_α = I_20.4°/I_18.4°) was computed for each composite composition. Further confirmation of β-phase enhancement is provided through Fourier Transform Infrared Spectroscopy (FTIR), which is discussed in the subsequent section.

XRD also allows for the quantitative determination of the degree of crystallinity using peak deconvolution methods. The crystalline and amorphous regions are separated mathematically from the diffractogram, and the crystallinity is calculated using the integrated area under the peaks:

$$X_c = \frac{A_{crystalline}}{A_{crystalline} + A_{amorphous}} \times 100\%$$

2.4.2 Fourier Transform Infrared Spectroscopy (FTIR) for Conformational Analysis

FTIR spectroscopy is widely used for identifying the molecular vibrations of chemical bonds within polymers. PVDF's various polymorphs are distinguished by their specific chain conformations, which give rise to distinct vibrational absorption bands in the mid-infrared range (400–4000 cm^{-1}). FTIR is particularly valuable because it is sensitive to molecular-level changes that may not be evident in XRD, especially in samples with low crystallinity or overlapping phases.

FTIR spectra were recorded for both pristine PVDF and its nanocomposites containing various concentrations of barium hexaferrite (BHF) nanoparticles. The resulting spectra, presented in Fig. 6, reveal a series of absorption bands characteristic of specific PVDF crystalline phases. Peaks observed at 485, 611, 760, 795, and 975 cm^{-1} are attributed to the α-phase. Notably, the bands at 611 cm^{-1} and 760 cm^{-1} correspond to the bending and wagging modes of CF_2 groups, and the rocking motion of the PVDF backbone, respectively [25].

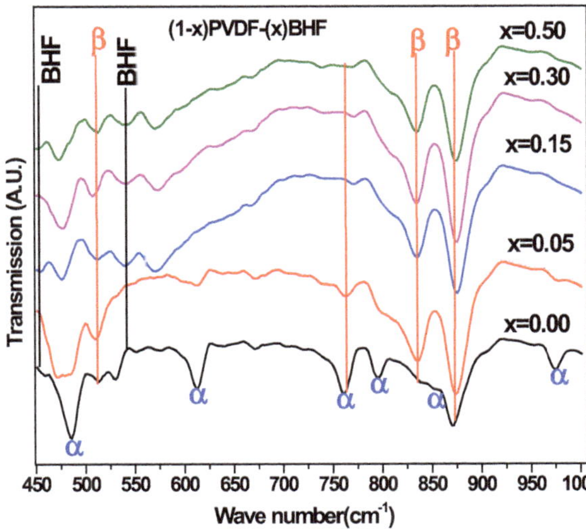

Fig. 6 FTIR spectra of pure PVDF and nanocomposites (1 − x) PVDF − (x) BHF with different weight fractions of BHF nanoparticle [16]

In contrast, the β-phase is identified by characteristic absorption bands at 511, 834, and 872 cm^{-1}. The peak at 872 cm^{-1}, in particular, is associated with the CH_2 rocking and CF_2 stretching vibrations, which are indicative of the all-trans (TTTT) chain conformation typical of the polar β-phase. An examination of the spectra reveals a notable trend: with increasing BHF content, the intensity of β-phase peaks rises consistently, while the α-phase peak intensities diminish. This shift suggests that the incorporation of BHF nanoparticles actively promotes the transformation of PVDF from the non-polar α-phase to the electroactive β-phase [3].

This phase transformation has significant implications, as the polar β-phase is crucial for achieving enhanced dielectric, ferroelectric, and piezoelectric properties. To quantify the extent of this transformation, the relative fraction of the β-phase in the samples was calculated using the empirical method proposed by Gregorio and Cestari. This method relies on the relative intensities of the absorption bands at 840 cm^{-1} (β-phase) and 760 cm^{-1} (α-phase), and is given by:

$$F(\beta) = \frac{A_\beta}{\left(1.26A_\alpha + A_\beta\right)}$$

where A_β and A_α represent the absorbance values at 840 cm^{-1} and 760 cm^{-1}, respectively. The calculated β-phase fractions for various BHF loadings are shown in Fig. 6. The data reveal that the β-phase content increases from 0.45 in pure PVDF to a peak value of 0.63 at 30 wt% BHF, followed by a slight decline with further addition of nanoparticles. Importantly, the increase in β-phase content aligns with the observed enhancement in dielectric properties of the nanocomposite system, as discussed in the subsequent sections. Thus, FTIR analysis serves as a vital tool for confirming phase transitions and establishing structure–property relationships in PVDF-based functional materials.

2.4.3 Differential Scanning Calorimetry (DSC) for Thermal Phase Transitions

Differential Scanning Calorimetry (DSC) is a powerful thermal analysis technique used to investigate thermal transitions in polymers by monitoring heat flow during controlled heating or cooling cycles. This method enables the precise determination of key thermal properties, including the glass transition temperature (T_g), melting temperature (T_m), and crystallization temperature (T_c), as well as associated enthalpic changes and specific heat capacity [26].

In the case of polyvinylidene fluoride (PVDF), each of its polymorphic crystalline phases—namely α, β, γ, and δ—exhibits distinct thermal behaviors that are influenced by polymerization conditions and processing parameters such as temperature and pressure [27]. Notably, the γ-phase displays a higher melting temperature compared to the α and β phases. Thermal studies reveal that when the α-phase is formed by crystallization below 140 °C, its melting point nearly overlaps with that of the β-phase. However, the γ-phase melts at a temperature approximately 8 °C higher than the α-phase, serving as a thermal signature for phase identification. An interesting characteristic of the δ-phase is its ferroelectric-to-paraelectric phase transition, observed around 154 °C. This transition suggests that the δ-phase maintains structural and ferroelectric Furthermore, DSC provides a means to evaluate the degree of crystallinity in PVDF samples. This is accomplished by comparing the measured melting enthalpy of a given sample to the reference enthalpy for 100% crystalline PVDF. Such an approach yields quantitative insight into how processing methods, filler incorporation, or thermal treatments influence crystalline content and, by extension, functional properties Representative DSC thermograms illustrating the endothermic melting behavior associated with the various PVDF polymorphs are presented in Fig. 7. These curves serve as valuable fingerprints for identifying crystalline phases and assessing thermal transitions critical for tailoring PVDF materials for advanced technological applications.

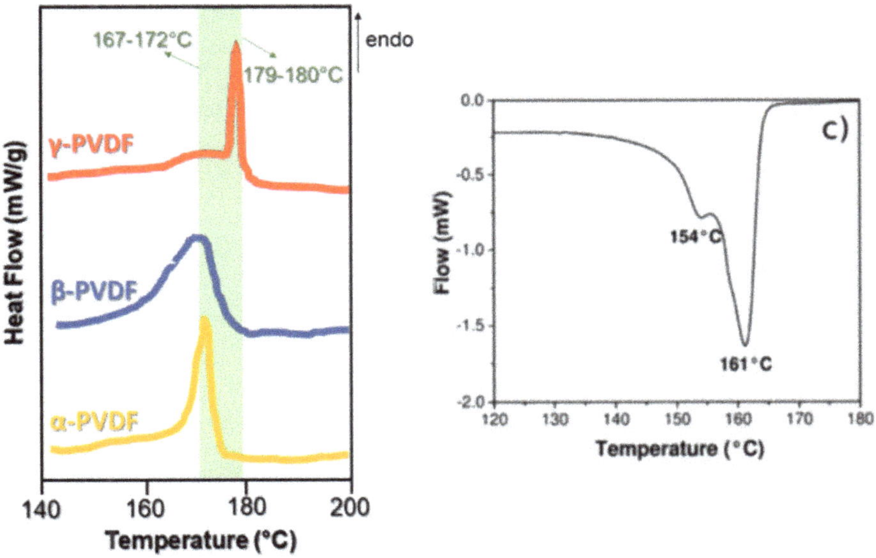

Fig. 7 Typical DSC curves presented by the PVDF polymorphs: α, β, and γ on the left-hand side and δ on the right hand [27]

2.4.4 Raman Spectroscopy

Raman spectroscopy has been accepted as an alternative method for discriminative determination of the crystal phases in PVDF [28]. The Raman fingerprints of the α- and β-phases are shown in Fig. 8, which shows clear differences in the spectra. The region of 700–900 cm^{-1} is the main fingerprint of the crystal phases. The α-phase PVDF shows a sharp peak at 794 cm^{-1},which is due to the rocking of the CH_2 bond, whereas the β-phase corresponding to the CH_2 rocking exhibits a shift to 839 cm^{-1}. Moreover, an additional peak at 812 cm^{-1} is responsible for the γ-phase [29].

2.5 *Phase Transformation Mechanisms in PVDF*

Poly(vinylidene fluoride) (PVDF) is a semi-crystalline fluoropolymer known for its ability to crystallize into multiple polymorphic phases, each characterized by distinct chain conformations and macroscopic properties. The most prominent phases—α (non-polar), β (highly polar), and γ (intermediate)—can interconvert under the influence of external stimuli such as mechanical stress, thermal treatment, electric field, and solvent environment [18]. Understanding the mechanisms of phase transformation in PVDF is essential to engineering materials for piezoelectric, ferroelectric, and optoelectronic applications.

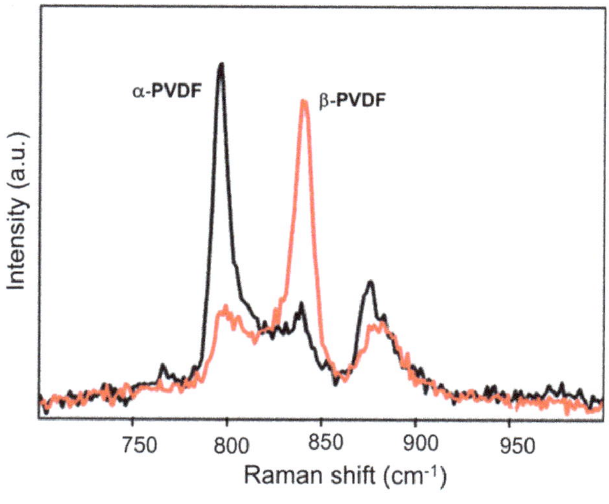

Fig. 8 Spectroscopic fingerprints of the α-, γ-, and β-phases of PVDF obtained by Raman spectra [28]

2.5.1 Molecular Conformational Theory

The Fig. 9, explain the torsional energy barrier (in kcal/mol) as a function of torsion angle (θ), high lighting two distinct conformations, A and B. The molecular dipole alignments associated with each conformation (μ_A and μ_B) influence the overall dipolar orientation and ferroelectric properties of the polymer [30]. Conformation A shows aligned dipoles leading to a stronger net polarization, whereas conformation B results in a tilted arrangement and reduced macroscopic dipole moment [31].

Poly(vinylidene fluoride) (PVDF) exhibits different molecular conformations depending on the torsional angles between adjacent monomer units in its polymer chain. These conformations play a crucial role in determining the crystalline phase, dipole orientation, and ultimately the ferroelectric and piezoelectric properties of the material. As shown in Fig. 9, energy profile plotted against the torsional angle (θ) shows a barrier between two local minima, representing stable conformations A and B. Conformation A corresponds to a lower energy state with a more ordered dipole alignment (μ_A), typically associated with the all-trans (TTTT) conformation, as found in the ferroelectric β-phase of PVDF. Conformation B, on the other hand, represents a slightly twisted or skewed molecular geometry, likely related to gauche defects (e.g., TGTG′ conformation), common in the non-polar α-phase. The net dipole moment μ_B is smaller and oriented off-axis compared to μ_A. The diagram emphasizes how small changes in torsional angles significantly impact the energy landscape and the macroscopic polarization behavior of PVDF. These conformational transitions are central to molecular conformational theory, which models the rotational freedom and energy preferences of the polymer backbone to explain the phase stability and electromechanical responses between these conformational states [31].

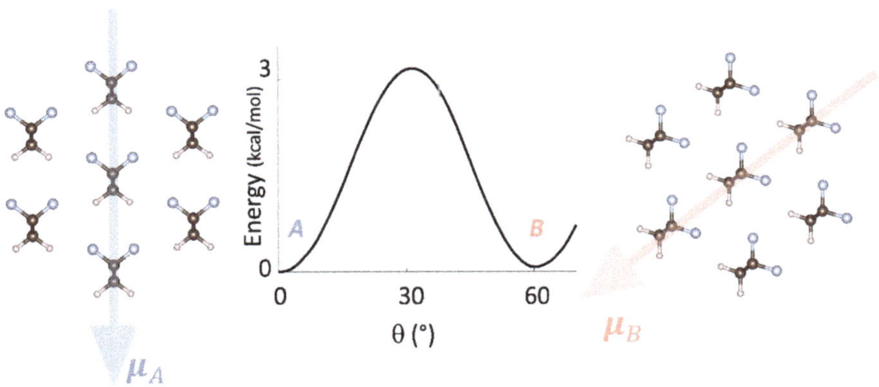

Fig. 9 Schematic representation of molecular conformational states in poly(vinylidene fluoride) (PVDF) [31]

2.5.2 Crystallographic Transformation Mechanisms

The transformation of α to β-phase is commonly achieved by uniaxial drawing of PVDF films at elevated temperatures (70–100 °C). At the crystal lattice level, this transformation changes the unit cell from monoclinic (α-phase) to orthorhombic (β-phase) symmetry. This involves a reorganization of the lattice parameters and hydrogen bonding networks as shown in Fig. 10.

Solvent-Mediated and Solution-Casting-Induced Transformations

Solvent evaporation methods can influence PVDF phase formation. Polar solvents such as DMF, DMSO, or acetone can stabilize the β-phase during film formation due to solvent–polymer dipole interactions and hydrogen bonding. The theoretical basis for this lies in solvent-induced dipole alignment and interchain electrostatic stabilization. The solution casting "freezes" the β-conformation before chains can relax into the more stable α-phase.

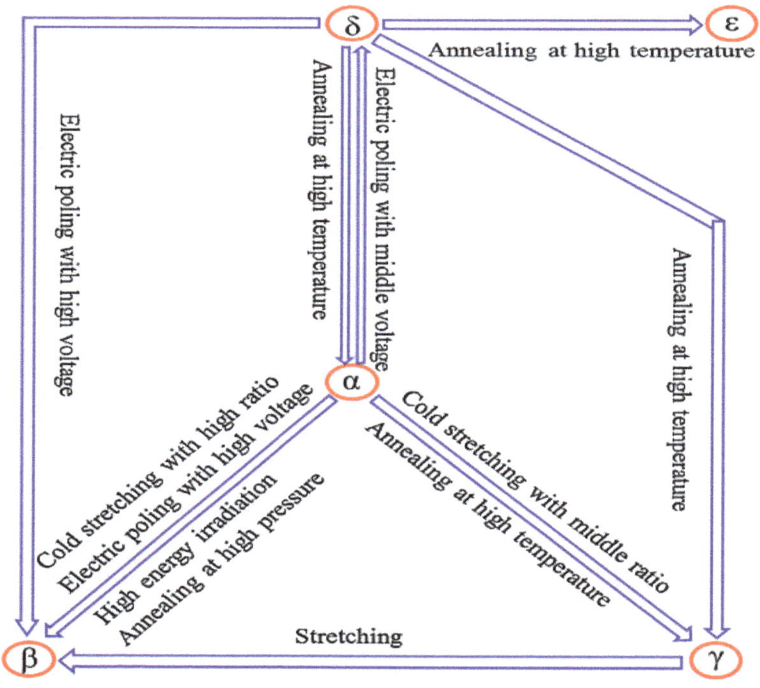

Fig. 10 Conversion between the PVDF polymorphs using thermal, mechanical, and electrical treatments [22]

Role of Additives and Nanofillers

Incorporation of nanomaterials (e.g., $BaTiO_3$, ZnO, graphene oxide) acts as heterogeneous nucleation sites, selectively favoring β-phase crystallization. Surface energy matching between filler and β-phase crystals. Electrostatic interactions that promote chain alignment. Mechanical constraints that induce local stress, mimicking stretching effects [32]. As a example, barium hexaferrite nanoparticles were taken as nanofiller to introduced in the PVDF matrixes as a model proposed in Fig. 11a–e. The most of the target to enhanced the β-phase in PVDF. In other words, another crystal phases have to convert into the β-phases for better electrical properties such as dielectric and ferrelectric. In the *gauche* configuration, substituent atoms are positioned approximately ±60° relative to each other around the central carbon atom, [3] as illustrated in Fig. 12a. The β-phase, in particular, is notable for its ferroelectric and dielectric properties. In this configuration, the highly electronegative fluorine atoms (F^-) and the electropositive hydrogen atoms (H^+) are aligned on opposite sides of the polymer chain, creating strong dipole moments across the molecule, as shown in Fig. 12b. When subjected to an alternating electric field, these aligned dipoles polarize coherently, resulting in a ferroelectric response. This behavior is akin to the action of microcapacitors embedded within the polymer matrix. The incorporation of nanofillers such as barium hexaferrite (BHF) significantly influences the crystalline phase behavior of PVDF. These nanoparticles not only act as nucleating agents that promote the crystallization of PVDF but also induce the transformation of the α-phase into the electroactive β-phase as shown in Fig. 12c. Experimental observations from both XRD and FTIR analyses (Fig. 13a, b) confirm that increasing the BHF content in the composite leads to a marked enhancement in β-phase content.

This phase transformation mechanism is attributed to interfacial interactions between PVDF chains and the surface of BHF nanoparticles. The nanoparticles possess localized surface plasmons, which generate a surrounding electric field. This localized field interacts with the charged constituents of PVDF, particularly the H^+ and F^- atoms, facilitating the nucleation and stabilization of the β-phase. Figure 13c schematically illustrates this interaction mechanism.

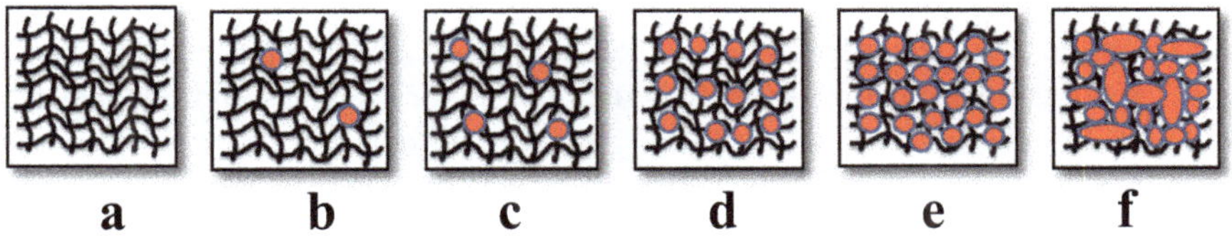

Fig. 11 Expected structure and BLCs model of (1 − x) PVDF (x) − BHF for x = (**a**) 0.00, (**b**) 0.025, (**c**) 0.05, (**d**) 0.15, (**e**) 0.30 and (**f**) 0.50 [16]

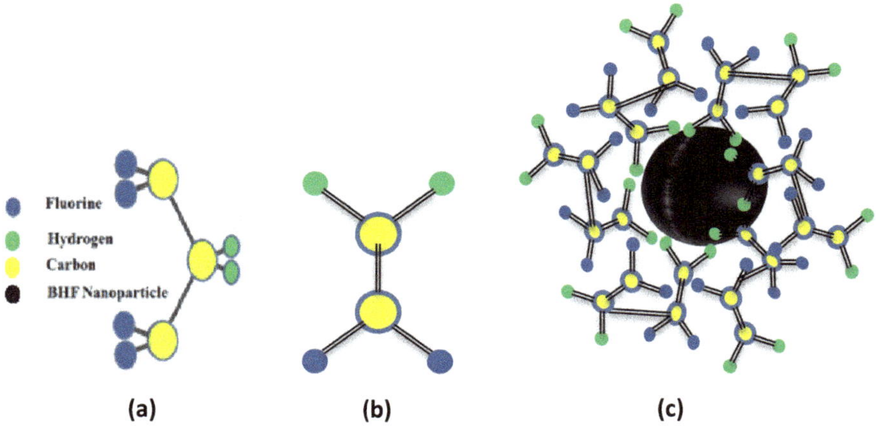

Fig. 12 The structure of PVDF; (**a**) α phase of PVDF, (**b**) β phase of PVDF and (**c**) proposed model of nanocomposite prepared from PVDF and BHF nanoparticles [16]

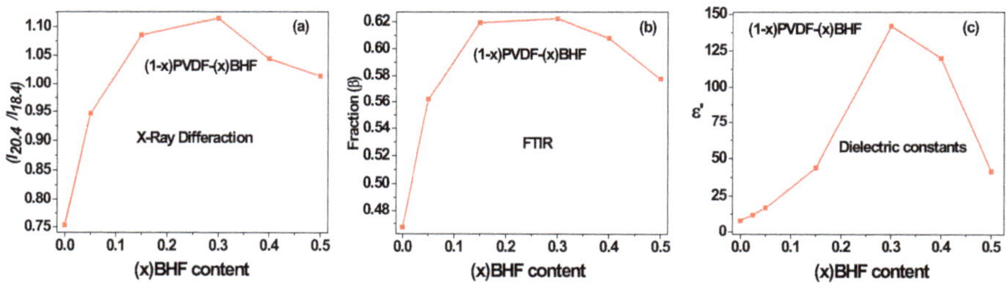

Fig. 13 The β phase of nanocomposites with increasing BHF obtained from (**a**) XRD, (**b**) FTIR analysis and, (**c**) variation of dielectric constants with the increase BHF in nanocomposites [16]

3 Properties of PVDF

The multifunctional characteristics of PVDF-based composites, and their outstanding electroactive properties, including high dielectric, piezoelectric, and ferroelectric response, being the best among polymer systems as shown in Fig. 14 [33]. Their superior ferroelectric properties, as evidenced by well-defined hysteresis loops, make them ideal for use in non-volatile memory devices and ferroelectric field-effect transistors (FeFETs). The high dielectric constant ensures their suitability for capacitors and energy storage systems, where stable charge accumulation is critical.

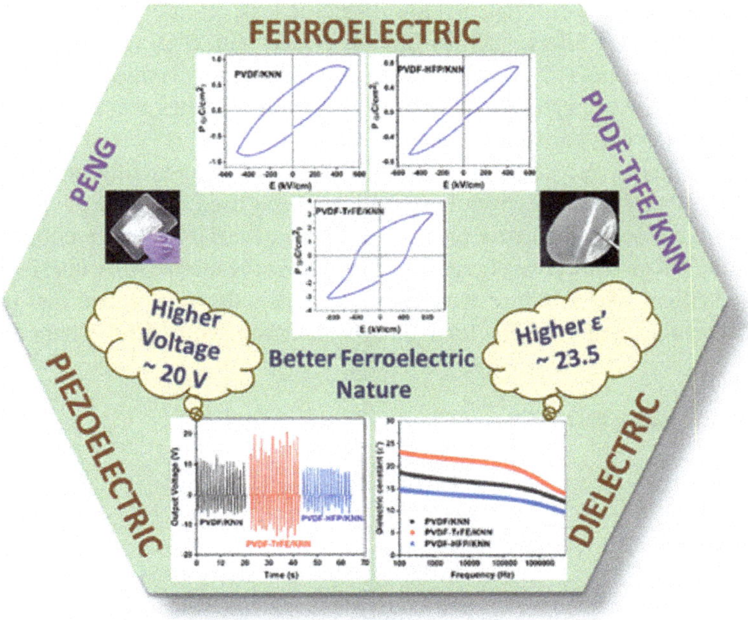

Fig. 14 Multifunctional characteristics of PVDF-based composites [33]

3.1 Dielectric Properties of PVDF and Its Composites

The dielectric behavior of PVDF arises from various polarization mechanisms, including electronic, ionic, dipolar, and interfacial (Maxwell–Wagner–Sillars) polarization. Among these, dipolar polarization is dominant in pure PVDF due to the strong permanent dipole moments in its molecular structure, while interfacial polarization becomes prominent in heterogeneous systems like composites [34]. To enhance its dielectric performance, researchers have incorporated a variety of fillers into the PVDF matrix, forming composites that leverage the synergistic effects of polymer flexibility and filler functionality. The inclusion of high-dielectric ceramic fillers such as barium titanate ($BaTiO_3$), titanium dioxide (TiO_2), and lead zirconate titanate (PZT) significantly increases the dielectric constant. For instance, PVDF/ $BaTiO_3$ composites with 50 wt% filler have reported dielectric constants exceeding 60, compared to ~10 for neat PVDF [35].

Conductive fillers, including carbon nanotubes (CNTs), graphene, and metallic nanoparticles like silver or copper, have also been employed to further boost dielectric properties through interfacial polarization. Near the percolation threshold, CNTs can significantly enhance the dielectric constant, although care must be taken to avoid high dielectric losses due to increased conductivity [36]. Graphene and graphene oxide, due to their large surface areas and tunable surface chemistry, facilitate better dispersion and interfacial compatibility, contributing to improved dielectric behavior with relatively low filler loading [37]. However, to address the trade-off between high dielectric constant and low dielectric loss, researchers have developed core–shell structured fillers, such as $BaTiO_3$ coated with SiO_2 ($BaTiO_3@SiO_2$) or CNTs encapsulated in TiO_2. These hybrid structures improve interfacial interaction and suppress charge carrier mobility, resulting in composites that exhibit high εr with minimal tan δ as enlisted in Table 3 [38].

Besides ceramic and conductive fillers, intrinsically conducting polymers like polyaniline (PANI) and polypyrrole (Ppy) have been used to prepare PVDF composites with enhanced dielectric properties. These materials not only contribute to polarization but also improve charge transport across the matrix-filler interface [36]. Bio-based fillers like cellulose nanofibers have also shown promise due to their renewable nature and capacity for hydrogen bonding with PVDF chains, which

Table 3 Dielectric constant and loss of PVDF and its composites

No.	Composite	Dielectric constant (ε_r)	Dielectric loss (tan δ)	Notes	References
1	PVDF (pristine)	7.4	0.056	Baseline for comparison	[35]
2	PVDF + 30 vol% $BaTiO_3$ (BT_IO)	18.8	0.088	154% Enhancement over pristine PVDF	[35]
3	PVDF + 50 vol% $BaTiO_3$@PVP	80.4	0.085	Improved dispersion and reduced defects	[41]
4	PVDF + 7 wt% MWCNT	43.6	0.079	Below percolation	[42]
5	PVDF + 20 vol% Fe_2O_3@$BaTiO_3$	31.7	~0.05	Satellite–core structure enhances	[36]
8	PVDF + 8 vol% c-MWCNTs/MG	448.0	2.35	High ε_r but increased loss	[43]
9	PVDF + 8 vol% MWCNTs-CuPc	45.0	0.15	Functionalization improves dispersion	[44]
10	PVDF + 6 wt% POS@CNT-B5	98.0	0.53	Enhances dielectric properties	[45]
11	PVDF + 8.3 vol% MEB/PVDF	260.0	1.3	Surface-functionalized MWCNTs	[43]
12	PVDF + 8 vol% MWCNTs (trifluorobromobenzene treated)	550.0	2.2	High ε_r with moderate loss	[35]
15	PVDF + 30 wt% TiO_2	25.0	Not specified	TiO_2 enhances dielectric properties	[46]
16	PVDF + 5 wt% CNTs	35.0	Not specified	Near percolation threshold	[42]
17	PVDF + $BaTiO_3@SiO_2$	60.0	<0.05	Core–shell structure reduces loss	[36]
19	PVDF + cellulose nanofibers	20.0	Not specified	Bio-based filler enhances ε_r	[39]
20	PVDF + $BaTiO_3@SrTiO_3$	50.0	Not specified	Core–shell nanoparticles	[47]

promotes better filler dispersion and matrix compatibility [39]. The dielectric performance of PVDF composites is influenced by several factors, including filler type, size, shape, and volume fraction, as well as the quality of dispersion and interfacial adhesion. Increasing filler content generally leads to higher dielectric constants but may also result in higher dielectric loss and brittleness. Therefore, an optimal balance must be achieved. Processing techniques such as solution casting, melt blending, and electrospinning also affect the morphology and crystalline phase of PVDF, which in turn influence the dielectric properties [40].

3.2 Ferroelectric Characteristics

Ferroelectric materials are characterized by their ability to exhibit spontaneous electric polarization that can be reversed by the application of an external electric field. While traditionally associated with inorganic ceramics like barium titanate ($BaTiO_3$), ferroelectricity has also been observed in certain polymers, with poly(vinylidene fluoride) (PVDF) being one of the most notable examples. PVDF combines the desirable mechanical properties of polymers such as flexibility and chemical resistance with electrical functionalities like ferroelectricity, piezoelectricity. The ferroelectric behavior in PVDF arises from the alignment of molecular dipoles within its crystalline regions, particularly the polar β-phase. Due to its molecular structure, PVDF exhibits unique phase-dependent electrical behavior, which is strongly influenced by its processing conditions. With advancements in nanotechnology and materials engineering, the ferroelectric performance of PVDF has been further enhanced by the incorporation of nanofillers and through structural modifications at the molecular level [16].

The theoretical stands for ferroelectricity in polymers such as PVDF can be described using the Landau–Ginzburg–Devonshire (LGD) theory, which models the free energy of a ferroelectric system as a function of polarization (P). The free energy landscape features two or more local minima, each corresponding to a stable polarization state. When an external electric field is applied, the system can transition between these states, resulting in hysteresis behavior. In PVDF, the origin of polarization lies in the difference in electronegativity between the hydrogen and fluorine atoms along the polymer backbone. This results in permanent dipoles on each monomer unit. When these dipoles are aligned through mechanical or electrical processing, the material gains a net polarization, which is reversible under an alternating electric field with a hallmark of ferroelectricity. The electrostatic interactions among molecular dipoles and their cooperative alignment in the crystalline domains underpin the macroscopic polarization observed in PVDF [31].

The ferroelectric (P–E) loops of PVDF and (0.70) PVDF – (0.30) BHF in different applied electric field at room temperature and 100 Hz frequency are shown in Fig. 15a, b respectively. The surface of both circular sides of the nanocomposite thick film was silver coated for electrode purpose. The maximum polarization (Pm) corresponds to the maximum applied field increased with increasing applied electric

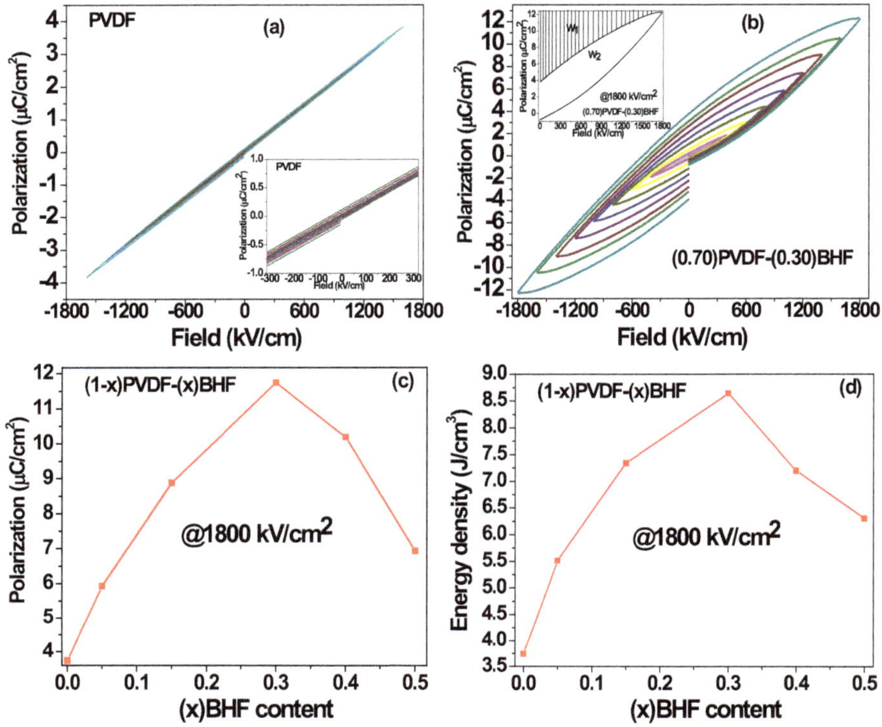

Fig. 15 (**a**) Ferro-electric (*P–E*) loops of PVDF with different applied electric field at room temperature, (**b**) polarization measure at different applied electric field of the (0.70) PVDF – (0.30) BHF sample, and inset shows the stored energy and release energy diagram and, (**c**) saturation polarization (**d**) energy storage density with different weight percentage of BHF nanoparticle in the PVDF matrix at 1800 kV/cm^2 respectively [16]

field. The maximum polarization (Pm) with weight percentage of BHF nanoparticles in PVDF matrix at 1800 kV/cm is shown in Fig. 15c. The nanocomposite with (0.70) PVDF – (0.30) BHF is showing maximum polarization. It could be due to presence of maximum β phase, which is confirmed from the XRD and FTIR analysis as shown in Fig. 13. The high polarization is due to the interaction between BHF nanoparticles filler and PVDF matrix on the interface. The energy storage density with different weight percentage of BHF nanoparticles in PVDF matrix at 1800 kV/cm is shown in Fig. 15d. The energy density of (0.70) PVDF – (0.30) BHF sample is 8.75 J/cm^3 while the PVDF energy density is only 3.85 J/cm^3 at higher applied electric field 1800 kV/cm. This energy storage density enhancement in (0.70) PVDF – (0.30) BHF due to having the maximum polarization. The Maxwell–Wagner–Sillars (MWS) interfacial polarization model is commonly used to explain the enhanced dielectric and ferroelectric response in heterogeneous systems. According to MWS theory, the accumulation of charges at the interface between the filler and matrix results in localized polarization, boosting the overall dielectric response.

3.3 Mechanical Properties

Poly(vinylidene fluoride) (PVDF) is not only known for its excellent electroactive properties but also for its remarkable mechanical performance, making it a highly versatile polymer in both structural and functional applications. PVDF exhibits a semi-crystalline morphology, where its mechanical strength is strongly influenced by its degree of crystallinity, crystal phase composition, and processing history. The polymer typically shows a tensile strength ranging from 40 to 55 Mpa, with a Young's modulus of approximately 1.0 to 2.5 Gpa, depending on the fabrication conditions and presence of additives [48]. The elongation at break is relatively high, often exceeding 50%, which contributes to its toughness and flexibility. These mechanical properties are retained over a broad temperature range due to PVDF's high thermal stability, with a melting temperature around 170–180 °C.

Importantly, the mechanical behavior of PVDF is also influenced by its crystalline phase. The non-polar α-phase offers higher mechanical stability, while the polar β-phase, though less thermodynamically stable, provides enhanced electroactive behavior with slightly reduced stiffness. The ability to process PVDF into various forms such as thin films, fibers, or membranes adds to its applicability in flexible electronics, sensors, and filtration devices. Furthermore, the addition of nanofillers (e.g., carbon nanotubes, graphene, ceramic nanoparticles) can significantly alter the mechanical behavior, often leading to enhanced modulus, strength, and toughness, though it may sometimes reduce flexibility depending on filler content and dispersion quality. The mechanical behaviour (tensile strength, Young's modulus and elongations at break) of PVDF and it composites with ceramics filler, carbon filler from the literatures are given in Table 4.

Table 4 Mechanical behaviour of PVDF and it composites

Filler type	Filler content (wt%)	Tensile strength (Mpa)	Young's modulus (Gpa)	Elongation at break (%)	References
None (pure PVDF)	0	42–55	1.2–2.1	50–200	[49]
ZnO	5	51	2.0	70	[50]
Carbon nanotubes (CNTs)	1	65	3.0	35	[51]
Graphene	0.5	63	3.2	40	[52]
rGO (reduced graphene oxide)	1	60	2.9	45	[53]
Montmorillonite (MMT) clay	5	54	2.2	65	[54]

4 Processing Techniques for Performance Enhancement

The physical properties of PVDF can be significantly enhanced through various processing techniques that influence its crystalline phase structure, orientation, and morphology. One of the most widely adopted methods is mechanical stretching, which transforms the non-polar α-phase into the highly polar β-phase by aligning the molecular chains in the direction of the applied force. This structural reorientation enhances the material's electroactive response. Another critical technique is electrical poling, where a high electric field is applied to orient the dipoles in a single direction, further stabilizing the β-phase and improving ferroelectricity. Solvent casting also plays a vital role; the choice of solvent and drying conditions can promote the formation of different crystalline phases, particularly the γ-phase. Additionally, thermal annealing is employed to control crystallinity and phase transformation; moderate heating can enhance the β-phase content and improve mechanical stability [20]. Electrospinning is a modern nanofabrication method used to produce PVDF fibers with high surface area and controlled orientation, which are ideal for energy harvesting and sensing applications. Spin coating and hot pressing techniques are also utilized to create thin films and dense pellets with controlled thickness and phase content, suitable for capacitors and piezoelectric devices. Moreover, additive manufacturing and solution blending with fillers or nanomaterials such as $BaTiO_3$ or graphene oxide can be used to form PVDF composites with enhanced dielectric and mechanical properties. Collectively, these processing strategies aim to maximize the β-phase content, enhance dipole alignment, and improve interfacial charge distribution.

The dielectric and ferroelectric properties of poly(vinylidene fluoride) (PVDF) are strongly influenced by its crystalline phase structure, molecular orientation, and interfacial characteristics, which can be significantly modified through various processing techniques as shown in Fig. 16. Among these, mechanical stretching is a widely adopted approach to induce the transformation of the non-polar α-phase to the electroactive β-phase. Uniaxial stretching at elevated temperatures (60–100 °C) aligns molecular dipoles along the stretching direction, enhancing ferroelectric polarization and increasing the dielectric constant. Additionally, electrical poling—applying a high DC electric field (~500–1000 kV/cm) to the polymer above its glass transition temperature—reorients dipoles within β-crystallites, resulting in increased remanent polarization and reduced coercive field, which are essential for memory and sensor applications [16].

Thermal annealing is another effective technique that influences crystallinity and phase composition. Controlled annealing around 90–130 °C helps enhance the β-phase fraction while stabilizing crystalline domains. Furthermore, solution processing using high-polarity solvents (such as DMF, DMSO, or acetone) promotes β-phase nucleation during solvent evaporation, leading to improved ferroelectric and dielectric behavior in thin films. The choice of solvent and casting conditions (temperature, humidity, drying rate) significantly affects the final morphology and interfacial polarization, especially when fillers are introduced.

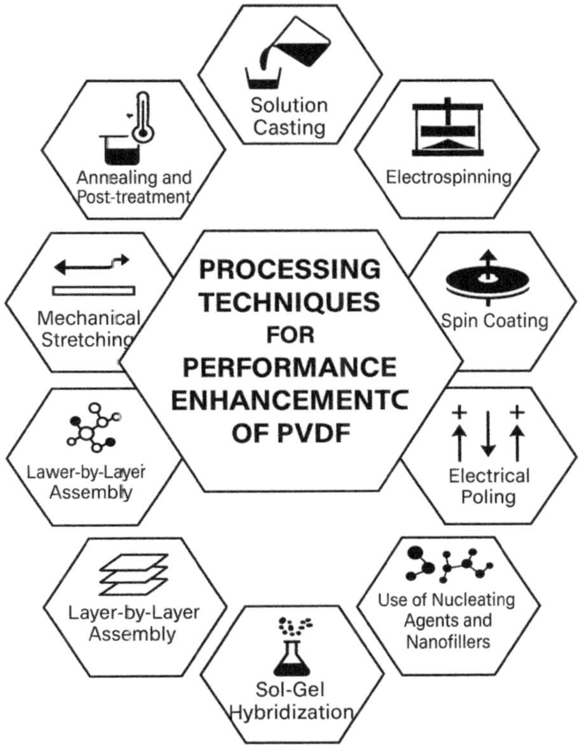

Fig. 16 Processing techniques for performance enhancement

The incorporation of functional nanofillers into PVDF through techniques like solution blending, in situ polymerization, or melt compounding further enhances its dielectric and ferroelectric properties. Fillers such as $BaTiO_3$, graphene oxide, or Mxene can act as nucleating agents for the β-phase and introduce interfacial polarization based on the Maxwell–Wagner–Sillars (MWS) effect. Uniform dispersion of these fillers, facilitated by techniques such as ultrasonication, surface functionalization, or the use of surfactants, ensures efficient dipole interaction and high dielectric permittivity. Recent advances also include electrospinning, which inherently aligns dipoles along fiber axes, creating nanofibrous PVDF mats with enhanced piezoelectric and ferroelectric responses [16].

5 Challenges and Future Directions

5.1 Challenges

Poly(vinylidene fluoride) (PVDF) exhibits multiple crystalline phases, with the electroactive β-phase being the most desirable for applications like sensors, actuators, and energy harvesters. However, achieving a high fraction of this phase is technically demanding. One of the primary challenges is that the α-phase, which is non-polar and less functional, tends to form more easily due to its thermodynamic stability under normal processing conditions. Transforming PVDF into the β-phase often requires precise mechanical stretching, electrical poling, or specific solvent systems, all of which are sensitive to environmental and operational conditions. Scaling these techniques for industrial production remains complicated due to variability in phase content, high processing costs, and energy requirements. Furthermore, when PVDF is blended with fillers or nanomaterials to enhance its properties, maintaining consistent phase formation and long-term stability becomes more complex. Another obstacle is the reliance on toxic solvents in some processing methods, which raises environmental and safety concerns. Additionally, accurately distinguishing and quantifying various PVDF phases in mixed-phase materials requires sophisticated instruments and methodologies, limiting easy characterization.

5.2 Future Prospects

To address these challenges, future efforts should focus on developing sustainable and efficient processing methods that promote electroactive phase formation. Green processing techniques using non-toxic or minimal solvents are gaining interest for their environmental benefits. Tailoring molecular structures through copolymerization or chemical modification could enable inherent stabilization of desirable phases like the β- or γ-phase without needing external stimuli. The use of real-time monitoring technologies, such as in-situ spectroscopy or advanced imaging, could enhance process control and reproducibility. Moreover, integrating artificial intelligence and machine learning models to predict optimal processing conditions may significantly improve efficiency and scalability. PVDF's role in next-generation applications can also be expanded through the design of smart composites and hybrid systems for flexible electronics, biomedical devices, and sustainable energy systems. Interdisciplinary collaboration and innovation in materials design, processing, and application engineering will be essential to fully exploit PVDF's polymorphic versatility and technological potential.

6 Conclusion

Polymorphism in polyvinylidene fluoride (PVDF) plays a pivotal role in defining its performance across a broad spectrum of advanced applications. Among the five known crystalline phases (α, β, γ, δ, and ε), the β-phase is the most technologically significant due to its all-trans (TTT) molecular conformation, which aligns molecular dipoles and imparts strong ferroelectric, piezoelectric, and pyroelectric properties. This phase transformation—from the non-polar α-phase to the polar β-phase—is influenced by processing methods such as mechanical stretching, electrospinning, annealing, solvent casting, and the incorporation of nanofillers like barium hexaferrite (BHF). Structural characterization techniques such as XRD and FTIR have confirmed that the inclusion of BHF nanoparticles enhances β-phase formation by inducing intra-molecular interactions and nucleation. The resulting changes in the crystal structure contribute to improved dielectric constant, polarization, and energy harvesting efficiency. These properties make PVDF-based materials highly suitable for cutting-edge applications including flexible electronics, non-volatile ferroelectric memory, biomedical sensors, piezoelectric nanogenerators, and smart coatings. The future of PVDF research lies in mastering the control over phase transformation and achieving stable, high β-phase content through optimized processing and hybrid material strategies, thereby unlocking its full potential in multifunctional and sustainable technologies.

References

1. Smith, T.F., Waterman, M.S.: Identification of common molecular subsequences. J. Mol. Biol. **147**, 195–197 (1981)
2. Nalwa, H.S.: Ferroelectric Polymers: Chemistry, Physics, and Applications. Marcel Dekker, New York (1995)
3. Martins, P., Lopes, A.C., Lanceros-Méndez, S.: Electroactive phases of poly(vinylidene fluoride): determination, processing and applications. Prog. Polym. Sci. **39**, 683–706 (2014)
4. Punetha, D., Kar, M., Pandey, S.K.: A new type low-cost, flexible and wearable tertiary nanocomposite sensor for room temperature hydrogen gas sensing. Sci. Rep. **10**, 2151 (2020)
5. Costa, C.M., et al.: Smart and multifunctional materials based on electroactive poly(vinylidene fluoride): recent advances and opportunities in sensors, actuators, energy, environmental, and biomedical applications. Chem. Rev. **123**(19), 11392–11487 (2023)
6. Tashiro, K.: Structure and phase transition of PVDF and related copolymers. Ferroelectrics. **115**, 317–344 (1991)
7. Lovinger, A.J.: Ferroelectric polymers. Science. **220**, 1115–1121 (1983)
8. Yoon, S.W., Park, Y.J., Kim, Y.S.: High-performance piezoelectric nanogenerators based on electrospun PVDF nanofibers. Nano Energy. **10**, 77–85 (2014)
9. Zhang, Y., Das, S., Yoshida, K.: Conformational changes in PVDF-based copolymers under mechanical stretching. Polymer. **56**, 450–458 (2015)
10. Punetha, D., Kumar, A., Pandey, S.K., Chakrabarti, S.: Tertiary nanocomposite-based self-powered E-skin as energy harvester and electronic nose. J. Mater. Sci. Mater. Electron. **35**(2), 160 (2024)
11. Kim, K., Park, J.: Atomistic insight into PVDF phase transformation by molecular dynamics simulation. Macromolecules. **45**, 9331–9338 (2012)
12. Liu, J., Wang, X., Li, Y.: Enhancing dielectric properties of PVDF composites via ceramic nanoparticle inclusion. J. Alloys Compd. **783**, 343–350 (2019)
13. Zhang, T., Tanaka, Y., Wang, H.: Interfacial polarization and dielectric behavior in PVDF/ $BaTiO_3$ nanocomposites. Compos. Sci. Technol. **129**, 68–75 (2016)
14. Das, A., Nandan, B., Jana, S.: Tin oxide (SnO_2) loaded PVDF composites for enhanced dielectric applications. J. Mater. Chem. C. **9**, 10125–10136 (2021)
15. Ahmad, S., Wang, J.: Machine-learning guided design of PVDF-based nanocomposites for high energy storage applications. ACS Appl. Mater. Interfaces. **14**, 25678–25688 (2022)

16. Kumar, S., Supriya, S., Kar, M.: Enhancement of dielectric constant in polymer–ceramic nanocomposite for flexible electronics and energy storage applications. Compos. Sci. Technol. **157**, 48–56 (2018)
17. Lovinger, A.J.: Crystallographic studies of ferroelectric poly(vinylidene fluoride). J. Polym. Sci. Polym. Phys. Ed. **16**, 753–775 (1978)
18. Balli, B., Şavk, A., Şen, F.: Graphene and Polymer Composites for Supercapacitor Applications, pp. 123–151. Woodhead Publishing (2019)
19. Tashiro, K., Kobayashi, M., Tadokoro, H.: Structural phase transitions of poly(vinylidene fluoride) and poly(vinylidene fluoride-trifluoroethylene) crystals. Macromolecules. **13**, 691–698 (1980)
20. Kim, K., Lee, J.H., Park, J.: Phase transitions and ferroelectric properties of PVDF studied by in-situ FTIR and XRD. Polymer. **65**, 45–53 (2015)
21. Mohammadi, B., Yousefi, A.A., Bellah, S.M.: Effect of tensile strain rate and elongation on crystalline structure and piezoelectric properties of PVDF thin films. Polym. Test. **26**, 42–50 (2007)
22. Gregorio, R.: Determination of the α, β and γ crystalline phases of poly(vinylidene fluoride) films prepared at different conditions. J. Appl. Polym. Sci. **100**, 3272–3279 (2006)
23. Salimi, A., Yousefi, A.A.: FTIR studies of β-phase crystal formation in stretched PVDF films. Polym. Test. **22**, 699–704 (2003)
24. Gregorio, R., Cestari, M.: Effect of crystallization temperature on the crystalline phase content and morphology of poly(vinylidene fluoride). J. Polym. Sci. Part B Polym. Phys. **32**, 859–870 (1994)
25. Cai, Z., Song, D., Lin, D.: Characterization of phase transition behavior in PVDF using FTIR and XRD. Macromolecules. **41**, 9345–9351 (2008)
26. Kepler, R.G.: Ferroelectric polymers. In: Ferroelectric Polymers: Chemistry, Physics, and Applications. Marcel Dekker, New York (1995)
27. Concha, V.O.C., Timóteo, L., Duarte, L.A.N., et al.: Properties, characterization and biomedical applications of polyvinylidene fluoride (PVDF): a review. J. Mater. Sci. **59**, 14185–14204 (2024)
28. Chatterjee, A., Das, A., Saha, K., Jeong, U.: Nanofiller-induced enhancement of PVDF electroactivity for improved sensing performance. Adv. Sens. Res. **2**, 2200080 (2023)
29. Horiuchi, D., Yamazaki, T.: Raman study of PVDF films with different phases. Spectrochim. Acta A. **58**, 1087–1091 (2002)
30. Lovinger, A.J.: Developments in the theory of ferroelectric polymers. Science. **220**, 1115–1121 (1983)
31. Serra, G., Arrigoni, A., Del Zoppo, M., Castiglioni, C., Tommasini, M.: Molecular mechanism of the piezoelectric response in the β-phase PVDF crystals interpreted by periodic boundary conditions DFT calculations. Materials. **16**, 6004 (2023)
32. Pan, H., Feng, X., Zhai, J., Shen, B.: Effects of nanofillers on phase transformation and piezoelectric properties of PVDF-based composites: a review. J. Mater. Sci. **56**, 13514–13542 (2021)
33. Tripathy, A., Maria Joseph Raj, N.P., Kim, S.J., Ramadoss, A.: Elucidating the piezoelectric, ferroelectric, and dielectric performance of lead-free KNN/PVDF and its copolymer-based flexible composite films. ACS Appl. Electron. Mater. **5**(10), 5422–5431 (2023)
34. Kumar, S., Supriya, S., Kar, M.: PVDF, barium hexaferrites and rGO nanocomposite for high energy density capacitor. In: 2017 IEEE 12th Nanotechnology Materials and Devices Conference (NMDC), pp. 99–100 (2017)
35. Pradhan, L.K., Pandey, R., Kumar, S., Kar, M.: Lead-free $Bi_{0.5}Na_{0.5}TiO_3$ (BNT) and polyvinylidene fluoride (PVDF) based nanocomposite for energy storage applications. AIP Conf. Proc. **1953**, 090036 (2018)
36. Li, Q., Wang, Q., Zhang, Y.: Surface modification of $BaTiO_3$ nanoparticles and their dielectric performance in polymer composites. Adv. Mater. **31**, 1804910 (2019)
37. Yao, S., Yu, S., Wang, X.: Core-shell structured dielectric nanocomposites for flexible energy storage. Nano Energy. **59**, 65–74 (2019)

38. Singh, B.P., Chandrasekhar, A., Mathur, R.B.: Dielectric properties of polyaniline composites with PVDF. Mater. Res. Bull. **64**, 97–104 (2015)
39. An, H., Park, H., Kim, Y., Lee, S.: Cellulose nanofiber reinforced PVDF composites for enhanced dielectric performance. Carbohydr. Polym. **264**, 118043 (2021)
40. Kumar, S., Sharma, R., Roy, P.K.: Processing strategies and phase transitions in PVDF-based nanocomposites: effects on dielectric performance. Polym. Eng. Sci. **60**, 1703–1714 (2020)
41. Kim, J.H., Lee, D.S., Park, C.: Enhanced dielectric performance in PVDF/$BaTiO_3$-PVP composites. Compos. Part B Eng. **140**, 57–64 (2018)
42. Wang, X., Liu, F., Zhang, Z.: Dielectric and electrical behaviors of PVDF/MWCNT composites near percolation. J. Appl. Polym. Sci. **137**, 48964 (2020)
43. Zhou, X., Yang, J., Xu, J.: High dielectric permittivity and low loss in PVDF composites with modified CNTs. ACS Appl. Mater. Interfaces. **10**, 19170–19178 (2018)
44. Zhang, Q., Zhang, H., Zhang, Y.: Functionalized carbon nanotubes for enhanced dielectric performance in PVDF nanocomposites. Compos. Sci. Technol. **164**, 75–82 (2018)
45. Sun, Y., Li, M., Wu, G.: POS@CNT hybrid fillers for dielectric enhancement of PVDF composites. J. Mater. Chem. C. **6**, 1653–1662 (2018)
46. Peng, M., Zhang, Y., Wang, L.: TiO_2-modified PVDF nanocomposites for energy storage. J. Mater. Sci. Mater. Electron. **33**, 465–475 (2022)
47. Zhao, H., Liu, Y., Xu, C.: Tailored core–shell $BaTiO_3$@$SrTiO_3$ structures for dielectric enhancement in PVDF. Adv. Compos. Hybrid Mater. **5**, 1015–1024 (2022)
48. Callister, W.D., Rethwisch, D.G.: Materials Science and Engineering: An Introduction, 10th edn. Wiley, Hoboken (2018)
49. Wang, X., Liu, F., Zhang, Z.: Mechanical properties of PVDF and its composites: a review. J. Appl. Polym. Sci. **137**, 48964 (2020)
50. Sencadas, V., Gregorio, R., Lanceros-Méndez, S.: Effects of filler content on the mechanical and dielectric properties of PVDF/ZnO nanocomposites. J. Non-Cryst. Solids. **355**, 1746–1750 (2009)
51. Li, L., Zhang, Z., Chen, L.: PVDF/CNT nanocomposites with improved mechanical and electrical properties. Compos. Sci. Technol. **80**, 1–7 (2013)
52. Liu, Y., He, J., Wang, W.: Enhanced mechanical properties of PVDF/graphene nanocomposites. Polym. Compos. **35**, 332–338 (2014)
53. Zhang, X., Jiang, J., Liu, H.: Effects of rGO on PVDF nanocomposites. Mater. Res. Express. **4**, 095033 (2017)
54. Lu, X., Wang, C., Zhang, Y.: Montmorillonite-reinforced PVDF nanocomposites. Appl. Clay Sci. **65–66**, 143–149 (2012)

PVDF-Based Piezoelectric and Pyroelectric Sensors

Sharmistha Roy and Debanjan Acharyya

Abstract Polyvinylidene fluoride, or PVDF, is a semicrystalline piezopolymer that has garnered significant attention due to its exceptional piezoelectric and pyroelectric properties, particularly in its β-phase. Furthermore, PVDF has significant mechanical rigidity, endurance, and chemical resistance. PVDF is an environmentally friendly polymer material and an excellent substitute for lead-based piezoceramics (PZTs) due to its lead-free composition, suitable for biomedical, structural, acoustic, and wearable sensor applications. PVDF can be fabricated into thin films or nanofibers according to specifications utilizing various facile fabrication techniques such as electrospinning, 3D printing, solution coating, and solution casting etc. Recent breakthroughs in co-polymerization and nanocomposite creation significantly boost the sensitivity of PVDF, its crystalline phase purity, and facilitate its seamless incorporation into next-generation improvements for piezo and pyroelectric device applications. This book chapter examines the physical, piezoelectric, and pyroelectric characteristics of PVDF material, its fabrication process, the adaptability of PVDF-based sensors, and the future roadmap towards the potential development route of high-performance piezoelectric and pyroelectric sensors.

Keywords Polyvinylidene-fluoride · Piezoelectric · Pyroelectric · Nanocomposite · Crystalline β-phase · Biomedical · Acoustic · Structural health monitoring

S. Roy
Department of Electronics and Instrumentation Engineering, National Institute of Technology Agartala, Agartala, India

Department of Electronics and Communication Engineering, Dream Institute of Technology, Kolkata, India

D. Acharyya (✉)
Department of Electronics and Instrumentation Engineering, National Institute of Technology Agartala, Agartala, India

D. Punetha (ed.), *Polyvinylidene Fluoride (PVDF)*, Engineering Materials,
https://doi.org/10.1007/978-981-95-6718-8_2

1 Introduction

The mounting demand of intelligent sensor technologies covering countless sectors-from biomedical, healthcare to infrastructure management, has initiate intensive researches with materials that are capable of physical transformation to electrical parameters conversion [1–3]. Among the polymer materials that have undergone investigations throughout the years, polyvinylidene fluoride (PVDF) $-(CH_2 - CF_2)n-$ stands out as an endowed material as it has a several properties like: a comparatively high melting temperature, strong chemical resistance, limited thermal conductivity, high mechanical strength with rising temperature, good resistance for ultraviolet radiation, and high piezoelectric as well as pyroelectric properties [2, 3]. The presence of piezoelectricity in a material was originally documented by curie brother back in 1880 [4]. The piezoelectric characteristics of polyvinylidene fluoride were observed in the late 1960s with several advancement till date [5]. Kawai et al. demonstrated the piezoelectric contributions of PVDF in 1969, which was interpreted as one of the best piezo-polymer compared to other polymer materials studies in that phase [5]. In early 1970s, the extensive researches on PVDF and its copolymers made a path of development of ultrasonic actuators [6].

PVDF (polyvinylidene fluoride) can crystallize into (i) α, (ii) β, (iii) γ, (iv) δ, and (v) ε phases. Each of the phases demonstrates distinct properties [7]. Non-polar molecular alignment is a significant attribute of α-phase PVDF which makes it more stable compare to other phases [7]. β-phase is the polar phase with all-transition configuration [8]. The β phase is the highly advantageous candidate due to its significant piezoelectric characteristics [9]. The γ [10, 11] and δ phases [11] shows intervening characteristics. γ-phase is partially polar [10]. The ability of PVDF to process into diverse architecture like: nano-fibers, thin films, composites, etc. place it as an outstanding material for advance sensor development, particularly in crucial domains such as biomedical engineering, acoustic sensing, health monitoring, radio engineering, medicine, pharmaceutical, metallurgical, food, paper, structural industries [12]. Copolymerization of PVDF with monomers such as trifluoro-ethylene (TrFE) [13], or hexafluoropropylene (HFP) [14], also enhance β-phase formation of PVDF. Observations reveal that PVDF-TrFE spontaneously generates the β-phase without any mechanical manipulation, which is advantageous for thin-film applications in sensors [15] and actuators [16]. These copolymers preserve PVDF's desirable properties—chemical resistance, thermal stability, while improving its pyroelectric and piezoelectric performance. These materials have flexibility, chemical resistivity, wide range of frequency response which make them suitable candidates for environmental applications: nuclear power [17], aerospace [18, 19], biomedical [20], industrial [21], automotive [22] sectors. PVDF is a piezopolymer with semi-crystalline arrangement with outstanding piezoelectric, and pyroelectric aspects mainly in its β-phase crystalline alignment, where dipoles of the PVDF align persistently on electrical poling [12]. This crystalline alignment of PVDF help to release electrical charges when temperature fluctuate, which make PVDF as an excellent candidate for a vast range of pyroelectric applications [18]. Its natural flexibility, low weight, chemical resistance, and mechanical robustness render PVDF an ideal candidate for being incorporated into wearable and flexible electronics [23]. PVDF thin films and nanostructures are also widely utilized in infrared sensors [24], thermal sensors [25], and motion-sensing devices [26], as well as in thermal imaging [27] and surveillance systems [28]. PVDF is used in pyroelectric energy harvesting technologies, which transforms waste or ambient heat into electrical energy suitable for powering low-energy devices. Processing of PVDF is simple as it uses techniques like spin coating, solution casting, electrospinning, etc. For this effectiveness, PDVF provides a scalable and cost-effective platform to produce pyroelectric sensors and harvesters [6]. Recent advancement in PVDF-based nanocomposites and multilayered architectures have also enhances pyroelectric coefficients and thermal sensitivity of PVDF based sensors which expands its application in smart sensing technologies [29].

One of the such areas is biomedicine where PVDF based piezoelectric or pyroelectric sensors may be utilized for tissue or membrane engineering, artificial muscle creation [30], medicine delivery system [31], health monitoring system [20] and many more. Due to its piezoelectric attributes, PVDF is used in development of biomedical instrumentation: smart skin [32], artificial muscle [30], power sensor for health monitoring [33], heart sound monitoring [34], bone tissue [35], etc. Membrane and fibers format of PVDF are commonly used for biomedical applications [33]. During COVID 19 outbreak, the global market of PVDF membrane remarkably reached at USD624.8 million in 2020 [36]. It is anticipated that the growth will reach around USD958.2 million by 2027. PVDF finds significant applications in medical sectors [37]. Some of the researches devotes attention concerning the piezoelectric attributes of PVDF and identifies considerable interest in exploring PVDF and its applications in biomedical sectors [38]. The utilization of PVDF underscores the role of fabrication process of piezoelectric sensor films, and also its use in the development in implantations with exceptional resilience, biological compatibility, and robust mechanical performances, as thoroughly as in the development of stem cells as nanoscaled membrane structures [39]. Many of the researches demonstrates the outstanding chemical attributes of PVDF, especially the relationship with the other polymers like chitosan, and shares certain similarities with PVDF. The advancement of PVDF based micro/nano structured composite membranes combined with graphene nanoplatelets, along with the requirement of the optical fiber hydrophone based on chitosan and PVDF for in-vivo ultrasonic measurement are also discussed [40]. Application of PVDF in tissue engineering are also studied. These studies concludes that PVDF can be used in polyfunctional electroactive scaffolds as a bone healing procedure and in the development of fibers that use the body's natural movement to generate consistent and dependable power for implanted medical devices and portable electronics [40]. The creation of polyvinylidene fluoride (PVDF) fibers and particles for nebulization employing electrohydrodynamic atomization (EHDA) in the pulmonary drug delivery systems are also investigated [41]. PVDF based piezoelectric and pyroelectric sensors offers lightweight, flexible, and chemically stable substitutes for conventional ceramic sensors in structural health monitoring (SHM) [42]. For concurrent vibration and strain detection in structures, buildings, bridges, airplanes, and many more [42]. PVDF based sensors are best for their versatility on dynamic surfaces. The creation of implantable sensors and energy harvesters that use body motion to power electronic appliances or to monitor cardiac or respiratory function, is made possible by biocompatibility and piezoelectricity of PVDF in the biomedical industry [43]. The ability of PVDF to process easily and as a non-toxic material, it is an appropriate candidate for novel technologies which requires soft, not so heavy, lead-free materials even when the piezoelectric coefficients of PVDF is much smaller compared to piezoceramic materials like lead zirconate titanate (PZT) [44]. PVDF provides an environmental-friendly alternatives lead-based piezoceramics for piezoelectric applications because environmental regulations and restriction on lead-based material usage [44]. PVDF is also at the forefront of energy harvesting technologies, particularly in piezoelectric nanogenerators (PENGs) [45]. These devices harness ambient mechanical energy to generate electricity, using sources like airflow, vibrations, and body movement [45]. By virtue of its subdued density and high durability, PVDF is ideal for wearable, implantable, and portable devices like self-sustaining Internet of Things (IoT) and biological sensing systems. PVDF is widely used in piezoelectric and pyroelectric acoustic sensing technologies as it can transduce mechanical and thermal stimuli into electrical output [46]. This feature makes PVDF a versatile material for sound and vibration detection [46]. In the β-phase crystalline arrangement, PVDF demonstrates strong piezoelectric characteristics to detect dynamic pressure change along with the mechanical vibrations produced by sound waves [47]. PVDF is remarkably suitable for microphones, hydrophones along with the structural health monitoring systems [46]. The pyroelectric properties of PVDF also enable its temperature sensing capabilities produced by acoustic wave induced heating by adding sensitivity into sensor's configurations [48]. Low acoustic impedance and compatibility with flexible surfaces enhances the integration of PVDF based sensors into acoustic engineering [49]. PVDF based acoustic sensors are chemically stable, comparatively light weighted. They can be fabricated in thin films. These features enable the application of PVDF based sensors in wearable technologies, biomedical devices, acoustic sensing technologies [47–49].

PVDF is a multipurpose material for piezoelectric and pyroelectric sensors with a variety of uses made possible by its ferroelectric, pyroelectric, and piezoelectric properties. The functional contribution of PVDF based sensors can be greatly augmented by using co-polymerization, and nano-composite techniques [37, 50]. PVDF is considered as the key component of sustainable technologies and next-generation smart devices as the need for flexible, lightweight, and environmentally friendly materials increases [43, 49]. This chapter aims to provide an invaluable resource for engineers and researchers by providing detailed perspective on the recent breakthroughs and prospective applications of the PVDF based sensors in various applications like biomedical, acoustic engineering, structural health monitoring and many more.

2 Working Mechanism of Piezoelectric and Pyroelectric Sensors

2.1 *Piezoelectric Effect*

Piezoelectricity is a fundamental quality that some of the crystals, and polymers demonstrates. These materials have the special features like they can generate electrical charges in response to mechanical stress, deformation or dynamic stimuli such as pressure, oscillation [2]. Piezoelectric materials facilitate the interconversion between electrical and mechanical (vibrational) motion. Several varieties of sensors, actuators, and transducers utilize the direct and reciprocal piezoelectric effects employed by piezoelectric material. Three modes of operation of the sensing mechanism: longitudinal, transverse, and shear are the main emphasis of the underlying principles of the piezoelectric method. There are variations in the charge coefficient (d_{ij}) values for each of the operation modes. The longitudinal mode, transverse mode, and shear mode are represented respectively based on the charge coefficients. Thus, the terms "d_{33}," "d_{31}" and "d_{15}" are also used as references to longitudinal, transverse, and shear modes accordingly [51]. A variety of charges are formed on all of the surfaces of a piezoelectric material when the d_{xy} values alter for different operation modes of the piezoelectric material [51] (Fig. 1).

Under defined condition, the inherent generation of electrical charges within designated crystalline structures, is the foundational mechanism of piezoelectricity. Spontaneous polarization in PVDF is induced by the relative shift of electronic charge distribution in response to the movement of positive ions corresponding to negative ions within crystalline structures. Dipoles are formed as the result of this phenomenon. One well known active piezoelectric material is polycrystalline in consists of minuscule crystallites arranged in a random manner [53]. Each of the

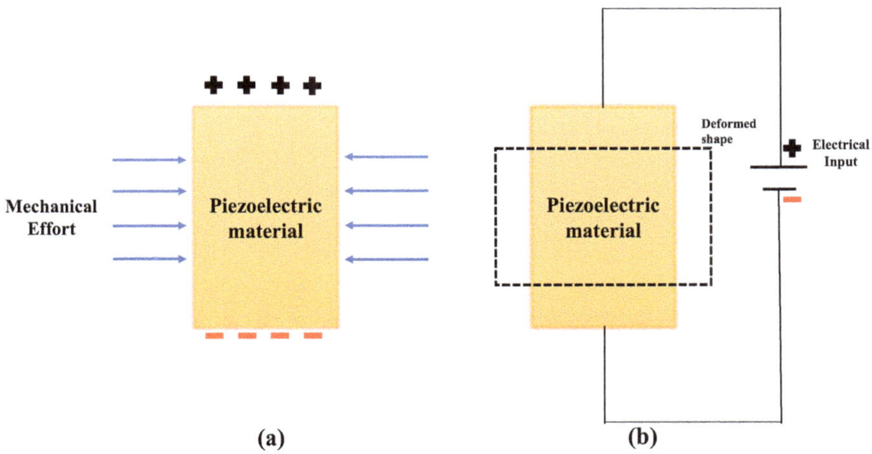

Fig. 1 Piezoelectric effect (**a**) Direct: change apparition (**b**) Indirect: shape deformation [52]

crystallites are again subdivided into smaller regions exhibiting uniform dipole orientations. Due to incorporation of electric field into piezoelectric material, piezoelectric polarization transpires. This situation is the result of the indirect piezoelectric effect that induces mechanical strain [2, 51, 53].

The stress-charge form is as follows:

$$T = c_E S + e^T E \tag{1}$$

$$D = eS + \varepsilon_0 \varepsilon_{rs} E \tag{2}$$

where ‘S’ stands for strain, ‘T’ means stress, ‘E’ denotes applied electric field, and ‘D’ defines electric displacement field. ′c_E′, ‘e’, and ‘ϵ_{rs}" are the stiffness coefficients, electromechanical coupling coefficients, and relative dielectric permittivity (strain) respectively at constant strain, whereas, ′ϵ_0′ is the permittivity (vacuum).

$$\begin{pmatrix} T_{xx} \\ T_{yy} \\ T_{zz} \\ T_{yz} \\ T_{xz} \\ T_{xy} \end{pmatrix} = \begin{pmatrix} c_{E11} & c_{E12} & c_{E13} & c_{E14} & c_{E15} & c_{E16} \\ c_{E21} & c_{E22} & c_{E23} & c_{E24} & c_{E25} & c_{E26} \\ c_{E31} & c_{E32} & c_{E33} & c_{E34} & c_{E35} & c_{E36} \\ c_{E41} & c_{E42} & c_{E43} & c_{E44} & c_{E45} & c_{E46} \\ c_{E51} & c_{E52} & c_{E53} & c_{E54} & c_{E55} & c_{E56} \\ c_{E61} & c_{E62} & c_{E63} & c_{E64} & c_{E65} & c_{E66} \end{pmatrix} \begin{pmatrix} S_{xx} \\ S_{yy} \\ S_{zz} \\ S_{yz} \\ S_{xz} \\ S_{xy} \end{pmatrix} + \begin{pmatrix} e_{11} & e_{21} & e_{31} \\ e_{12} & e_{22} & e_{32} \\ e_{13} & e_{23} & e_{33} \\ e_{14} & e_{24} & e_{34} \\ e_{15} & e_{25} & e_{35} \\ e_{16} & e_{26} & e_{36} \end{pmatrix} \begin{pmatrix} E_x \\ E_y \\ E_z \end{pmatrix} \tag{3}$$

$$\begin{pmatrix} D_x \\ D_y \\ D_z \end{pmatrix} = \begin{pmatrix} e_{11} & e_{12} & e_{13} & e_{14} & e_{15} & e_{16} \\ e_{21} & e_{22} & e_{23} & e_{24} & e_{25} & e_{26} \\ e_{31} & e_{32} & e_{33} & e_{34} & e_{35} & e_{36} \end{pmatrix} \begin{pmatrix} S_{xx} \\ S_{yy} \\ S_{zz} \\ S_{yz} \\ S_{xz} \\ S_{xy} \end{pmatrix} + \varepsilon_0 \begin{pmatrix} \varepsilon_{rs11} & \varepsilon_{rs12} & \varepsilon_{rs13} \\ \varepsilon_{rs21} & \varepsilon_{rs22} & \varepsilon_{rs23} \\ \varepsilon_{rs31} & \varepsilon_{rs32} & \varepsilon_{rs33} \end{pmatrix} \tag{4}$$

The stain-charge form is given as:

$$S = s_E T + d^T E \tag{5}$$

$$D = dT + \varepsilon_0 \varepsilon_{rT} E \tag{6}$$

Where, ‘s_E’, ‘d’, and ‘ϵ_{rT}’are the compliance matrix coefficients, coupling properties, and relative dielectric permittivity (stress) respectively at constant stress.

$$\begin{pmatrix} S_{xx} \\ S_{yy} \\ S_{zz} \\ S_{yz} \\ S_{xz} \\ S_{xy} \end{pmatrix} = \begin{pmatrix} s_{E11} & s_{E12} & s_{E13} & s_{E14} & s_{E15} & s_{E16} \\ s_{E21} & s_{E22} & s_{E23} & s_{E24} & s_{E25} & s_{E26} \\ s_{E31} & s_{E32} & s_{E33} & s_{E34} & s_{E35} & s_{E36} \\ s_{E41} & s_{E42} & s_{E43} & s_{E44} & s_{E45} & s_{E46} \\ s_{E51} & s_{E52} & s_{E53} & s_{E54} & s_{E55} & s_{E56} \\ s_{E61} & s_{E62} & s_{E63} & s_{E64} & s_{E65} & s_{E66} \end{pmatrix} \begin{pmatrix} T_{xx} \\ T_{yy} \\ T_{zz} \\ T_{yz} \\ T_{xz} \\ T_{xy} \end{pmatrix} + \begin{pmatrix} d_{11} & d_{21} & d_{31} \\ d_{12} & d_{22} & d_{32} \\ d_{13} & d_{23} & d_{33} \\ d_{14} & d_{24} & d_{34} \\ d_{15} & d_{25} & d_{35} \\ d_{16} & d_{26} & d_{36} \end{pmatrix} \begin{pmatrix} E_x \\ E_y \\ E_z \end{pmatrix} \quad (7)$$

$$\begin{pmatrix} D_x \\ D_y \\ D_z \end{pmatrix} = \begin{pmatrix} d_{11} & d_{12} & d_{13} & d_{14} & d_{15} & d_{16} \\ d_{21} & d_{22} & d_{23} & d_{24} & d_{25} & d_{26} \\ d_{31} & d_{32} & d_{33} & d_{34} & d_{35} & d_{36} \end{pmatrix} \begin{pmatrix} S_{xx} \\ S_{yy} \\ S_{zz} \\ S_{yz} \\ S_{xz} \\ S_{xy} \end{pmatrix} + \varepsilon_0 \begin{pmatrix} \varepsilon_{rT11} & \varepsilon_{rT12} & \varepsilon_{rT13} \\ \varepsilon_{rT21} & \varepsilon_{rT22} & \varepsilon_{rT23} \\ \varepsilon_{rT31} & \varepsilon_{rT32} & \varepsilon_{rT33} \end{pmatrix} \begin{pmatrix} E_x \\ E_y \\ E_z \end{pmatrix} \quad (8)$$

2.2 *Principle of Piezoelectricity*

2.2.1 Mechanism of Piezoelectricity in Crystalline Material

The piezoelectric effect originates form the non-centrosymmetric nature of a crystalline structure of a material which enables the generation of electric polarization. Quartz and Berlinite are inherently occurring piezoelectric materials characterized by crystallographic structure devoid of a center of symmetry, thereby allowing them to exhibit piezoelectric traits [54]. Conversely, formulated materials like barium titanate (BT) and lead zirconate titanate (PZT) require a poling procedure to manifest piezoelectricity. Perovskite materials, with crystal orientation (ABX_3),also demonstrate outstanding responsiveness to poling which significantly enhances their piezoelectric characteristics [54, 55].

2.2.2 Poling

Poling is used to escalate material's piezoelectric features, subjected to elevated thermal condition and strong electric field [56]. Primarily, the material's dipoles are distributed in a disordered arrangement, resulting in no noticeable piezoelectric effect. With the application of an electric field exceeding the saturation point but operating under the critical breakdown field is applied in the crystal structure, the dipoles are oriented. The temperature is increased to sustained in the reorientation of the dipoles, through constraint to maintain temperature below curie point, to prevent any adverse effects. This alignment of dipoles establishes orientational dependency of the structure. To maintain the alignment, the substance is exposed to a high electric field during cooling [56, 57].

2.2.3 Bending Modes (d_{31} & d_{33})

The operation of piezoelecgric materials can be catagorised in two fundamental modes. The d_{33}– mode or longitudinal mode where the polarization direction aligns parallelly with the stress direction [56 yields a higher voltage when bending compared to d31, making it the preferred choice for applications like piezoelectric energy harvesting [56, 58]. Whereas the d_{31}–mode or transverse mode where the polarization and stress directions are orthogonal. The d33 mode (Fig. 2).

2.2.4 Figures of Merit (FOM)

In the selection of piezoelectric materials, it is crucial to evaluate the figure-of-merit (FOM) as a fundamental element contingent upon the designated applications [56, 58]. The three FOMs that holds significance for piezoelectric energy harvesters and sensors are enumerated as follows:

***Piezoelectric Charge/Strain Constant** (d)* Piezoelectric charge/strain coefficient (d) is defined as the interrelation between mechanical stress (T) and induced charge density (x). It also indicates the linkage between the strain (S) due to externally applied electric field (E). this figure of merit is an essential consideration in energy harvester applications [56–58].

$$d = \frac{X}{T} \tag{9}$$

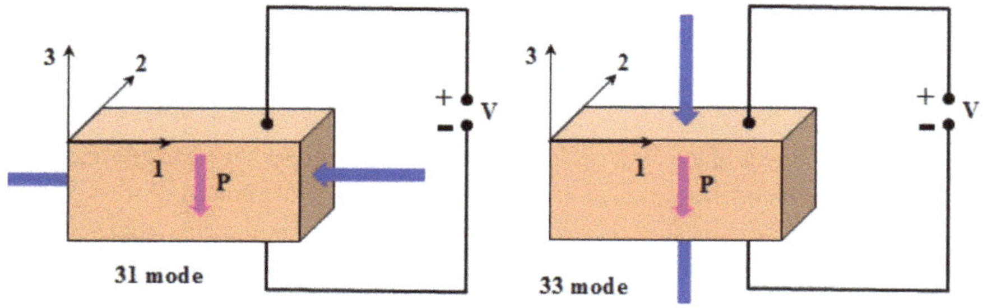

Fig. 2 Bending modes of piezoelectric material (reproduced from Ref [57] with permission from Elsevier)

$$d = \frac{S}{E} \tag{10}$$

Piezoelectric Voltage Constant (g) Piezoelectric voltage constant (g) is characterized by the relationship $\frac{E}{T}$, where E is the induced electric field and T is applied mechanical stress. This figure of merit is particularly significant when evaluating and comparing the performance of piezoelectric sensors [56, 58].

$$g = \frac{E}{T} \tag{11}$$

Electrotechnical Coupling Factor (k^2) The electromechanical coupling factor is formulated as a fraction of stored energy to the energy dissipated per cycle in a resonating system, both for the direct and indirect piezoelectric effect [58].

$$k^2 = \frac{\text{stored mechanical energy}}{\text{input electrical energy}} \tag{12}$$

$$k^2 = \frac{\text{stored electrical energy}}{\text{input mechanical energy}} \tag{13}$$

Mechanical Quality Factor The figure-of-merit (FOM) of any piezoelectric material utilized for high-power applications, which operate in proximity to their natural frequencies (eigenfrequencies), is defined as product between piezoelectric strain coefficient (d_{ij}) and quality factor (Q_m). This relationship exhibits a direct proportionality to both the eigenfrequency and the resultant output power [56]. Furthermore, the temporally combined thermal power loss per unit volume of the piezoelectric material is directly related to the dielectric and mechanical loss factors. The dielectric loss factor is predominantly significant at zero resonance, whereas the mechanical loss factor operates as the principal contributor to power dissipation at resonance. Material characterized by elevated mechanical quality factor and reduced mechanical losses yield minimal dielectric losses. Therefore, quality factor is an effective FOM applicable to high-load power system [58].

2.3 *Pyroelectric Effect*

Pyroelectric effect occurs in some of the polar materials due to change of spontaneous polarization caused by temperature variation [59]. It is a phenomenon under which pyroelectric materials produce electrical charges when subjected to thermal fluctuations. Some of the electrical dipoles are coinciding in the pyroelectric materials and display spontaneous polarization emerging perpendicularly from the untextured surface. The sustained intrinsic polarization of the pyroelectric materials attracts light weighted particles of positive and negative charges which are roaming nearby [60]. Due to electrostatic induction, when conductive electrodes are attached on the opposite interfaces of the pyroelectric surfaces, the internal spontaneous electric field will cause two electrodes to acquire equal charges separately with opposing electrical polarity through the external circuitry. During a positive rate of temperature change with time (i.e.; $\frac{dT}{dt} > 0$), the intensity of oscillation exhibited by the electrical dipoles are also increased which will dominate the fundamental electric polarization with the pyroelectric material and that causes the migration of electrical diploes to restore electrostatic equilibrium through charge redistribution in the external circuitry. In contrast, when $\frac{dT}{dt} < 0$, the spontaneously formed polarization will be increased and break the electrical equilibrium condition that leads to inverse electron migration in the pyroelectric material [59, 60] (Fig. 3).

2.3.1 Pyroelectric Coefficient

In pyroelectric materials, the pyroelectric coefficients demonstrate the capability of the pyroelectric materials to vary polarity and intensity of the dielectrics with respect to thermal fluctuations. The dielectric displacement (D) [60, 61] is established as:

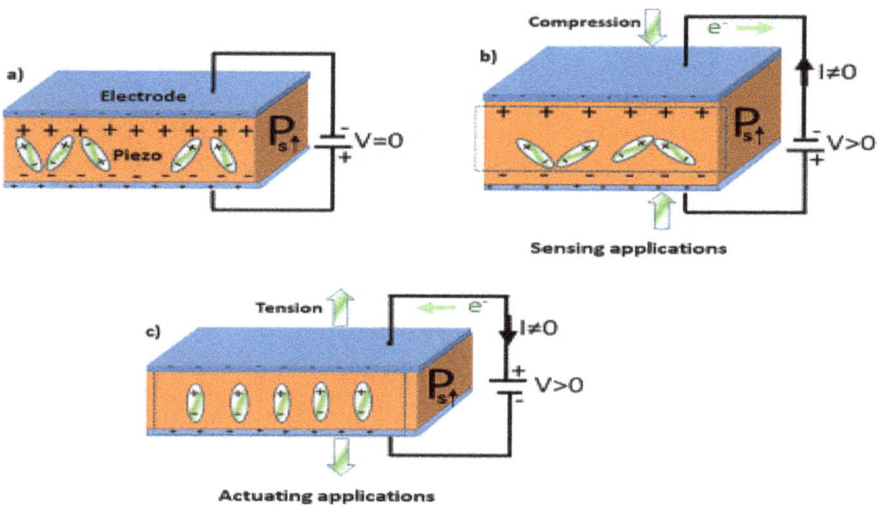

Fig. 3 Functional mechanism based on pyroelectric response (**a**) Spontaneous polarization arises due to internal electrical dipoles (**b**) sensing application (**c**) actuating application (reproduced from Ref. [59] with permission from Elsevier)

$$D = \frac{Q}{S} = \varepsilon E + P \tag{14}$$

Here, 'Q', 'S', 'E', 'P' and 'ε' represents pyroelectric charges, surface area of the electrodes, external electric field, polarization and permittivity respectively.

During simultaneous application of constant stress and uniform electric field, pyroelectric coefficient [60, 61] can also be represented as:

$$p = \frac{dQ}{dS} = \left(\frac{dP}{dT}\right)_{\sigma,E} \tag{15}$$

Where (σ, E) represents the constant conditions. We have considered the situation where the material's temperature denoted by 'T' is homogeneously distributed with stable pyroelectric properties. Again, pyroelectric current [60, 61] is given by:

$$i_p = \frac{dQ}{dT} = pS\left(\frac{dT}{dt}\right) \tag{16}$$

Where $\frac{dT}{dt}$ demonstrate the rate of temperature variation. Equation (16) is suitable for short circuit condition. In addition to that, the direction of the polarisation should be vertical to the surface of the electrodes. It is evident from Eq. (16) that the current generated due to pyroelectricity, maintains a linear relationship with the external surface area of the structure, and remain unaffected by variation in thickness and volume of the material (pyroelectric) as the pyroelectric current arises due to temporal variation of pyroelectric charge (Q), which is only distributed at the interface of the pyroelectric material. In accordance with the above situation, pyroelectric coefficient [61] can also express as:

$$p = i_p\left(\frac{dT}{dt}\right)^{-1} / S \tag{17}$$

So, it is clear from the discussion that pyroelectric coefficients are obtained by calculating the pyroelectric current, rate of temperature variation and surface area.

2.3.2 Figure of Merit (FOM)

Generally, there are majorly three different categories of figure-of-merit (FOMs), associated with the pyroelectric sensors based on their applications and requirements. Figure of merit [60, 61], using the current sensitive measurement, is given as:

$$FOM_i = \frac{p}{c} = p / \rho c_p \tag{18}$$

Where ρ and c_p are the density and specific heat capacity of the pyroelectric material respectively.

For voltage output response, FOM [60, 61] of the pyroelectric material is:

$$FOM_v = p / \rho c_{p\varepsilon_0\varepsilon_r} \quad (19)$$

Where, 'ε_0', and 'ε_r' stands for dielectric constant of vacuum and that of the pyroelectric material respectively. For any pyroelectric detector, its detection efficiency is a significant aspect in performance assessment. The conforming figure of merit [60, 61] is given as:

$$FOM_D = \frac{p}{\rho c_p \sqrt{\varepsilon_0 \varepsilon_r \tan \delta}} \quad (20)$$

Where tanδ indicates the dielectric loss. It is also evident from Eq. (20) that a situation with higher pyroelectric coefficient, low dielectric constant, low specific heat, and low dielectric loss are preferable to get highly sensitive pyroelectric detector. Along with these parameters, Curie temperature, cost etc. these factors should also be considered while designing pyroelectric detectors during manufacturing process [59–61].

3 PVDF-Material Properties

Material properties of PVDF have been the focus of widespread research throughout the years to improve the functional behaviors to achieve better performance. We will explore the fundamental properties of PVDF in this part of our chapter, and multiple processing method to understands the advantages as well as the disadvantages of optimizing PVDF performance (Fig. 4).

3.1 Physical Properties

PVDF (poly(1,1-difluoroethylene)) is a thermoplastic belonging to fluoropolymer family [62]. PVDF exhibits piezoelectric characteristics, meaning it can physically deform when an electric field is present. PVDF exhibits semicrystalline nature, characterized by crystallites that are comprised of elongate molecular chains that are arranged to facilitate dipoles, which are pivotal for its functional properties [63]. In comparison to other polymer materials, PVDF not only demonstrates extremely high piezoelectric coefficient, but it also emerges as an exceptional and effective candidate within piezoelectric polymers and delivers outstanding performance matrices in numerous applications [5]. PVDF showcases an outstanding

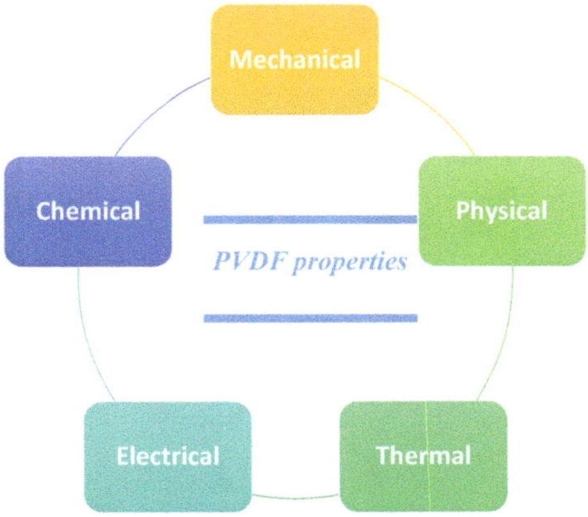

Fig. 4 Different properties of PVDF

electromechanical response and maintain relatively higher degree of stability. PVDF also possesses low acoustic impedance. These are crucial attributes of PVDF for its utilization in diverse technological contexts. PVDF exhibits piezoelectric coefficient in the range − 24 to −34 pCN^{-1} when subjected to applied electric field, which indicates a specific behaviour that is basic requirement for its application as a piezoelectric material in energy harvesters and sensing technologies. The negative nature of the piezoelectric coefficient justifies the fact that PVDF undergoes compression when exposed to applied electric field, irrespective of other piezoelectric polymers [64]. Despite the fact that PVDF displays significant level of crystallinity, estimated to be between 50–60%, it is noteworthy that the material continues to undergo expansion prior to be subjected to tensile force, therefore indicating a complicated relationship between amorphous and crystalline phases of PVDF [65]. PVDF is distinguished by its high melting point, which ranges between 170° to 180 °C, which signifies a thermal attribute of PVDF that permits the piezopolymer to endure substantial temperature fluctuations without experiencing significant degradation of functionality [64, 65] (Fig. 5, Table 1).

3.2 *Crystaline Phases*

The key factors that drive variation in the crystalline orientation pertaining to the functional group are intrinsic configurations and conformational rotation of the single bond of the molecules [7]. PVDF demonstrates polymorphism into five distinct phases: alpha (α) phase [7], beta (β) phase [8, 9, 12, 68], gamma (γ) phase [10, 11], delta (δ) phase [11], and epsilon (ε) phase (Fig. 3) [66]. Out of these phases α, β, and γ, are most prominent phases owning to their thermal, electrical, and mechanical properties [68]. Delta (δ) phase, and epsilon (ε) phase are harder compared to other phases and cannot produce through conventional methods [10, 11, 66]. The observable properties of PVDF are enumerated in Table 2 (Figs. 6 and 7).

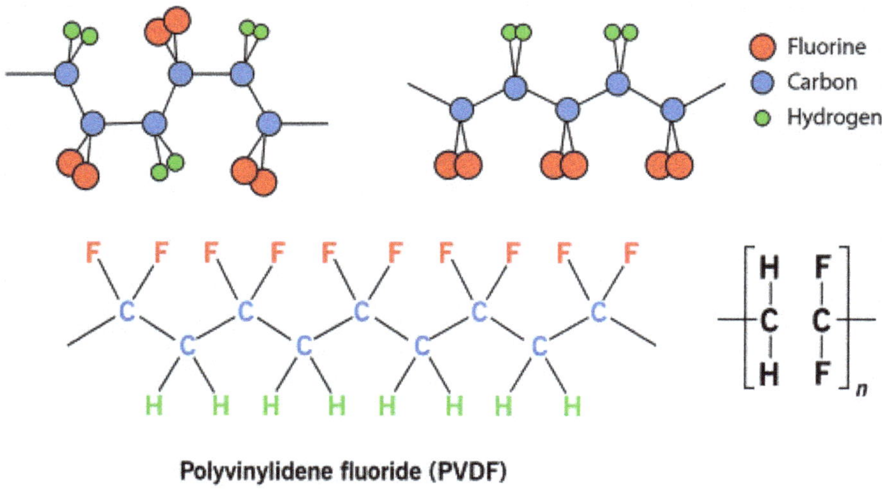

Fig. 5 Chemical structure of PVDF (reproduced from Ref. [63] with permission from Elsevier)

Table 1 Physical Properties of PVDF

Parameters	Value	Reference
Density (g/cm^3)	1.78	[62, 67]
Material color	Transparent without color	[62, 67]
Water absorption (%)	0.02–0.07	[62, 67]
Refraction index	1.40–1.42	[62, 67]
Oxygen index (%)	42–44	[62, 67]

Table 2 Properties of different crystalline phases of PVDF [7–12, 66, 68]

Crystalline phases	α phase	β phase	γ phase	δ phase
Polarity	Non-polar	All-trans polar	Partially polar	Polar (related to α phase)
Dipole orientation	Dipoles cancel out	Dipoles aligned in same direction	Partial dipole alignment	Dipoles aligned under electric field
Piezoelectricity	–	High	Low to moderate	Moderate
Ferroelectricity	–	Strong	Weak	Moderate
Dielectric constant	~8–10	~12–13	~10–11	~12–13
Crystal structure	Trans-gauche TGTG′	All trans polar	Trans-gauche TTTT, TTTG-TTTG′	Polar (related to α phase)

3.2.1 α Phase

When the polymer transits from its molten state to a solidification form, it attains most scientifically recognized as its most resilient structural phase, designated as α-phase. Nevertheless, despite having reached this advanced stage of solidity, the polymer retains a dipole moment that is intrinsically lined to its molecular

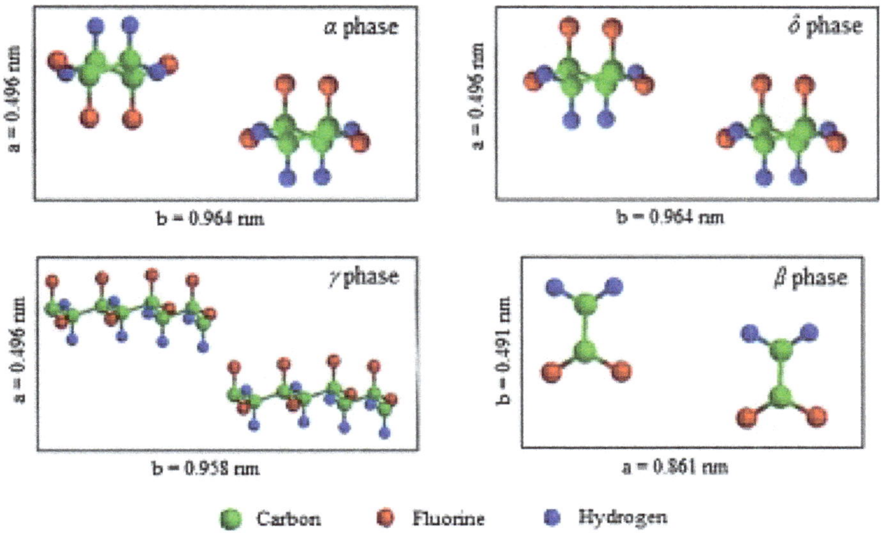

Fig. 6 Crystalline phases of PVDF (**a**) α-phase (**b**) β-phase (**c**) γ-phase (**d**) δ-phase (reproduced from Ref. [66] with permission from Elsevier)

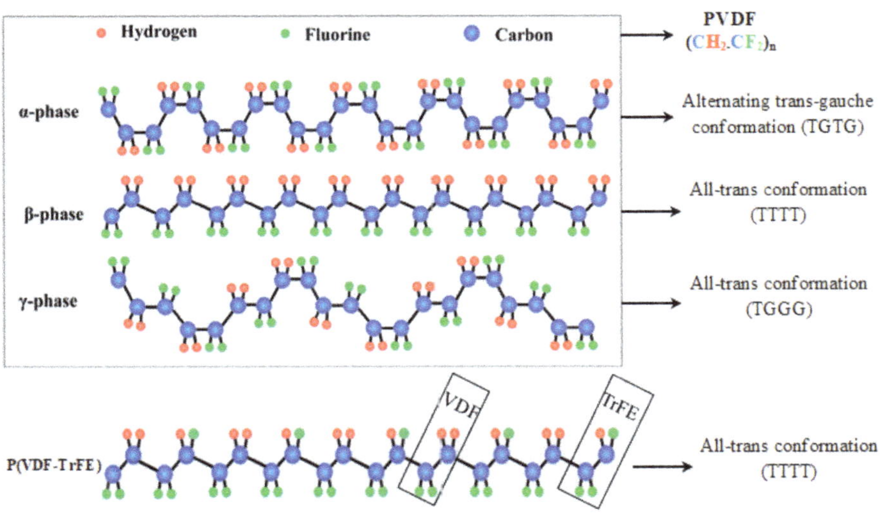

Fig. 7 Visualisation of α, β, γ phases of PVDF with their crystal structure and PVDF-TrFE (reproduced from Ref. [57] with permission from Elsevier)

architecture, a property arises from the sequential placement of fluorine and hydrogen atoms on the carbon backbone [69]. α-phase of PVDF is non-polar and resulting from the absence of dipole moment, it exhibits non-piezoelectric behaviour [70]. Fluorine atoms are arranged in an anti-parallel mannerism throughout the chains which neutralized the entire configuration. Also, the α-phase lacks the ability to demonstrate electrical behaviour caused by unaligned dipolar configuration, thus it is non-polar. This phase is prevailing when the PVDF is crystalized up to temperature 160 °C [7, 69].

3.2.2 β Phase

β-phase of polyvinylidene fluoride (PVDF) is among the foremost and technically relevant crystalline form of PVDF for its all-transitional molecular orientation (TTTT), which gives it remarkable piezoelectric and ferroelectric abilities [67, 71]. The β phase is strongly polar due to the net polarization that occurs as the result of dipole moments of the $-CH_2-$ and $-CF_2-$ groups aligned in the same direction along the polymer chain axis of PVDF [72]. The β phase is metastable. This phase of PVDF mainly arises through post processing method such as solvent-induced crystallization, electric field poling, and mechanical stretching [7, 72]. However, it is imperative to achieve a considerable degree of crystallization at an earlier stage in order to achieve β-phase transformation to commence effectively and subsequently proceed to its desired state. In order to initiate the β-phase transformation successfully, it is essential that PVDF undergoes a poling process prior to the commencement of the aforementioned phase transition [57]. The incorporation of various group imperfections within the molecular structure of PVDF, beyond improving the piezoelectric attributes, contributes as a catalyst that facilitates the transition to the β-phase. A widely recognised technique employed to include transformation from α to β phase, involves the stretch-induced PVDF films, which is optimally performed at elevated temperature from 70–100 °C, while applying a stretch ratio that typically falls within the range of 3–5 [65, 72, 73]. Higher stretch ratio (>5) implies greater conversion efficiency but temperature beyond 120–140 °C prevent the stabilization of β phase [74]. β phase development is also favoured during solvent evaporation in lower temperature (<70 °C) when dimethylformamide (DMF) or dimethylacetamide (DMA) is used for solvent processing [74]. Evaporation at temperature above 110 °C combine the α and β phase or strong for α phase [72]. Induced electric field of 0.5–0.8 MV/cm can also produce β phase with the help of corona discharge or heating which predominates in setting the electric field strength needed for polarization [5, 75]. In implementations like sensors, actuators, and energy harvesters, PVDF with the β-phase content is essential for better functional performance [5, 7, 68]. To maximize its development, temperature, mechanical stress, solvent evaporation, and electric field conditions must be carefully controlled [5, 62, 73–75] (Fig. 8).

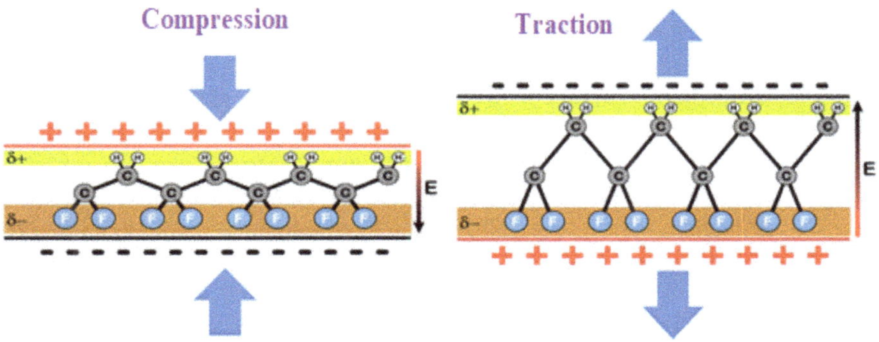

Fig. 8 Polarization in β-phase PVDF (reproduced from Ref. [57] with permission from Elsevier)

3.2.3 γ Phase

The γ phase of PVDF results due to delayed cooling of molten PVDF closely to the slightly below the melting temperature of α phase of the same, or from overheating the remaining phases (α, β, and δ phases) [73]. This particular phase can also be created by annealing α phase of PVDF at 175–185 °C [7, 73]. Solution crystallization or from molten state near 160 °C, γ phase of PVDF can be produced [76]. An alternative method is to melt PVDF at 175 °C and dissolve it in dimethylformamide (DMF) to create thin films and then continue by heating that film at 200 °C for 10–15 min to remove thermal impurities [76]. The γ phase of PVDF can maintain significant polarization while exhibiting several processing flexibilities compared to β phase, which demonstrates variety of application in piezoelectric sensors, capacitors, along with the optoelectronic devices [75, 76].

3.2.4 δ Phase

The δ phase of polyvinylidene fluoride (PVDF) is a polar crystalline form similar to non-polar α phase, but it occurs when the α-phase is exposed to strong electric field inducing dipole movement without exchanging the basic crystal structure [77]. This phase can be engineered through several processing method where nanoprecipitation is one of the common methods. In this technique, PVDF is dissolved in a polar solvent (i.e.; acetone) under continuous stirring along with another solvent which is injected from the bottom of the phase separation, forming δ phase at the interface [11, 77]. These δ phase particles are then collected by centrifugation followed by heat treatment at 80°C for almost 6 hours. The δ phase of PVDF demonstrate linear polarization electric filed behaviour at low induced electric field and effective polarization at high electric field. These characteristics of the δ phase of PVDF like structural stability, and polar response make it appropriate fir MEMS application, precision dielectric device applications [78].

3.3 Chemical Properties

How well a substrate dissolves is determined by its chemical and physical traits along with some of the external variables (pressure, temperature, etc.) and interactions with other substances in the solution medium. PVDF dissolves in a limited number of organic solvents: dimethylformamide (DMF), dimethyl sulfoxide (DMSO), N-methyl pyrrolidone (NMP), dimethylacetamide (DMAc) [17, 79]. Also, PVDF is insoluble in alcohol, strong acids, halogens, aromatic compounds, and chlorine solution [80]. PVDF copolymers are more soluble than pure PVDF due to their low crystallinity [80]. While PVDF is renowned for its chemical stability in extreme environments, it is incompatible with ketone, esters, and concentrated alkalis such as NaOH, which can result in discoloration of PVDF membranes [81]. This degradation can cause dehydrofluorination that forms carbon-carbon double bond that causes weakening of polymer chains. In addition to NaOH, the solubility of PVDF has also been tested with sodium hypochlorite (NaOCl) and potassium hydroxide (KOH) [82]. It is clear from the various researches that PVDF demonstrates prominent chemical stability, it is essential to practice caution when exposed to strong alkaline solutions. The chemical characteristics of PVDF are enlisted in Table 3.

Table 3 Solvent compatibility of PVDF [17, 38, 67, 68, 79–83]

Chemical type	Chemical	20 °C	50 °C
Organic solvent resistance	Acetone	–	–
	Benzene	Yes	Limited
	Chloroform	Yes	Yes
	Dimethylformamide	–	–
	Trichloroethene	Yes	Yes
	Xylol	Yes	Yes
Alcohol resistance	Ethanol (30%)	Yes	Yes
	Methyl alcohol (10%)	Yes	Yes
	Phenol (10%)	Yes	Yes
	Methyl alcohol (pure)	Yes	Yes
	Phenol (100%)	Yes	Yes
	Methanol	Yes	Yes
Food product resistance	Glucose	Yes	–
	Milk	Yes	Yes
	Olive oil	Yes	Yes
	Wine	Yes	Yes
	Vinegar	Yes	Limited
	Coconut oil	Yes	Yes
Acids and bases resistance	Formic acid (10%)	Yes	Yes
	Acetic acid (10%)	Yes	Yes
	Sulfuric acid (10%)	Yes	Yes
	Nitric acid (10%)	Yes	Yes
	Hydrochloric acid	Yes	Yes
	Lactic acid	Yes	Limited

3.4 Mechanical Properties

Owning to its versatile mechanical properties, PVDF is widely employed in various industrial and scientific implementations, that encompasses high tensile strength, moderate stiffness, and significant elongation at break impact strength, etc. Escalated tensile strength stipulates that PVDF has better tearing and breaking resistance compared to other piezoelectric polymers. PVDF can withstands significant amount of pulling force before breaking [83]. This property of PVDF makes it suitable for functional areas demanding resistance to significant mechanical deformations (e.g., in structural assemblies or surface potential layers) of the film. PVDF also exhibits greater deformation sufferance prior to rupture that means it is a flexible polymer that maintains structural integrity when subjected to mechanical strain [84]. This ability makes PVDF suitable for applications which require bending and flexibility [67]. Young modulus of a material defines the measurement of stiffness of that material. PVDF is stiffer material with comparatively high young modulus [83]. Additionally, PVDF exhibits outstanding impact resistance so that it is capable for withstanding shock. PVDF also possesses better fatigue resistance which also justifies that PVDF can resist a cycle of stress for a longer time without noticeable degradation [62]. PVDF also demonstrates good creep resistance which means it can maintain its shape and structure under continuous load. This mechanical property is required for environments demanding for long-time consistency of shape and size under mechanical stress [85]. PVDF has high melting point so it can withstand temperature variation without significant amount of deformation in its shape, size, and dimensions [83–85]. The factors that have impact on the stress-strain responses of PVDF are: polymer chain length, manufacturing approaches, post-processing treatments, etc. The mechanical characteristics of PVDF piezopolymer are summarized in Table 4.

3.5 Thermal Properties

Thermal stability of any polymer suggests its high thermal resistance while retaining mechanical strength, dimensional stability, structural durability and flexibility. The glass-transition threshold and phase transition temperature are two key thermal

Table 4 Mechanical characteristics of PVDF [62, 67]

Parameters	Value
Tensile strength (MPa) [*at 23 °C]	35–55
Hardness	50–80
Young modulus (MPa) [*at 23 °C]	1340–2000
Permittivity in vacuum (F/m)	$\varepsilon_0 = 8.854 \times 10^{-12}$
Elastic constant (N/m^2)	$C_{11} = 2.184 \times 10^9$ $C_{12} = 0.633 \times 10^9$

Table 5 Thermal properties of PVDF [62, 67]

Parameters	Unit	Value
Melting point	°C	160–190
Defection temperature	°C	114–118
Decomposition temperature	°C	375
Flammability	°C	V-O
Glass transition temperature	°C	−20 to 60
Thermal expansion coefficient	μm/m-°C	10^{-4}
Thermal conductivity	W/m-K	0.144–0.2

properties that influence its thermal stability. The glass-transition temperature falls between −20 to 60 °C making it a slurry polymer above 0 °C [64]. The phase transition temperature (solid-to-liquid) of PVDF are between 160–190 °C which depends on the crystalline phases and processing methods. Within −30 to 170 °C, PVDF has remarkable thermal stability. The temperature of heat-induced deterioration of PVDF also remain constant in the range 400–450 °C which varies as per the crystalline phases along with the processing techniques. Curie temperature lies between 195 °C to 197 °C as per the previous researches. Reduced heat transfer diffusivity of PVDF makes it applicable for insulation [64, 83] (Table 5).

3.6 Electrical Properties of PVDF

See Fig. 9.

3.6.1 Piezoelectric Properties

In order to exemplify piezoelectric properties, it is essential for a polymer that it incorporates molecular dipoles within its structural framework [56]. A piezoelectric structure may demonstrate either an amorphous or a semicrystalline configuration [58]. Semicrystalline piezoelectric polymer like PVDF and its copolymers, consist of ferroelectric crystalline domain that are embedded within an amorphous matrix [86]. Upon the application of poling to these polymers, their dipolar configurations undergo a recognition and alignment process that facilitates the manifestation of piezoelectric qualities [87]. Amorphous polymer can enhance the presence of dipoles but they do not maintain these dipoles across a higher order range within the structure. Piezoelectric material generates an electrical charge as a result of applying mechanical stress (Fig. 1). This charge may be harnessed in different applications like sensors, transducers, and energy harvesters [52]. Whereas, in inverse piezoelectric effect phenomenon, the material undergoes mechanical deformation due to applied electric field. This inverse effect can be utilized as the basis for an electromechanical actuator [52, 55] (Table 6).

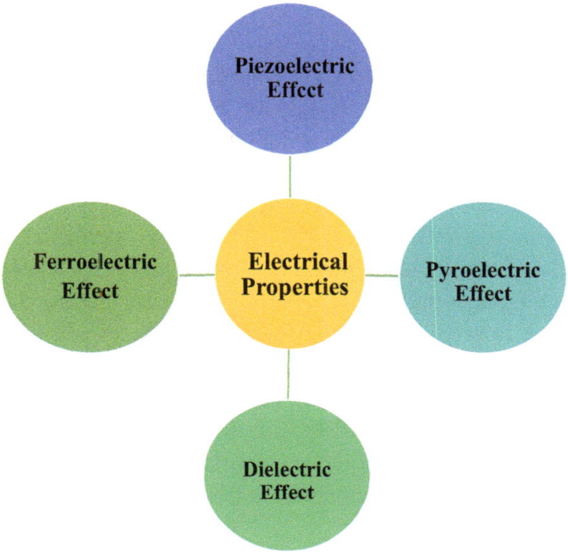

Fig. 9 Different electrical properties of PVDF

Table 6 Piezoelectric properties of PVDF [62, 67]

Parameters	Value
Curie temperature (°C)	80
Strain constant (10^{-12}C/N)	$d_{31} = 20 - 30$ $d_{33} = -30$
Stress constant (10^{-3}mV/N)	$g_{33} = 335$

3.6.2 Pyroelectric Properties

The pyroelectric phenomenon is attributed to the intrinsic polarization of the polymeric architecture of PVDF, predominantly in β-phase, wherein the dipolar entities are oriented along the molecular chains [59, 60]. This characteristic is pertinent for applications in thermal detection and energy-harvesting systems [61]. The pyroelectric response of PVDF is significantly augmented by phase transition methods like mechanical stretching or electrical poling which precisely alter the dipolar orientation within the molecular structure. This enhances the sensitivity of the PVDF polymer to the fluctuation of temperature [88].

3.6.3 Dielectric Properties

The rate of dissipated energy of PVDF is called dielectric loss denoted by tanδ. This rate quantifies the amount of lost energy in dielectric material under applied electric field. The dielectric loss is impacted by the frequency and integration of nanofillers, and lies within the range 0.04–0.11 depending on the samples [89].

3.6.4 Ferroelectric Properties

The ferroelectric effect within PVDF occurs due to dipole moments within the molecular structure mainly in β-phase PVDF induced by strong electronegative fluorine atoms [84]. These dipoles align along the molecular chain in response to the induced electric field so that PVDF can create hysteresis loop which is a key characteristic of ferroelectric effect. The ferroelectric features of PVDF were discovered in the year 1974 [90]. The ferroelectric properties of PVDF also enhances the piezoelectric attributes within it by enabling it to retain polarization even the electric field is no longer present. PVDF films those are produced from molten or dissolved state, have fewer ferroelectric properties as the dipole moments are removed due to the crystal structure. Mechanical stretching or copolymers of PVDF [polyvinylidene fluoride-trifluoro ethylene (PVDF-TrFE)] are important to induce ferroelectricity within PVDF. When PVDF is heated over curie temperature, the material transits from β (ferroelectric) to α (paraelectric) phase. PVDF maintain spontaneous polarization below this temperature [91].

4 PVDF Fabrication Methods

PVDF based piezopolymer membranes, or thin films are widely recognized for their versatile materials which are used in multiple applications. These films can form piezoelectric structure quite easily with inducing β phase formation through low stress mechanical stretching and comparatively lower voltage to retain permanent electric polarization [67]. These PVDF films are suitable for a wide range of electromechanical devices because of their lower cost of fabrication and operation. PVDF may be synthesized using multiple techniques to get α and β phase PVDF materials. Some of the widely used conventional polymer fabrication processes are: (a) solution casting, (b) spin coating, (c) solution coating, (d) hot pressing, (e) electrospinning, and (f) 3D printing. Moreover, solution casting techniques like: electrospinning, solvent evaporation, etc. have received much attention due to their requirements such as nano-scale thickness of PVDF films in numerous applications [62, 76].

4.1 Solution Casting

PVDF and its copolymer films can be fabricated effectively using solution casting fabrication process. This fabrication method produces remarkable films with outstanding mechanical, physical and optical attributes through coplanar distribution. Dimethylformamide (DMF), methyl ethyl ketone (MEK), acetone, dimethyl sulfoxide (DMSO), dimethylacetamide (DMAc) can diffuse PVDF and its copolymers [92]. These dissolution of the PVDF and its copolymers are processed in room temperature but sometime high temperature requirements (<70 °C) have been observed. Ultra-sonication-added mixing techniques are often used in dissolution. Ultra-sonication time is less than 30 min in major cases but sometimes it requires operation time of 3 h. After evaporation process, the PVDF along with the mixed organic-based dispersion is overlaid with the aluminum (Al) substrate or glass to fabricate the initial PVDF layer. Temperature of the evaporation process lies in the range of 50–120 °C [93]. Some of the advantages of solution casting are: (a) solution casting is a cost-effective process of fabrication; (b) this technique can control the thickness of the PVDF films; (c) preparation process is quite simple compared to other fabrication process. Although, solution casting method has some drawbacks over the technique as: (a) critical solvent selection, (b) limited scalability, (c) post treatment steps are required in this fabrication method [92, 93]. However, solution casting of PVDF, and PVDF-TrFE (copolymer of PVDF) is among the most commonly employed methods of creating films or membrane due to its control throughout the physical properties and crystalline morphology of the synthesized material [92, 93].

4.2 Solution Coating

Solution coating is commonly employed to engineer composite membranes through material depositing a very thin polymer solution layer onto a porous base layer. In order to maintain a compatible structure that exhibits higher porosity compared to resistance of the membranes to electrical transport, measures can be taken before selecting the coating layer. Before the coating process, a lower evaporation temperature of solvent which does not dissolve in coating formulation, is deposited onto the substrate to reduce the invasion [94]. Upon solvent evaporation, a membrane is formed onto the substrate. Solution coating method is compatible with variety of substrates. This method allows precious control over deposition techniques. This technique requires optimization to achieve uniform film thickness. Also in this method, the physical or other properties of the selected material get affected because of solvent selection and drying process. Solution coating method also requires post treatment step like solution casting technique [95].

4.3 Electrospinning

Electrospinning is another famous fabrication technique which create PVDF ultrathin fiber-based films. Electrospinning comprises of the following steps: preparation of the polymer solution, application of the high voltage field, jet formation, jet initiation and elongation, and lastly removal of solvent through evaporation [2, 3]. Imposition of an elevated electrical potential to the polymer solution induces the establishment of jet, then solidifies into nano-fibers due to the interaction with the

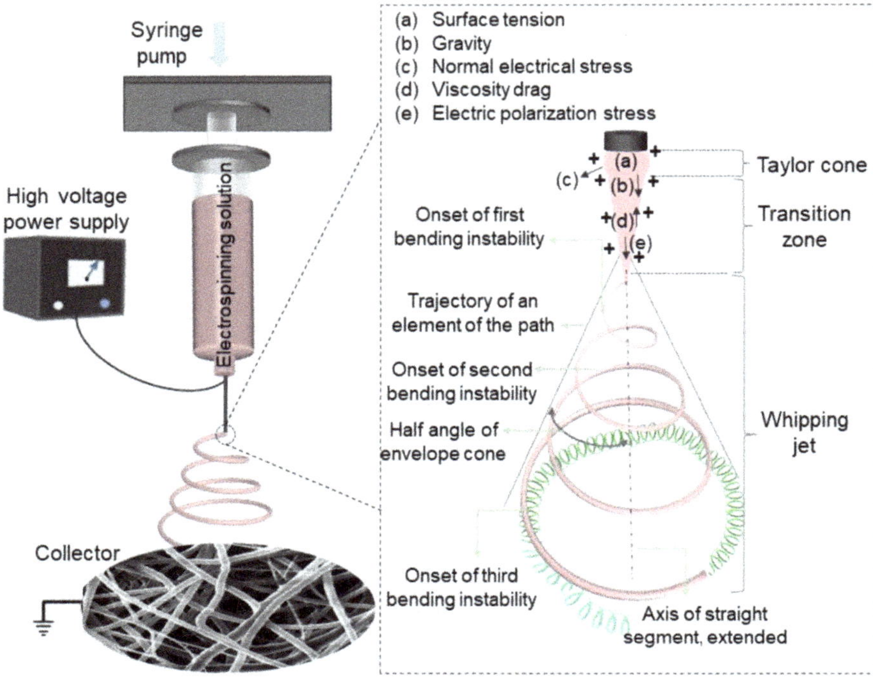

Fig. 10 Diagrammatic representation of the electrospinning setup and working principle (reproduced from ref. [94], under the Creative Commons Attribution license [CC BY], Elsevier)

collector [94]. Electrostatic force is generated across the electrically charge polymer solution and the collector surface (grounded) which creates nano-to-microscale fibers. Electrospinning has some advantages over other fabrication methods of PVDF polymers because in this method, PVDF molecular main elongates oriented along the length of the fiber-axis creates the β-phase PVDF structure. PVDF can exhibits piezoelectric features without post-poling due to elongational forces and whipping instability in polymer contribute to the poling of the piezoelectric polymer [68, 94]. Electrical force can conclude the dipole alignment as well as the stretching in electrospinning method. Thermal treatment should be administered to PVDF during or post electrospinning process to tailor the material's structure and properties. Mechanically robust cross-linked membranes can also be fabricated through this fabrication approach [95] (Fig. 10).

4.4 Spin Coating

The spin coating fabrication method is a productive, cost-effective and swift technique which is used to prepare thin film of PVDF. In this technique, uniform polymer film is processed from dilute polymer solution deposited on a flat substrate

[96]. Spin coating is appropriate for preparing well-ordered polymer coating for microelectronic device integrations. The final width of the resulting film governed by the basic features of the solvent-based solution and the spinning speed. Humidity or rotational motion, can be utilized to analyze the surface profile of the thin films [97]. These thin films are non-poled α-phase PVDF. These PVDF films requires further polarization to elevate the piezoelectric properties of PVDF. Spin coating methods have more favorable attributes when they are incorporated with microscale devices built on silicon wafer substrates. The scalability of this technique is limited owning the limited scale of spinning apparatus. This technique is appropriate for research-oriented or laboratory scaled applications [96, 97].

4.5 3D Printing

3D printing fabrication technique (Fig. 6) allows to create precise shape of the PVDF film using layer by layer deposition. PVDF samples of various forms can be produced using Fused Deposition Modelling (FDM) by raising the thermal environment of the nozzle above the melting point of PVDF material. Fused Deposition Modelling (FDM) can incorporate β-phase content via stress. Many of the previous researches undertook strategies to promote β-phase by optimizing the printing parameters. Polymer of PVDF, PVDF-TrFE with elevated concertation of ferroelectric phase, can be added to the optimization process. Post-processing of poling is important to maintain dipole alignment to achieve microscopic piezoelectricity after retaining higher β-phase content. Sebastian et al. [98] engineered Integrated 3D Printing and Corona Poling (IPC) which enhanced the β-phase content up to 56.8%. Lee et al. [99] developed electric-poling-assisted-additive-manufacturing-process (EPAM) which employed electrical poling with mechanical stretching to convert α-phase PVDF into β-phase PVDF. This approach elevates the content of β phase of PVDF materials but it still a challenge to improve piezoelectric characteristics. Substitution of PVDF with PVDF-TrFE improve crystallinity around 85.8% and d_{33} of $-18pcN^{-1}$ (in-situ polarized PVDF-TrFE). Yuan et al. [100] demonstrated six-layered PVDF-TrFE with d_{33} of $-130\,pcN^{-1}$. 3D printing fabrication techniques are promising candidate for producing cost-effective PVDF based piezoelectric material with increasing β-phase content [62] (Fig. 11).

4.6 Hot Pressing Method

Hot pressing method of PVDF fabrication is gaining attention due to its ability to improve crystalline phases of PVDF mainly the transition from the α-phase to β-phase. This technique can produce denser film of PVDF with towering β-phase content that can improve piezoelectric attributes of the polymer with superior mechanical performance [101]. The hot-pressing method of PVDF film fabrication

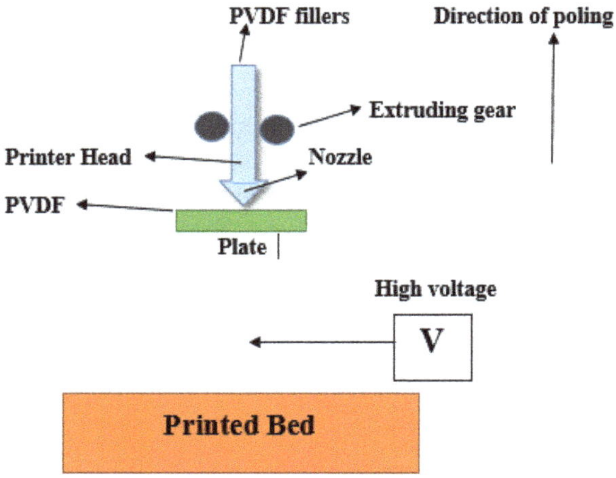

Fig. 11 3D printing mechanism of PVDF film (redrawn from ref. [62])

starts with preparation of PVDF powder substrates. The material is melted at the temperature from 150° to over 200 °C with maintaining the melting of the material without decomposition [101]. High pressure around 5–40 MPa is applied to the PVDF powder to compress it and transform it in to continuous film without void places and with increased density [101]. After the temperature and pressure treatment, the PVDF films are cooling down and solidification of the polymer into the required crystalline phase are done. Stretching and electrical poling may be further integrated into the process to enhance the piezoelectric behaviors of the polymer [101, 102]. Previous researches have demonstrated that the rising temperature during hot pressing fabrication technique regulates the crystal size of the PVDF polymer. It has been manifested in a study that the average crystallite dimension of PVDF has grown (7.2 to 20.54 nm) with increasing consolidation temperature (170 °C) [102]. The hot-pressing method can also enhance the dielectric features of PVDF polymer. At 150 °C, films can achieve high energy densities near $19.24 J/cm^3$ with 68.99% efficiency. The temperature and pressure play very important role in the hot-pressing technique to engender PVDF films with high crystallinity and phase composition. Huge amount of great-quality commercial PVDF films are produced using hot-pressing fabrication techniques which is amalgamated with elevated temperature pressing to synthesize the PVDF film, mechanical elongation to enhance the β phase composition and subsequent annealing to augment the crystallinity of PVDF [101, 102] (Table 7).

Table 7 various techniques for the fabrication of PVDF along with their fundamental parameters [100]

Fabrication method	Parameters
Solution casting	Concentration of PVDF: 10–20 wt% Processing temperature: 25–80 °C
Solution coating	Concentration of PVDF: 5–15 wt% Processing temperature: 25–60 °C
Electrospinning	Output voltage: 10–30 kV Rate of discharge: 0.5–2.0 mL/h
Spin coating	Speed: 1000–5000 rpm Processing time: 30–60 s
3D printing	Height of each layer: 0.1–0.4 mm Speed: 10–100 mm/s

5 Application of PVDF-Based Sensor

5.1 Biomedical Applications

PVDF exhibits outstanding piezoelectric characteristics, mechanical rigidity, thermal stability, and impermeability to multiple elements like acid, base, oil, and fats [67, 68, 83, 84]. These properties make PVDF applicable for far reaching supplications in medicine, optics, nanotechnology and many more. PVDF based sensors can be used to manufacture artificial ligaments, multifilament for vascular grafts, artificial corneas. Alternatively, PVDF membranes can be utilized in immunoblotting to neutralize protein compositions because of how well they bind to amino acid. PVDF has been utilized in the preparation of membranes that has been incorporated in mask during COVID 19. PVDF has multiple application depending on its crystalline or copolymerised configurations. PVDF based sensors can be used to simulate extracellular environment. These sensors are used in processing of physiological activities to generate electrical signals to generate artificial bone [103], nerve [104], muscle [105]. PVDF based sensors have great utilization in tissue engineering also. A PVDF-TrFE/BaTiO_3 nano-particle composite film was prepared by Genchi et al., [106] which demonstrate better piezoelectric characteristics than pure PVDF-TrFE. The piezoelectric activities pertaining to the composite-film assemblies can also be enhanced with ultrasonic stimulation. The length of axon and the quantity of β3-tibuline-positive cells of PVDF-TrFE/BNTP composite cells could also be elevated, so these films are applicable in wireless neuron stimulation devices. To optimize the interface compatibility between BTO nano-particle and PVDF-TrFE composites, polydopamine was incorporated within BTO nano-particles which delivered better piezoelectric attributes than PVDF-TrFE/BaTiO_3 (barium titanate) nano-particle composite film [107]. It has been reported that these composite films are used in -vitro-osteogenic behaviour of bone-marrow derived mesenchymal stem cells (BM-MSCs). These can further felicitate the constant preservation for extensive healing of bone imperfections. Damaraju et al. [108] prepared three-dimensional fiber scaffold using electrospinning for tissue formation within mesenchymal stem cells (MSCs). Cheng et al. [104] prepared a cellular nerve tissue engineered scaffold using degradable material such as polycaprolactone (PCL) and PVDF by annealing solvent replacement method. Scaffold with high elasticity modulus prepared by electrospinning is comparatively exhibit strong mechanical strength. PVDF composite prepared with PVDF (10 wt%) and $Bi_4Ti_3O_{12}$ with dimension of 40 μm, provide maximum generated potential (41 V) and highest power density (4.3 Wm^{-2}), can be used as portable and wearable self-generated biochemical sensor in health care monitoring industries [109]. PVDF and polyethylene glycol (PEG), ZnO nanoparticle composite works in large relative humidity (19–92%) with high sensitivity (0.7 pF/RH%), has a very quick response time (2S) [110]. This composition is used in wearable humidity sensor coupled with mask for breath monitoring. PVDF-$CaTiO_3$ (calcium titanate-8 wt%) composite has output voltage of 20 V, current in the range 250 nA, and power density of 0.19 μWcm^{-2} at 108 Ω, is utilized in PVDF nanogenerators (PENG) for human health monitoring that prevent major chronic attack condition [111]. The stacked layer of polyamide (nylon 66) and PVDF nanofibers demonstrate self-charging because of the triboelectric interactions and piezoelectric attributes of PVDF. The study of this composite showed an optimal filtration performance for triple-layered mask [112]. PVDF-chitosan (CS-2 wt%)-hydroxyapatite (Hap-8 wt%) nanocomposites, which has great mechanical strength (young modulus: 889 MPa), is used as smart implantation for veterinary orthopedic area [113]. Composite prepared with PVDF membrane and single walled carbon nanohorns (CNHs), treated with plasma which enhanced the hemocompatibility of the PVDF nanocomposite, is used as implants for cardiological and circulatory system [114]. PVDF-ZnO nanocomposite prepared with acetone and DMF solution was subjected to magnetic stirring at 55 °C for 12 hours, has enhanced antibacterial activities and stimulated osteoblastic development which is majorly incorporated with orthopedic application [115]. The nanocomposite produced with PVDF, caesium (CS), silica nanoparticles (MCM41), and polyvinyl alcohol (PVA), has great potential for wound dressing in response to electrical voltage, antibacterial activities, cell-adhesion, and hydrophilicity [116]. The addition of PVDF with polyhydroxybutyrate (PHB) and caesium (CS) nanocomposite, improves tensile strength which is used for wound dressing of post-surgical ulcers [117] (Table 8).

5.2 *Acoustic Sensing*

The construction of Sound detection and recognition sensors (SDR) basically depends on the electromagnetic, electrostatic, electrical methods. SDRs have drawbacks due to their huge dimensions, integrated power source, and low electro-acoustic conversion efficiency [118]. Whereas, PVDF based sensor designs have better features like smaller size, flexibility, high electro-acoustic conversion efficiency, self-powered mechanism, inflated signal-to-noise ratio (SNR) which can be great selection for the latest SDR requirements [118]. Acoustic sensors require the

Table 8 Biomedical application of PVDF-based sensors

Composition	Application	Ref.
PVDF-TrFE + $BaTiO_3$	Wireless neuron stimulation devices with enhanced piezoelectric response	[107]
PVDF-TrFE + $BaTiO_3$+ polydopamine +BTO	Enhanced osteogenic behaviour of stem cells, bone healing and regeneration	[105]
PVDF electrospun 3D fiber scaffold	Tissue formation of mesenchymal stem cells (MSC)	[108]
PVDF + Polycaprolactone (PCL)	Cellular nerve tissue engineering scaffolds	[104]
PVDF-TrFE + BNTP	Wireless neuron stimulation devices	[106]
PVDF (10 wt%) + $Bi_4Ti_3O_{12}$	Self-powered portable and wearable biochemical sensor	[109]
PVDF+ PEG+ ZnO	Wearable humidity sensor coupled with mask to measure breath	[110]
PVDF-$CaTiO_3$ (8 wt%)	PVDF nanogenerators in human health monitoring applications that prevent major chronic attack	[111]
Nylon 66 + PVDF	Self-charging filters for triple-layered masks	[112]
PVDF-chitosan (CS-2 wt%)-hydroxyapatite (8 wt%)	Smart implantation for orthopaedic areas (veterinary)	[113]
PVDF+ (CNHs)	Cardiological and circulatory implants with enhanced hemocompatibility	[114]
PVDF+ ZnO	Antibacterial orthopedic implants, enhanced osteoblast growth	[115]
PVDF + CS + silica (MCM-41) + PVA	Electrically-responsive wound dressing with antibacterial and cell-adhesion properties	[116]
PVDF + PHB + CS	Wound dressing for post-surgical ulcers with improved tensile strength	[117]

sensing material that can deliver better flexibility and higher sensitivity. PVDF nanofibers fabricated using electrospinning method can vibrate easily under sound wave. These nanofibers can differentiate sound waves of different frequencies [119] (Fig. 12).

These kinds of sensors demonstrate high sensitivity to sound above 100 dB. When sound frequency is 220 Hz, the maximum voltage output of these sensors can be utilized for noise detection. A nano-resonant acoustic sensor is designed using electrospinning fabrication approach. This acoustic sensor has a wider frequency spectrum, great resolution, and better fidelity. This sensor may be utilised in voice recognition and recording system. The frequency ranges from 200 to 5000 KHz which is the most prevalent audible acoustic spectrum observed in our daily life, can be recognised by the acoustic sensor [118, 119]. Due to high permeability of the nanofibers, this sensor is getting recombined with optical as well as gas sensors to gather multidirectional data within the same crystalline arrangement. To obtain highly sensitive PVDF-TrFE film, ionic liquid in-situ electric field polarization is incorporated, which may identify as well as differentiate complicated sound waves, such as musical notes of different frequencies [120]. These sensors have numerous applications in microphones, biometrics, artificial intelligence and environmental

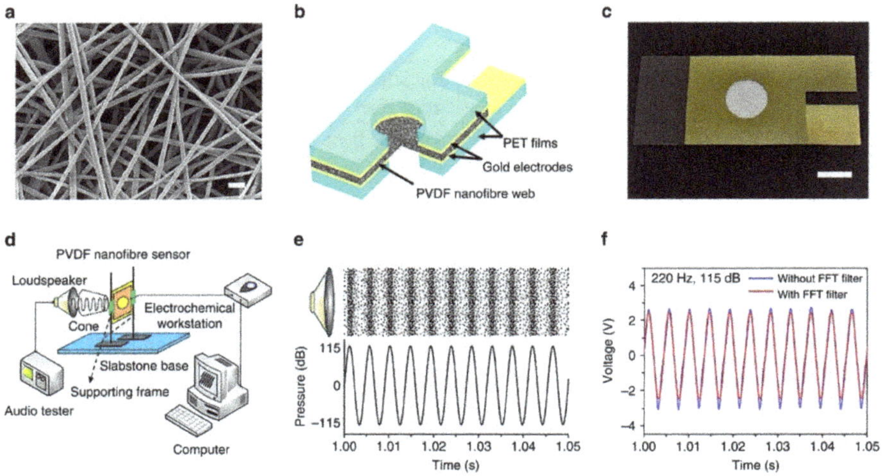

Fig. 12 (**a**) Scanning electron microscopic visualization of the PVDF nanofibrous structure with 1 mm scaling, (**b**) structural representation of the nanofiber, (**c**) digital image of the structure with 1 mm scale bar (**d**) functional block diagram of testing device (**e**) sound wave illustration (**f**) output voltages of the device under (reproduced from ref. [119], under the Creative Commons Attribution license [CC BY], Elsevier)

protection systems. PVDF film sensors were used to detect acoustic emission (AE) signals to locate fatigue damage growth in graphite-epoxy laminates [121]. Despite of having lower signal-to-noise ratio (SNR) compared to lead zirconate titanate (PZT) based sensors, low cost and disposability of PVDF based sensors made them appropriate for permanent mounting on composites [120, 121].

5.3 Structural Health Monitoring (SHM)

The flexibility, durability and high sensitivity of PVDF make it suitable for various structural health monitoring applications. PVDF films are prepared using solution blow-spinning method which exhibits high β-phase content (83%). These films are used to create flexible piezoelectric sensors with excellent strain sensitivity (low response close to 23 μ∈ and high response accuracy near 1.11 mV/μ∈) [121]. These sensors are demonstrated better performance compared to commercial strain gauges and useful for low-cost structural health monitoring (SHM) field. PVDF sensors embedded on glass fiber reinforced polymer (GERP) composite used for in-situ monitoring of damage and fatigue has been studied [121, 122]. These sensors have monitored structural health and integrity without compromising the mechanical behaviour of the piezocomposite. To identify the deterioration in polymer matrix composites (PMC), non-linear vibration methods like non-linear resonance (NLR) or single frequency excitation (SFE) were incorporated with integrated PVDF sensors. This study demonstrated different approaches that work to find the post-production degradation and errors in manufacturing [123]. An accelerometer was used to examine the result to verifying the robustness of the PVDF based sensor for structural health monitoring. PVDF-graphene nanoplatelet (GNP) based nanocomposite membranes were investigated for application in fiber reinforcement polymer composites [124]. Glass fabrics are mixed with PVDF and GNP fibers to prepare self-sensing piezocomposites. PVDF nanofibers with 0.05% GNP exhibited the highest consistency and β-phase content which additionally demonstrated better mechanical and sensing capabilities [124]. For self-powered for structural health monitoring system, a permeable piezoelectric nanogenerator (PENG) based on PVDF was designed. As compared to pure PVDF based nanogenerators, this PENG with embedded ZnO nanoparticles demonstrated enhanced output voltage as well as current [125]. It was perfect choice for aviation structural health monitoring since it produced enough electrical energy from environmental vibrations to run wireless sensor unit. PVDF strips were utilized as actuators and sensors in an innovative structural impedance measurement technique where the capacitance impedance of the PVDF sensor was separated from structural impedance [126]. Signals with weak impedance were recovered using phase detection method which allowed the actual resonant frequency measurement to monitor structural health [124]. A wireless experimental framework was created employing PVDF sensors for observing structural health of bridges [127]. The efficiency of the PVDF sensors in documenting structural impact responses and recognizing damages were the major attributes of the research [226]. In order to determine specific damage in steel moment-resisting frames, PVDF based piezoelectric films were employed as dynamic strain sensors [128]. A damage detection system was created to track the variations in the distribution of the bending moments. To replicate seismic damage, a 5-story steel testbed was incorporated with an array of wireless strain sensors [128]. A PVDF film-based sensor approach was originated in order to track the growth of cracks due to fatigue in technological structure. Through the solution of non-linear inverse equations, the sensor enabled the in-situ estimation to track fractural properties [129]. The electric potential of PVDF sensors array demonstrated the applicability of these sensors in real-time fatigue crack growth monitoring by stress intensity factor calculations [130]. Another approach for crack identification in beam structure was created using functional bending deflection shapes detected by PVDF based sensors. This method used frequency analysis as a function of damage structure to calculate damage indices. Again, for aircraft structural health monitoring, 1D PVDF array sensor was prepared for lamb wave reception in aircraft [131]. Experimental proofs were reported by the sensor to identify and demonstrate flaws using modal analysis in the wavenumber-frequency domain [131]. A multielement PVDF sensor was also reported in structural health monitoring field which determined the mode content without sensor movement. This sensor was fixed in the waveguide surface and detected both the symmetric and asymmetric lamb wave modes [131]. This sensor is suitable for fixed location structural health monitoring system due to its low mass, and conformability. A multi-operational transducer was prepared on a single PVDF film with features like lamb wave generation, impact detection, and temperature sensing. This sensor can function as transducer, sensor, and as a resistance

Table 9 Utilization of PVDF based sensors in structural health monitoring (SHM)

Sensor type	Detection capabilities	Ref.
Embedded PVDF sensors	Monitored damage without compromising mechanical properties	[121]
PVDF sensors with NLR and SFE methods	Detected manufacturing and post-manufacturing defects	[122]
PVDF/GNP micro-nanocomposite membranes	Detected damage through piezoelectric response changes	[123]
Porous PVDF-based PENG	Generated energy from vibrations for wireless SHM systems	[124]
PVDF patches as actuators and sensors	Precisely measured resonance frequencies and mass changes	[125]
Wireless PVDF sensor system	Monitored local damage in bridge structures	[127]
PVDF film sensors	Detected local damage through dynamic strain monitoring	[128]
PVDF sensor Array	Determined crack tip position and stress factors	[129]
PVDF sensors	Identified cracks using operating curvature deflection	[130]
PVDF Array sensor	Detected and characterized defects through modal analysis	[131]
PVDF single film	Integrated lamb wave generation, reception, and impact detection	[131]
PVDF sensors	Monitored bolted joint conditions through impedance changes	[126]
PVDF films	Provided high sensibility for structural vibration monitoring	[124]
PVDF sensors	Impact damage is detected in composite laminate with deceleration in velocity	[132]
PVDF film sensors	Located crack fronts in composite laminates	[133]

temperature detector (RTD) [124]. PVDF based sensors were also examined based on their vibration measurement attributes across numerous fields including structural health monitoring (SHM). Some of the features like flexibility, chemical resistance, low acoustic impedance made these PVDF based sensors suitable for structural vibration detection [124] (Table 9).

6 Recent Advancements in PVDF-Based Materials

Advancement in sensors formulated with polyvinylidene fluoride concentrate on optimizing the piezoelectric and pyroelectric performances to improve their sensing efficacy through microstructural optimization and material combinations, among other factors [5–8, 10, 11]. Currently, such advancements are significantly influential in flexible and wearable application, particularly within biomedical sectors [134]. The self-powered piezoelectric sensors present a promising platform for a variety of wearable devices [67]. PVDF exhibits distinct polymorphic phases namely α [7], β [8, 9, 68], γ [10, 11], δ [11], and ε [66]. Among these phases, α, β, γ, and δ crystalline forms are predominantly utilized [134, 135]. The α-phase composition PVDF is defined as exhibiting paraelectric properties, whereas the δ-phase of PVDF is identified as ferroelectric, and the γ-phase of PVDF is noted for generating non-zero dipole moments and polarity [136]. The β -phase is distinguished by achieving the highest dipole movement (8×10^{-30} *cm*) [137]. Within the β-phase composition, the polymer chains are organized such that the dipole arrangement is superimposed. To attain superior performance from PVDF, it is imperative to enhance the β-phase constant. As illustrated, the *TGTG·* configuration possesses the lowest free energy, thus numerous researchers are engaged in the transition of the *TGTG·* configuration [138] to the metastable all-trans configuration by applying external forces through mechanical and interfacial interactions [139]. Electrospinning has emerged as prevalent technique to achieve a high content β-phase when an external electric field incorporates on the surface of the electrodes. Till date, researchers have successfully attained approximately 85.2% β-phase fraction [139]. An additional method that has emerged to augment the β-phase fraction within PVDF, involves interfacial interaction, utilizing N, N-dimethylformamide (DML), Dimethyl-sulfoxide (DMSO) and water. This approach facilitates the production of thin films under low temperature. Nanofiller integration contributes to an increased β-phase content. Content of PVDF, however an increased concentration of nanofillers may lead to heightened dielectric loss and diminished flexibility, given the crucial role of dielectric properties in PVDF [134]. Presently, barium di-titanate ($BaTi_2O_5$) is employed to facilitate the generation of β-phase in PVDF [140]. Furthermore, PVDF material is increasingly utilized in the fabrication of nanofibers, particularly in electrospun format. In this context, researchers are exploring the β-phase crystallization within electrospun materials [94]. The electrospinning of PVDF has been conducted using EIG-DIG apparatus, equipped with a climate control system [141]. The resultant materials are utilized for the transmission electron microscopy (TEM) [142] studies and for X-ray photoelectron spectroscopy (XPS) with 10 nm gold-coated wafer of silicon [143] and piezo-response force microscopy (PFM) [144]. Scanning electron microscopy (SEM) imaging [53], differential scanning calorimetry (DSC), and wide-angle X-ray scattering (WAXS) [38] are performed on fibers electrospun directly onto aluminium foil. Fourier-transformed infrared spectroscopy (FTIR) [145] samples are prepared by placing them on the paper frame directly connected to the steel drum. The mesh generating piezoelectricity, in the form of electrospun fibers, is directly affixed to aluminium foil, which serves as the electrode in aforementioned experiments [141]. The results indicate that the electrospun fibers exhibit a 60% human interface, with variations including PVDF30+, PVDF30-, PVDF60+, and PVDF60-[146]. Some of the recent investigations focuses on the generation of randomly oriented fibers to mitigate the mitigate the mechanical stretching effects inherent in polyvinylidene fluoride (PVDF) fibers. Additionally, PVDF is employed in the fabrication of nanogenerator through the incorporation of various lead zirconate titanate (PZT) or alternative materials to enhance operational efficacy [121]. A novel Ytterbium-assisted porous PVDF composite film incorporates a ferroelectric nanogenerator (FTNG) [44], completely circumventing traditional poling treatments. The pioneering material manifests a progressive absorption of atmospheric moisture linked to the polar aprotic and hydrophilic traits of PVDF. Consequently, water molecules envelop the fluoride (F^-) ions present in PVDF. The pronounced hydrogen dipole-dipole interplay between the water and the fluoride (F^-) ions within PVDF induce a unidirectional alignment of the $-(CH_2 - CF_2)$ dipoles [147]. The methodology provides superior techniques for achieving a self-polarized β-phase in PVDF when compared to spin coating, instead of utilizing the Langmuir-Schafer method [148]. Recent advancement in PVDF materials have led to their application in charge storage devices, specifically in advanced level capacitors referred to as supercapacitor [149]. The function of electrolyte during the fabrication of supercapacitors is of paramount importance. Polymer-based electrolytes are considered as favorable alternatives for integrating supercapacitors with lightweight configurations [150]. These types of capacitors exhibit thin film structures, and demonstrates excellent ionic conductivity [149, 150]. The design can be modified according to specific requirements. Furthermore, these capacitors are environmentally friendly and effectively prevents the leakage of liquid electrodes [149].

7 Summary and Outlook

PVDF based piezoelectric and pyroelectric sensors have demonstrated major applications in biomedical industries, power generation system, structural health monitoring systems etc. owning to their superior mechanical attributes along with their chemical resistivity (low toxicity), and cost effectiveness. PVDF showcases mainly four polymorphs, namely a) α, b) β, c) γ, and d) δ, out of which β-phase is predominantly desirable structure due to its all-trans molecular orientation. In β-phase, PVDF demonstrates outstanding piezoelectric activities and highest dipole moment. These properties have escalated numerus studies of PVDF structure-property relationship and the processing techniques. Developing multidimensional integrated system by combining the piezoelectric capabilities of the materials with their other physiological attributes like acoustic, optics, and thermal properties, is becoming a significant area of research these days. PVDF and its copolymer-based sensors are being explored for applications in biomedical devices, energy harvesters, drug delivery systems, nanogenerators, etc. PVDF membranes are outstanding candidates for artificial muscles, smart skin, masks, underwater transducers, due to its piezoelectric properties and as auricular prostheses which can respond to pressure and temperature stimuli using the pyroelectric properties of PVDF. PVDF sensors are also incorporated as drug carriers which release drugs on the targeted locations like; tissue, bone, cell, organ, etc. Recent advancement of PVDF sensors have been reviewed to understand their role in structural health monitoring applications like; civil, aviation. Although there are plenty of methods or approaches for PVDF based polymer films available in the market, generation of pure PVDF is tedious and necessitates stretching the post-processing polarization steps. Consequently, one of the subsequent investigations focuses on improving the preparation procedure and the piezoelectric performance. Furthermore, there are several ongoing researches on PVDF-based composite piezoelectric materials. The attributes of PVDF-based composites would be enhanced by investigating intriguing composite films made of material with elevated piezoelectric and pyroelectric abilities. The comparable piezoelectric magnitude of the PVDF films and their relevant functionality of the accompanying device will be strengthened, if compatible organic or inorganic materials are employed for constructing new structure. Owing the challenging structure of PVDF and the juxtaposition of the crystalline (ordered) and amorphous (non-ordered) regions, the crystallization mechanism is complicated. Fundamental and extensive study on the inherent qualities of pure PVDF is rare, notwithstanding the fact that the majority of the studies so far concentrate on the advancement and implementation of associated technologies. Additional investigation is required to understand the primary factors like which factor is responsible for the piezoelectricity and how the transformation between different configurations take place to adopt PVDF in next-generation technologies.

PVDF and its copolymers have attracted as engrossing interest in various sectors like wearable sensors, biosensors, acoustic sensors, etc. based on their piezoelectricity, pyroelectricity, along with their biocompatibility. The future directives of PVDF-based piezoelectric and pyroelectric sensors are poised for significant advancements driven by the requirements of innovative solutions in smart technology integration such as smart sensors, wearable technologies, and health monitoring system. The unique mechanical properties combined with piezoelectric and pyroelectric attributes of PVDF, open avenues for developing flexible sensor arrays that can monitor physiological signals in real-time applications, thus revolutionizing health monitoring systems. The implementation of self-sufficient PVDF thin films in sensors demonstrates optimal sensitivity, biocompatibility, which is critical parameter for capturing the human body movement accurately, thereby facilitating applications in area supervision and traffic monitoring. Ongoing researches on the PVDF fabrication processes such as bilayer structure design that remarkably improves the signal-to-noise ratio (SNR), demonstrate that the future of these sensors will not only be more efficient but also more accessible and cost effective for wider spectrum of applications. While advancements in PVDF-based sensors are promising, but the obstacles are there for the optimization of the performance of the sensors and assurance of long-term stability in numerous environmental conditions. Continued research is required to address these issues to realise the full potential of the PVDF-based technologies. Although considerable advancements have been incorporated in the investigations to understand the enhanced polarization methods but the achievement of PVDF-based sensor with high dielectric constants and high electrical conductivity along with enhanced piezoelectric coefficients, is still a critical aspect of explorations.

References

1. Smith, M., Kar-Narayan, S.: Piezoelectric polymers: theory, challenges and opportunities. Int. Mater. Rev. **67**(1), 65–88 (2021)
2. Ramadan, K.S., Sameoto, D., Evoy, S.: A review of piezoelectric polymers as functional materials for electromechanical transducers. Smart Mater. Struct. **23**(3), 033001 (2014)
3. Gao, X., Yang, J., Wu, J., Xin, X., Li, Z., Yuan, X., Dong, S.: Piezoelectric actuators and motors: materials, designs, and applications. Adv. Mater. Technol. **5**(1), 1900716 (2020)
4. Curie, J., Curie, P.: Développement par compression de l'électricité polaire dans les cristaux hémièdres à faces inclinées. Bull. Soc. Minéral. Fr. **3**(4), 90–93 (1880)
5. Kawai, H.: The piezoelectricity of poly (vinylidene fluoride). Jpn. J. Appl. Phys. **8**(7), 975 (1969)
6. Tamura, M., Yamaguchi, T., Oyaba, T., Yoshimi, T.: Electroacoustic transducers with piezoelectric high polymer films. J. Audio Eng. Soc. **23**(1), 21–26 (1975)
7. Hu, X., You, M., Yi, N., Zhang, X., Xiang, Y.: Enhanced piezoelectric coefficient of PVDF-TrFE films via in situ polarization. Front. Energy Res. **9**, 621540 (2021)
8. Barrau, S., Ferri, A., Da Costa, A., Defebvin, J., Leroy, S., Desfeux, R., Lefebvre, J.M.: Nanoscale investigations of α-and γ-crystal phases in PVDF-based nanocomposites. ACS Appl. Mater. Interfaces. **10**(15), 13092–13099 (2018)
9. Gebrekrstos, A., Muzata, T.S., Ray, S.S.: Nanoparticle-enhanced β-phase formation in electroactive PVDF composites: a review of systems for applications in energy harvesting, EMI shielding, and membrane technology. ACS Appl. Nano Mater. **5**(6), 7632–7651 (2022)
10. Zhang, L., Li, S., Zhu, Z., Rui, G., Du, B., Chen, D., Zhu, L.: Recent progress on structure manipulation of poly (vinylidene fluoride) -based ferroelectric polymers for enhanced piezoelectricity and applications. Adv. Funct. Mater. **33**(38), 2301302 (2023)
11. Si, S.K., Paria, S., Karan, S.K., Ojha, S., Das, A.K., Maitra, A., Khatua, B.B.: In situ-grown organo-lead bromide perovskite-induced electroactive γ-phase in aerogel PVDF films: an efficient photoactive material for piezoelectric energy harvesting and photodetector applications. Nanoscale. **12**(13), 7214–7230 (2020)
12. Wang, F., Tanaka, M., Chonan, S.: Development of a PVDF piezopolymer sensor for unconstrained in-sleep cardiorespiratory monitoring. J. Intell. Mater. Syst. Struct. **14**(3), 185–190 (2003)
13. Boyer, C., Ameduri, B., Boutevin, B., Dolbier, W.R., Winter, R., Gard, G.: Radical terpolymerization of1,1,2-trifluoro-2-pentafluorosulfanylethylene and pentafluorosulfanylethylene in the presence of vinylidene fluoride and hexafluoropropylene by iodine transfer polymerization. Macromolecules. **41**(4), 1254–1263 (2008)
14. Mao, D., Gnade, B.E., Quevedo-Lopez, M.A.: Ferroelectric properties and polarization switching kinetic of poly (vinylidene fluoride-trifluoroethylene) copolymer. In: Ferroelectrics-physical effects, vol. 23, (2011)
15. Behera, R., Elanseralathan, K.: J. Energy Storage. **48**, 103788 (2022)
16. Kekana, L.: Applications of Heterogeneous Multicationic Inorganic Perovskites in Organic Transformations. University of Johannesburg (South Africa) (2023)
17. Ameduri, B.: From vinylidene fluoride (VDF) to the applications of VDF-containing polymers and copolymers: recent developments and future trends. Chem. Rev. **109**(12), 6632–6686 (2009)
18. Ahbab, N., Xu, T.B.: Multifunctional polyvinylidene fluoride (PVDF)-based polymer, copolymer, and compo-sites for aerospace applications: a comprehensive technical review. In: ASME Aerospace Structures, Structural Dynamics, and Materials Conference, vol. 87745, p. V001T02A001. American Society of Mechanical Engineers (2024)
19. Xu, T.B., Naz, S.: Review of Cryogenic Piezoelectric Ultrasonic Motors for Aerospace Applications, p. 1090. AIAA SCITECH 2025 Forum (2025)
20. Drobny, J.G.: Fluoropolymers in automotive applications. Polym. Adv. Technol. **18**(2), 117–121 (2007)

21. Costa, C.M., Cardoso, V.F., Martins, P., Correia, D.M., Gonçalves, R., Costa, P., Lanceros-Méndez, S.: smart and multifunctional materials based on electroactive poly (vinylidene fluoride): recent advances and opportunities in sensors, actuators, energy, environmental, and biomedical applications. Chem. Rev. **123**(19), 11392–11487 (2023)
22. Naz, S., Xu, T.B.: A comprehensive review of piezoelectric ultrasonic motors: classifications, characterization, fabrication, applications, and future challenges. Micromachines. **15**(9), 1170 (2024)
23. Dong, W., Xiao, L., Hu, W., Zhu, C., Huang, Y., Yin, Z.: Wearable human–machine interface based on PVDF piezoelectric sensor. Trans. Inst. Meas. Control. **39**(4), 398–403 (2017)
24. Pullano, S.A., Mahbub, I., Islam, S.K., Fiorillo, A.S.: PVDF sensor stimulated by infrared radiation for temperature monitoring in microfluidic devices. Sensors. **17**(4), 850 (2017)
25. Liu, L., Xu, Y., Zhu, J., Liu, T., Hou, T., Li, Y., Zhou, X.: A flexible thermal sensor based on PVDF film for robot finger skin. Integr. Ferroelectr. **201**(1), 23–31 (2019)
26. Li, J., Zhou, G., Hong, Y., He, W., Wang, S., Chen, Y., Zhu, Y.: Highly sensitive, flexible and wearable piezoelectric motion sensor based on PT promoted β-phase PVDF. Sensors and Actuators A: Physical. **337**, 113415 (2022)
27. Gusarov, B.: PVDF piezoelectric polymers: characterization and application to thermal energy harvesting (Doctoral dissertation, Université Grenoble Alpes) (2015)
28. Dange, E., Hamsa, R., Madhuri, M., Jain, A.: PVDF-based piezoelectric alarm system. ISSS Journal of Micro and Smart Systems. **10**(1), 83–86 (2021)
29. Xin, Y., Qi, X., Qian, C., Tian, H., Ling, Z., Jiang, Z.: A wearable respiration and pulse monitoring system based on PVDF piezoelectric film. Integr. Ferroelectr. **158**(1), 43–51 (2014)
30. Haines, C.S., Li, N., Spinks, G.M., Aliev, A.E., Di, J., Baughman, R.H.: New twist on artificial muscles. Proc. Natl. Acad. Sci. U. S. A. **113**(42), 11709–11716 (2016)
31. Khandelwal, G., Raj, N.P.M.J., Kim, S.J.: Triboelectric nanogenerator for healthcare and biomedical applications. Nano Today. **33**, 100882 (2020)
32. Gao, C., Liu, Y., Gu, Z., Li, J., Sun, Y., Li, W., Xu, W.: Hierarchical structured fabrics with enhanced pressure sensing performance based on orientated growth of functional bacterial cellulose. Adv. Fiber Mater. **6**(5), 1554–1568 (2024)
33. Lei, K.F., Hsieh, Y.Z., Chiu, Y.Y., Wu, M.H.: The structure design of piezoelectric poly (vinylidene fluoride) (PVDF) polymer-based sensor patch for the respiration monitoring under dynamic walking conditions. Sensors. **15**(8), 18801–18812 (2015)
34. Wei, J., Wang, Z., Xing, X.: A wireless high-sensitivity fetal heart sound monitoring system. Sensors. **21**(1), 193 (2020)
35. Lang, S.B., Tofail, S.A.M., Kholkin, A.L., Wojtaś, M., Gregor, M., Gandhi, A.A., Wang, Y., Bauer, S., Krause, M., Plecenik, A.: Ferroelectric polarization in nanocrystalline hydroxyapatite thin films on silicon. Sci. Rep. **3**, 2215 (2013)
36. Leung, W.W.F., Sun, Q.: Charged PVDF multilayer nanofiber filter in filtering simulated airborne novel coronavirus (COVID-19) using ambient nano-aerosols. Sep. Purif. Technol. **245**, 116887 (2020)
37. iu, H., Luo, S., Yu, S., Ding, S., Shen, Y., Sun, R., Wong, C.P.: Flexible BaTiO3nf-Ag/PVDF nanocomposite films with high dielectric constant and energy density. IEEE Trans. Dielectr. Electr. Insul. **24**(2), 757–763 (2017)
38. Concha, V.O., Timóteo, L., Duarte, L.A., Bahú, J.O., Munoz, F.L., Silva, A.P., Souto, E.B.: properties, characterization and biomedical applications of polyvinylidene fluoride (PVDF): a review. J. Mater. Sci. **59**(31), 14185–14204 (2024)
39. Fan, X., Sun, M., Li, Z., Chen, Z., Zhou, X., Lu, Q., Zhang, Z.: Tuning piezoelectric properties of P (VDF-TFE) films for sensor application. React. Funct. Polym. **180**, 105391 (2022)
40. Cheng, W.H., Wu, P.L., Huang, H.H.: Polyvinylidene fluoride piezoelectric fiber glass/carbon hybrid self-sensing composites for structural health monitoring. Sensors. **23**(8), 3813 (2023)
41. Mir, A., Kumar, A., Riaz, U.: Synthesis and characterization of conducting polymer/alginate composite hydrogels: effect of conducting polymer loading on the release behaviour of metformin drug. J. Mol. Liq. **372**, 121193 (2023)

42. Ahbab, N., Xu, T.: B.: multifunctional polyvinylidene fluoride (PVDF)-based polymer, copolymer, and compo-sites for aerospace applications: a comprehensive technical review. In: ASME Aerospace Structures, Structural Dynamics, and Materials Conference, vol. 87745, p. V001T02A001. American Society of Mechanical Engineers (2024)
43. Parlato, S., Esposito, D., Centracchio, J., Andreozzi, E., Gragnaniello, M., Riccio, M., Bifulco, P.: A new, simple wrist-mounted PVDF sensor for continuous heart rate monitoring. In: 2024 IEEE Sensors Applications Symposium (SAS), pp. 1–6. IEEE (2024)
44. Pei, H., Jing, J., Chen, Y., Guo, J., Chen, N.: 3D printing of PVDF-based piezoelectric nanogenerator from programmable metamaterial design: promising strategy for flexible electronic skin. Nano Energy. **109**, 108303 (2023)
45. Zhao, J., Li, F., Wang, Z., Dong, P., Xia, G., Wang, K.: Flexible PVDF nanogenerator-driven motion sensors for human body motion energy tracking and monitoring. J. Mater. Sci. Mater. Electron. **32**(11), 14715–14727 (2021)
46. Shehata, N., Hassanin, A.H., Elnabawy, E., Nair, R., Bhat, S.A., Kandas, I.: Acoustic energy harvesting and sensing via electrospun PVDF nanofiber membrane. Sensors. **20**(11), 3111 (2020)
47. Xu, F., Yang, J., Dong, R., Jiang, H., Wang, C., Liu, W., Zhu, G.: Wave-shaped piezoelectric nanofiber membrane nanogenerator for acoustic detection and recognition. Adv. Fiber Mater. **3**, 368–380 (2021)
48. Gupta, S., Lorenzelli, L., Dahiya, R.: Multifunctional flexible PVDF-TrFE/$BaTiO_3$ based tactile sensor for touch and temperature monitoring. In: 2017 IEEE sensors, pp. 1–3. IEEE (2017)
49. Bloomfield, P.E., Lo, W.J., Lewin, P.A.: Experimental study of the acoustical properties of polymers utilized to construct PVDF ultrasonic transducers and the acousto-electric properties of PVDF and P (VDF/TrFE) films. IEEE Trans. Ultrason. Ferroelectr. Freq. Control. **47**(6), 1397–1405 (2000)
50. Fortunato, M., Bidsorkhi, H.C., Chandraiahgari, C.R., De Bellis, G., Sarto, F., Sarto, M.S.: PFM characterization of PVDF nanocomposite films with enhanced piezoelectric response. IEEE Trans. Nanotechnol. **17**(5), 955–961 (2018)
51. Tressler, J.F., Alkoy, S., Newnham, R.E.: Piezoelectric sensors and sensor materials. J. Electroceram. **2**, 257–272 (1998)
52. Ruiz-Gonzalez, A., Wang, M., Haseloff, J.: Sustainable approaches to incorporate plant-based biomaterials in power generation. Solids. **4**(2), 133–155 (2023)
53. Hao, J., Li, W., Zhai, J., Chen, H.: Progress in high-strain perovskite piezoelectric ceramics. Mater Sci Eng R: Rep. **135**, 1–57 (2019)
54. Dargaville, T.R., Celina, M.C., Elliott, J.M., Chaplya, P.M., Jones, G.D., Mowery, D.M., Martin, J.W.: Characterization. In: Performance and Optimization of PVDF as a Piezoelectric Film for Advanced Space Mirror Concepts, pp. 1–40. Sandia National Laboratories (SNL), Albuquerque (2005)
55. Hoshyarmanesh, H., Maddahi, Y.: Poling process of composite piezoelectric sensors for structural health monitoring: a pilot comparative study. IEEE Sens. Lett. **2**(1), 1–4 (2018)
56. Xu, R., Kim, S.G.: Figures of merits of piezoelectric materials in energy harvesters. In: Proceedings of the PowerMEMS, vol. 12, (2012)
57. Veved, A., Ejuh, G.W., Djongyang, N.: Review of emerging materials for PVDF-based energy harvesting. Energy Rep. **8**, 12853–12870 (2022)
58. Habib, M., Lantgios, I., Hornbostel, K.: A review of ceramic, polymer and composite piezoelectric materials. J. Phys. D Appl. Phys. **55**(42), 423002 (2022)
59. Bassani, F., Liedl, G.L., Wyder, P.: Encyclopaedia of Condensed Matter Physics. Elsevier, Amsterdam (2005)
60. Bowen, C.R., Taylor, J., LeBoulbar, E., Zabek, D., Chauhan, A., Vaish, R.: Pyroelectric materials and devices for energy harvesting applications. Energy Environ. Sci. **7**(12), 3836–3856 (2014)
61. He, H., Lu, X., Hanc, E., Chen, C., Zhang, H., Lu, L.: Advances in lead-free pyroelectric materials: a comprehensive review. J. Mater. Chem. C. **8**(5), 1494–1516 (2020)

62. Ahbab, N., Naz, S., Xu, T.B., Zhang, S.: A comprehensive review of piezoelectric PVDF polymer fabrications and characteristics. Micromachines. **16**(4), 386 (2025)
63. Balli, B., Şavk, A., Şen, F.: graphene and polymer composites for supercapacitor applications. In: Nanocarbon and Its Composites, pp. 123–151. Woodhead Publishing (2019)
64. Sappati, K.K., Bhadra, S.: Piezoelectric polymer and paper substrates: a review. Sensors. **18**(11), 3605 (2018)
65. Uchino, K.: The development of piezoelectric materials and the new perspective. In: Advanced Piezoelectric Materials, pp. 1–92. Woodhead Publishing (2017)
66. Zhu, Y., Jiang, P., Zhang, Z., Huang, X.: Dielectric phenomena and electrical energy storage of poly (vinylidene fluoride) based high-k polymers. Chin. Chem. Lett. **28**(11), 2027–2035 (2017)
67. Saxena, P., Shukla, P.: A comprehensive review on fundamental properties and applications of poly (vinylidene fluoride) (PVDF). Adv. Compos. Hybrid Mater. **4**(1), 8–26 (2021)
68. Sencadas, V., Gregorio Jr., R., Lanceros-Méndez, S.: α to β phase transformation and microestructural changes of PVDF films induced by uniaxial stretch. J. Macromol. Sci. B. **48**(3), 514–525 (2009)
69. Bormashenko, Y., Pogreb, R., Stanevsky, O., Bormashenko, E.: Vibrational spectrum of PVDF and its interpretation. Polym. Test. **23**(7), 791–796 (2004)
70. Kozdra, S., Wójcik, A., Możdżonek, M., Florczak, Ł., Opaliński, I., Michałowski, P.P.: Poly (vinylidene fluoride) solid polymer electrolyte structure revealed by secondary ion mass spectrometry. Polymer. **259**, 125364 (2022)
71. Wang, Y., Zhu, L., Du, C.: Progress in piezoelectric nanogenerators based on PVDF composite films. Micromachines. **12**(11), 1278 (2021)
72. Gregorio, R., Ueno, E.M.: Effect of crystalline phase, orientation and temperature on the dielectric properties of poly (vinylidene fluoride) (PVDF). J. Mater. Sci. **34**, 4489–4500 (1999)
73. Gregorio Jr., R.: Determination of the α, β, and γ crystalline phases of poly (vinylidene fluoride) films prepared at different conditions. J. Appl. Polym. Sci. **100**(4), 3272–3279 (2006)
74. Ruan, L., Yao, X., Chang, Y., Zhou, L., Qin, G., Zhang, X.: Properties and applications of the β phase poly (vinylidene fluoride). Polymers. **10**(3), 228 (2018)
75. Nakamura, K., Sawai, D., Watanabe, Y., Taguchi, D., Takahashi, Y., Furukawa, T., Kanamoto, T.: Effect of annealing on the structure and properties of poly (vinylidene fluoride) β-form films. J. Polym. Sci. B Polym. Phys. **41**(14), 1701–1712 (2003)
76. Wang, H., Yang, X., Zhao, Y., Yan, C., Wang, S., Yang, H., Schultz, J.M.: Preparation of gamma-PVDF with controlled orientation and insight into phase transformation. Polymer. **123**, 282–289 (2017)
77. Mishra, H.K., Gupta, V., Roy, K., Babu, A., Kumar, A., Mandal, D.: Revisiting of δ− PVDF nanoparticles via phase separation with giant piezoelectric response for the realization of self-powered biomedical sensors. Nano Energy. **95**, 107052 (2022)
78. Malik, P., Gupta, V., Mishra, H.K., Kumar, A., Mandal, D.: Long term retention in δ-PVDF thin film prepared by rapid ice quenching technique. J. Appl. Polym. Sci. **140**(15), e53714 (2023)
79. Hashim, N.A., Liu, Y., Li, K.: Stability of PVDF hollow fibre membranes in sodium hydroxide aqueous solution. Chem. Eng. Sci. **66**(8), 1565–1575 (2011)
80. Shinohara, H.: Fluorination of polyhydrofluoroethylenes. II. Formation of perfluoroalkyl carboxylic acids on the surface region of poly (vinylidene fluoride) film by oxyfluorination, fluorination, and hydrolysis. J. Polym. Sci. Polym. Chem. Ed. **17**(5), 1543–1556 (1979)
81. Wegener, M., Küunstler, W., Gerhard-Multhaupt, R.: Poling behavior and optical absorption of partially dehydrofluorinated and uniaxially stretched polyvinylidene fluoride. Ferroelectrics. **336**(1), 3–8 (2006)
82. Puspitasari, V., Granville, A., Le-Clech, P., Chen, V.: Cleaning and ageing effect of sodium hypochlorite on polyvinylidene fluoride (PVDF) membrane. Sep. Purif. Technol. **72**(3), 301–308 (2010)
83. Uchino, K.: Ferroelectric devices. CRC Press (2018)

84. Dallaev, R., Pisarenko, T., Sobola, D., Orudzhev, F., Ramazanov, S., Trčka, T.: Brief review of PVDF properties and applications potential. Polymers. **14**(22), 4793 (2022)
85. Zhang, Q.M., Cheng, Z.Y., Bharti, V., Xu, T.B., Xu, H., Mai, T.X., Gross, S.J.: Piezoelectric and electrostrictive polymeric actuator materials. In: Smart Structures and Materials 2000: Electroactive Polymer Actuators and Devices (EAPAD), vol. 3987, pp. 34–50. SPIE (2000)
86. Song, J., Zhao, G., Li, B., Wang, J.: Design optimization of PVDF-based piezoelectric energy harvesters. Heliyon. **3**, e00377 (2017)
87. Naz, S., Raja, M.A.Z., Mehmood, A., Jaafery, A.Z.: intelligent predictive solution dynamics for dahl hysteresis model of piezoelectric actuator. Micromachines. **13**(12), 2205 (2022)
88. Xu, T.B., Kang, J.H., Siochi, E.J., King, G.C.: Pyroelectric sandwich thermal energy harvesters (No. Patent Application Number: US-Patent-Appl-SN-14/879,506) (2018)
89. Cauda, V., Stassi, S., Bejtka, K., Canavese, G.: Nanoconfinement: an effective way to enhance PVDF piezoelectric properties. ACS Appl. Mater. Interfaces. **5**(13), 6430–6437 (2013)
90. Naber, R.C., Asadi, K., Blom, P.W., De Leeuw, D.M., De Boer, B.: Organic nonvolatile memory devices based on ferroelectricity. Adv. Mater. **22**(9), 933–945 (2010)
91. Lines, M.E., Glass, A.M.: Principles and applications of ferroelectrics and related materials. Oxford University Press (2001)
92. Thomas, P., Varughese, K.T., Dwarakanath, K., Varma, K.B.R.: Dielectric properties of poly (vinylidene fluoride)/$CaCu_3$Ti4O12 composites. Compos. Sci. Technol. **70**(3), 539–545 (2010)
93. Xia, Q., Zhao, X.J., Chen, S.J., Ma, W.Z., Zhang, J., Wang, X.L.: Effect of solution-blended poly (styrene-co-acrylonitrile) copolymer on crystallization of poly (vinylidene fluoride). Express Polym. Lett. **4**(5), 284–291 (2010)
94. Mokhtari, F., Samadi, A., Rashed, A.O., Li, X., Razal, J.M., Kong, L., Zhao, S.: Recent progress in electrospun polyvinylidene fluoride (PVDF)-based nanofibers for sustainable energy and environmental applications. Prog. Mater. Sci. **148**, 101376 (2024)
95. Wu, C.M., Chou, M.H.: Polymorphism, piezoelectricity and sound absorption of electrospun PVDF membranes with and without carbon nanotubes. Compos. Sci. Technol. **127**, 127–133 (2016)
96. Tyona, M.D.: A theoretical study on spin coating technique. Adv. Mater. Res. **2**(4), 195 (2013)
97. Yin, Z., Tian, B., Zhu, Q., Duan, C.: Characterization and application of PVDF and its copolymer films prepared by spin-coating and Langmuir–Blodgett method. Polymers. **11**(12), 2033 (2019)
98. Sebastian, T., Bach, M., Geiger, A., Lusiola, T., Kozielski, L., Clemens, F.: Investigation of electromechanical properties on 3-D printed piezoelectric composite scaffold structures. Materials. **14**(20), 5927 (2021)
99. Lee, C., Tarbutton, J.A.: Electric poling-assisted additive manufacturing process for PVDF polymer-based piezoelectric device applications. Smart Mater. Struct. **23**(9), 095044 (2014)
100. Yuan, X., Gao, X., Shen, X., Yang, J., Li, Z., Dong, S.: A 3D-printed, alternatively tilt-polarized PVDF-TrFE polymer with enhanced piezoelectric effect for self-powered sensor application. Nano Energy. **85**, 105985 (2021)
101. Jaffari, G.H., Khan, M.S.I., Mumtaz, F., Wang, Y., Khan, N.A.: Manipulation of crystallization and dielectric relaxation dynamics via hot pressing and copolymerization of PVDF with Hexafluoropropylene. J. Polym. Res. **30**(1), 11 (2023)
102. Dhumrash, S., Gautam, A., Kumar, A., Jaglan, N., Uniyal, P.: Appreciable amelioration in the dielectric and energy storage behavior of the electrospun fluoropolymer PVDF-HFP thick films: effect of hot-pressing. J. Energy Storage. **103**, 114337 (2024)
103. Rodrigues, M.T., Gomes, M.E., Mano, J.F., Reis, R.L.: β-PVDF membranes induce cellular proliferation and differentiation in static and dynamic conditions. Mater. Sci. Forum. **587**, 72–76 (2008)
104. Cheng, Y., Xu, Y., Qian, Y., Chen, X., Ouyang, Y., Yuan, W.E.: 3D structured self-powered PVDF/PCL scaffolds for peripheral nerve regeneration. Nano Energy. **69**, 104411 (2020)

105. Martins, P.M., Ribeiro, S., Ribeiro, C., Sencadas, V., Gomes, A.C., Gama, F.M., Lanceros-Méndez, S.: Effect of poling state and morphology of piezoelectric poly (vinylidene fluoride) membranes for skeletal muscle tissue engineering. RSC Adv. **3**(39), 17938–17944 (2013)
106. Genchi, G.G., Ceseracciu, L., Marino, A., Labardi, M., Marras, S., Pignatelli, F., Ciofani, G.: P (VDF-TrFE)/BaTiO3 nanoparticle composite films mediate piezoelectric stimulation and promote differentiation of SH-SY5Y neuroblastoma cells. Adv. Healthc. Mater. **5**(14), 1808–1820 (2016)
107. Zhang, X., Zhang, C., Lin, Y., Hu, P., Shen, Y., Wang, K., Chen, L.: Nanocomposite membranes enhance bone regeneration through restoring physiological electric microenvironment. ACS Nano. **10**(8), 7279–7286 (2016)
108. Damaraju, S.M., Shen, Y., Elele, E., Khusid, B., Eshghinejad, A., Li, J., Arinzeh, T.L.: Three-dimensional piezoelectric fibrous scaffolds selectively promote mesenchymal stem cell differentiation. Biomaterials. **149**, 51–62 (2017)
109. Ryu, C., Hajra, S., Sahu, M., Jung, S.I., Jang, I.R., Kim, H.J.: PVDF-bismuth titanate based self-powered flexible tactile sensor for biomechanical applications. Mater. Lett. **309**, 131308 (2022)
110. Shen, D., Liu, Y., Yu, M., Kong, F., Xin, B., Liu, Y.: Bioinspired flexible and highly responsive PVDF-based humidity sensors for respiratory monitoring. Polymer. **254**, 125103 (2022)
111. Panda, S., Hajra, S., Jeong, H., Panigrahi, B.K., Pakawanit, P., Dubal, D., Kim, H.J.: Biocompatible CaTiO3-PVDF composite-based piezoelectric nanogenerator for exercise evaluation and energy harvesting. Nano Energy. **102**, 107682 (2022)
112. Kang, D.H., Kim, N.K., Kang, H.W.: Electrostatic charge retention in PVDF nanofiber-nylon mesh multilayer structure for effective fine particulate matter filtration for face masks. Polymers. **13**(19), 3235 (2021)
113. Husain, M., Singh, R., Pabla, B.S.: On PVDF composite as partially absorbable smart implants. Proc. Inst. Mech. Eng. H J. Eng. Med. **237**(4), 517–526 (2023)
114. Zięba, M., Rusak, T., Misztal, T., Zięba, W., Marcińczyk, N., Czarnecka, J., Terzyk, A.P.: Nitrogen plasma modification boosts up the hemocompatibility of new PVDF-carbon nanohorns composite materials with potential cardiological and circulatory system implants application. Biomater. Adv. **138**, 212941 (2022)
115. Xi, Y., Pan, W., Xi, D., Liu, X., Yu, J., Xue, M., Ye, X.: Optimization, characterization and evaluation of ZnO/polyvinylidene fluoride nanocomposites for orthopedic applications: improved antibacterial ability and promoted osteoblast growth. Drug Deliv. **27**(1), 1378–1385 (2020)
116. Mohseni, M., Delavar, F., Rezaei, H.: The piezoelectric gel-fiber-particle substrate containing short PVDF-chitosan-gelatin nanofibers and mesoporous silica nanoparticles with enhanced antibacterial activity as a potential of wound dressing applications. J. Macromol. Sci. A. **58**(10), 694–708 (2021)
117. Amini, F., Semnani, D., Karbasi, S., Banitaba, S.N.: A novel bilayer drug-loaded wound dressing of PVDF and PHB/chitosan nanofibers applicable for post-surgical ulcers. Int. J. Polym. Mater. Polym. Biomater. **68**(13), 772–777 (2019)
118. Guo, Z., Liu, S., Hu, X., Zhang, Q., Shang, F., Song, S., Xiang, Y.: Self-powered sound detection and recognition sensors based on flexible polyvinylidene fluoride-trifluoroethylene films enhanced by in-situ polarization. Sensors Actuators A Phys. **306**, 111970 (2020)
119. Lang, C., Fang, J., Shao, H., Ding, X., Lin, T.: High-sensitivity acoustic sensors from nanofiber webs. Nat. Commun. **7**(1), 11108 (2016)
120. Caneva, C., De Rosa, I.M., Sarasini, F.: Monitoring of impacted aramid-reinforced composites by embedded PVDF acoustic emission sensors. Strain. **44**(4), 308–316 (2008)
121. Tuloup, C., Harizi, W., Aboura, Z., Meyer, Y.: Integration of piezoelectric transducers (PZT and PVDF) within polymer-matrix composites for structural health monitoring applications: new success and challenges. Int. J. Smart Nano Mater. **11**(4), 343–369 (2020)
122. Wang, Z.: Manufacturing and Characterization of Blended PVDF/PVDF-TrFE Polymers: a Study towards Future Sensor Applications. University of Alberta (2022)

123. Fukunaga, H., Hu, N., Chang, F.K.: Structural damage identification using piezoelectric sensors. Int. J. Solids Struct. **39**(2), 393–418 (2002)
124. Rana, M.M., Khan, A.A., Huang, G., Mei, N., Saritas, R., Wen, B., Ban, D.: Porosity modulated high-performance piezoelectric nanogenerator based on organic/inorganic nanomaterials for self-powered structural health monitoring. ACS Appl. Mater. Interfaces. **12**(42), 47503–47512 (2020)
125. Zhang, X., Xu, J., Yan, R.: A structural impedance measurement method by using polyvinylidene fluoride as actuator and sensor. Rev. Sci. Instrum. **91**(8), 085111 (2020)
126. Pavelko, I., Pavelko, V., Kuznetsov, S., Ozolinsh, I.: Bolt-joint structural health monitoring by the method of electromechanical impedance. Aircraft Eng. Aerosp. Technol. **86**(3), 207–214 (2014)
127. Yu, Y., Zhao, X., Wang, Y., Ou, J.: A study on PVDF sensor using wireless experimental system for bridge structural local monitoring. Telecommun. Syst. **52**, 2357–2366 (2013)
128. Kurata, M., Li, X., Fujita, K., He, L., Yamaguchi, M.: PVDF piezo film as dynamic strain sensor for local damage detection of steel frame buildings. In: Sensors and Smart Structures Technologies for Civil, Mechanical, and Aerospace Systems 2013, vol. 8692, pp. 100–109. SPIE (2013)
129. Gupta, S., West, G., Wilson, M.A., Grutzik, S.J., Warner, D.H.: Identifying crack tip position and stress intensity factors from displacement data. Int. J. Fract. **243**(1), 47–63 (2023)
130. do Nascimento Oliveira, J. E.: Deformation and damage analysis of composite beams equipped with polyvinylidene fluoride film sensors (Doctoral dissertation, Cape Peninsula University of Technology) (2008)
131. Fotouhi, M., Xiao, B., Pozegic, T., Jalalvand, M., Hamerton, I., Wisnom, M.R.: High performance piezo electric nanocomposite sensor nodes for structural health monitoring. In: 21st international conference on composite materials. ICCM (2017)
132. Kim, I.G., Lee, H.Y., Kim, J.W.: Impact damage detection in composite laminates using PVDF and PZT sensor signals. J. Intell. Mater. Syst. Struct. **16**(11–12), 1007–1013 (2005)
133. Bae, J.H., Lee, S.W., Chang, S.H.: Characterization of low-velocity impact-induced damages in carbon/epoxy composite laminates using a poly (vinylidene fluoride–trifluoroethylene) film sensor. Composites Part B: Engineering. **135**, 189–200 (2018)
134. Yu, L., Yi, Y., Yao, T., Song, Y., Chen, Y., Li, Q., Sun, J.: All VN-graphene architecture derived self-powered wearable sensors for ultrasensitive health monitoring. Nano Res. **12**, 331–338 (2019)
135. Cao, X., Xiong, Y., Sun, J., Zhu, X., Sun, Q., Wang, Z.L.: Piezoelectric nanogenerators derived self-powered sensors for multifunctional applications and artificial intelligence. Adv. Funct. Mater. **31**(33), 2102983 (2021)
136. Li, Q., Zhao, J., He, B., Hu, Z.: Solution processable poly (vinylidene fluoride)-based ferroelectric polymers for flexible electronics. APL Mater. **9**(1), 010902 (2021)
137. Zhu, L., Wang, Q.: Novel ferroelectric polymers for high energy density and low loss dielectrics. Macromolecules. **45**(7), 2937–2954 (2012)
138. Yan, J., Liu, M., Jeong, Y.G., Kang, W., Li, L., Zhao, Y., Yang, G.: Performance enhancements in poly (vinylidene fluoride)-based piezoelectric nanogenerators for efficient energy harvesting. Nano Energy. **56**, 662–692 (2019)
139. Martins, P., Lopes, A.C., Lanceros-Mendez, S.: Electroactive phases of poly (vinylidene fluoride): determination, processing and applications. Prog. Polym. Sci. **39**(4), 683–706 (2024)
140. Ye, L., Wei, J., Ran, C., Zhang, J., Zhang, Y., Ma, S.: PVDF/PI/BT nanocomposites with excellent charge/discharge efficiency obtained by inserting linear polymer to adjust the ferroelectricity. Colloids Surf. A Physicochem. Eng. Asp. **688**, 133653 (2024)
141. He, Z., Rault, F., Lewandowski, M., Mohsenzadeh, E., Salaün, F.: Electrospun PVDF nanofibers for piezoelectric applications: a review of the influence of electrospinning parameters on the β phase and crystallinity enhancement. Polymers. **13**(2), 174 (2021)
142. Miao, J., Reneker, D.H., Tsige, M., Taylor, P.L.: Molecular dynamics simulations and morphology analysis of TEM imaged PVDF nanofibers. Polymer. **125**, 190–199 (2017)

143. Wang, Y., Lin, H., Xiong, Z., Wu, Z., Yu, X., Wang, Y., Liu, F.: Investigation of abnormal thermoresponsive PVDF membranes on casting solution, membrane morphology and filtration performance. RSC Adv. **6**(33), 27485–27493 (2016)
144. Ravi Sankar, M.S., Pramod, K., Gangineni, R.B.: Local ferroelectric studies on interconnected PVDF nano-dot thin films using piezo force microscopy. J. Mater. Sci. Mater. Electron. **30**(23), 20716–20724 (2019)
145. Vasic, N., Steinmetz, J., Görke, M., Sinapius, M., Hühne, C., Garnweitner, G.: Phase transitions of polarised PVDF films in a standard curing process for composites. Polymers. **13**(22), 3900 (2021)
146. Szewczyk, P., Gradys, A., Kim, S., Persano, L., Marzec, M., Kryshtal, A., Busolo, T., Toncelli, A., Pisignano, D., Bernasik, A., Kar-Narayan, S., Sajkiewicz, P., Stachewicz, U.: Enhanced piezoelectricity of electrospun polyvinylidene fluoride (PVDF) fibers for energy harvesting. ACS Appl. Mater. Interfaces. **12**, 13575 (2020). https://doi.org/10.1021/acsami.0c02578
147. Jana, S., Garain, S., Sen, S., Mandal, D.: The influence of hydrogen bonding on the dielectric constant and the piezoelectric energy harvesting performance of hydrated metal salt mediated PVDF films. Phys. Chem. Chem. Phys. **17**(26), 17429–17436 (2015)
148. Kazak, A.V., Nikitin, K.S., Marchenkova, M.A., Savelyev, M.S., Vasilevsky, P.N., Gerasimenko, A.Y., Tolbin, A.Y.: Langmuir-Schaefer films based on cyclotriphosphazene-substituted phthalocyanines: supramolecular organization, UV/Vis study, and laser-induced nonlinear absorption. Appl. Surf. Sci. **638**, 158077 (2023)
149. Rajeevan, S., John, S., George, S.C.: Polyvinylidene fluoride: a multifunctional polymer in supercapacitor applications. J. Power Sources. **504**, 230037 (2021)
150. Lu, Y., Jiang, Y., Lou, Z., Shi, R., Chen, D., Shen, G.: Wearable supercapacitor self-charged by P (VDF-TrFE) piezoelectric separator. Prog. Nat. Sci. Mater. Int. **30**(2), 174–179 (2020)

PVDF as a Flexible Photovoltaic Material

Poonam Subudhi, Praveena Balasubramaniyan, A. Rahul Narasimhan, and Maimur Hossain

Abstract Polyvinylidene fluoride (PVDF) has emerged as a multifunctional polymer in the field of flexible photovoltaics, offering critical benefits for both organic and perovskite solar cells. Its ferroelectric and piezoelectric properties enhance internal electric fields, promote exciton dissociation, and facilitate efficient charge transport. PVDF effectively passivates surface and grain boundary defects, suppresses ion migration, and improves film morphology, thereby boosting power conversion efficiency and device stability. Its hydrophobic nature protects devices from moisture and oxygen ingress, enhancing operational longevity. Compatible with low-temperature, solution-based processing techniques like spin coating and roll-to-roll fabrication, PVDF supports scalable and cost-effective device manufacturing. Additionally, its mechanical strength, flexibility, and environmental resilience make it an ideal substrate or interfacial layer for wearable and bendable solar applications. PVDF thus plays a pivotal role in advancing next-generation, high-efficiency, and durable flexible solar technologies.

Keywords Photovoltaics · Organic · Perovskite solar cells · PVDF · Flexible film · Piezoelectric

P. Subudhi · P. Balasubramaniyan
School of Advanced Sciences, Vellore Institute of Technology, Chennai, India

A. Rahul Narasimhan
School of Electronics Engineering, Vellore Institute of Technology, Chennai, India

M. Hossain (✉)
Aiiso Yufeng Li Family Department of Chemical and Nano Engineering, University of California San Diego, San Diego, CA, USA
e-mail: mahossain@ucsd.edu

D. Punetha (ed.), *Polyvinylidene Fluoride (PVDF)*, Engineering Materials,
https://doi.org/10.1007/978-981-95-6718-8_3

1 Introduction

1.1 Need for Solar Energy Harvesting

According to the economic principle that "human wants are unlimited, but resources are limited," the current global energy scenario starkly illustrates this truth [1]. Non-renewable sources, such as coal, oil, and natural gas, are being rapidly depleted, prompting an urgent transition towards sustainable and renewable energy alternatives [2]. These sources are considered cleaner and more environmentally benign, leading to growing interest across academic and industrial sectors. Prominent among renewable resources are solar, wind, and hydropower [3]. The dual challenge of diminishing fossil fuel reserves and increasing environmental threats due to greenhouse gas emissions and global warming has compelled global stakeholders to seek cleaner energy pathways. In this regard, expanding the use of renewable energy technologies is critical for ensuring sustainable development and securing the planet's environmental future [4].

Solar energy, derived from the sun's light and heat, stands out due to its abundance and accessibility. The process of converting sunlight into usable electrical energy is known as the photovoltaic (PV) effect, and the devices employed for this conversion are called photovoltaic or solar cells [5]. Various approaches exist to harness solar power, including PV systems, solar water heating systems, and solar thermal power plants. PV systems operate on the principle that incident sunlight on a semiconductor material leads to the generation of electron–hole pairs, resulting in electric current when a load is connected. Commercial PV cells generally achieve power conversion efficiencies (PCE) of over 15%. For optimal energy output, PV modules comprising arrays of solar cells are typically installed facing south, either in a fixed-tilt configuration or with solar tracking systems that adjust orientation to maximize sunlight capture [6].

1.2 Generations of Solar Cells

It is estimated that the amount of solar energy received by Earth's surface in just 3 days equals the total energy stored in all fossil fuel reserves. This immense potential makes solar energy a highly attractive option for addressing future energy demands. The modern photovoltaic era began in the 1950s with the development of the first practical solar cells at Bell Laboratories. Since then, solar cell technology has advanced significantly. Once limited to small-scale applications, PV systems are now integral to global energy strategies and are projected to contribute around 35% of the new electricity generation capacity worldwide by 2040.

Solar cell technologies are typically classified into three generations based on the materials used and the structural complexity of the devices (see Fig. 1). First-generation solar cells are based on crystalline silicon. Monocrystalline silicon cells,

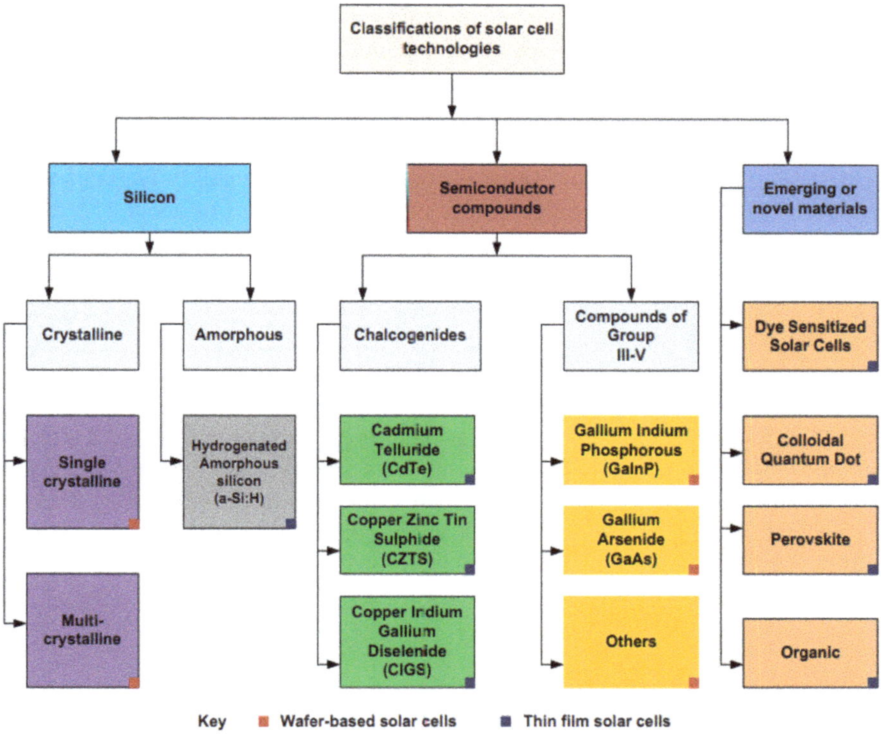

Fig. 1 Generations of solar cells [7]

initially created for aerospace use, were later adapted for terrestrial power generation. Although more efficient, they are costlier to manufacture. Multicrystalline (polycrystalline) silicon cells, produced through casting methods, are somewhat less efficient but are more economical and energy-efficient to fabricate. Currently, approximately 85–90% of all PV modules in the market are based on first-generation technologies, which can achieve efficiencies of up to 25% in laboratory settings.

Second-generation solar cells, commonly referred to as thin-film technologies, utilize much thinner layers of semiconducting materials, typically only a few micrometres thick. This category includes amorphous silicon, cadmium telluride (CdTe), and copper indium gallium selenide (CIGS) cells. Though amorphous silicon cells tend to have lower efficiencies around 10.2% in lab settings their manufacturing requires less energy and material. Microcrystalline variants and multi-junction thin films have shown improved performances up to 13.6%. CdTe cells, while simpler and less costly to produce, raise environmental concerns due to cadmium content, though they can reach efficiencies of 21% in controlled conditions. Among the most promising of the second-generation class are CIS and CIGS technologies, with lab-scale efficiencies nearing 21%. However, their fabrication processes are comparatively complex.

The third generation encompasses emerging PV technologies, many of which are still under research or early-stage development. This group includes both high-efficiency multi-junction solar cells such as those made by Spectrolab, which have achieved record-breaking efficiencies of 38.8% and low-cost solutions like organic photovoltaics (OPVs), dye-sensitized solar cells (DSSCs), and perovskite solar cells (PSCs) [8]. OPVs, with their low fabrication costs and suitability for large-scale roll-to-roll (R2R) manufacturing, currently reach efficiencies of about 11.5% [9]. While their efficiency lags behind other technologies, their scalability and economic feasibility hold significant promise. PSCs have shown an exceptionally fast improvement trajectory, reaching efficiencies over 22% within just a few years of development [10].

Considering the rising demand for lightweight, stretchable, and mechanically robust photovoltaic platforms, flexible substrates and polymers have emerged as critical components in advancing third-generation PV technologies. Polyvinylidene fluoride (PVDF) has garnered attention due to its excellent ferroelectric, piezoelectric, chemical-resistant, and mechanical properties, making it a promising material for integration in flexible OSCs and PSCs. The subsequent sections of this chapter will explore in detail the role and advantages of PVDF in the fabrication, performance enhancement, and long-term stability of organic and perovskite solar cells.

2 Solution Processed Photovoltaic Cells

2.1 Organic Photovoltaics

The origins of the photovoltaic (PV) effect date back to the early 1800s when Alexandre Edmond Becquerel found that light would cause an electrochemical cell to produce a photocurrent. Very late, and far later than the initial discoveries, and advancement of photovoltaics, particularly inorganic devices such as silicon (Si), was the development of organic materials for photovoltaics. Research on photoconductivity in organic materials provided the scientific basis for organic solar cells (OSCs). Pochetin (1906) and Volmer (1913) found that anthracene was one of the first organic materials that displayed photoconductivity. This interest was rekindled when a number of artificial dyes such as methylene blue, were found to be semiconducting and light absorbing in the 1950s and 1960s. Hence, the initial organic solar cells employing thin sheets of organic semiconductors were developed [11]. The initial organic solar cells (OSCs) consisted of a singular organic layer positioned between two electrodes with varying work functions, so charge carrier separation had to occur via the intersection of junction sharp Schottky barriers' as illustrated in Fig. 2a. The energy level band diagram of single layer OSC was shown on Fig. 2b.

In those early days, efficiencies were very low, in the range of 10^{-3} to 10^{-2}%. A notable upgrade was a bilayer heterojunction that was based on an electron acceptor and electron donor materials and allowed charge separation and charge transport

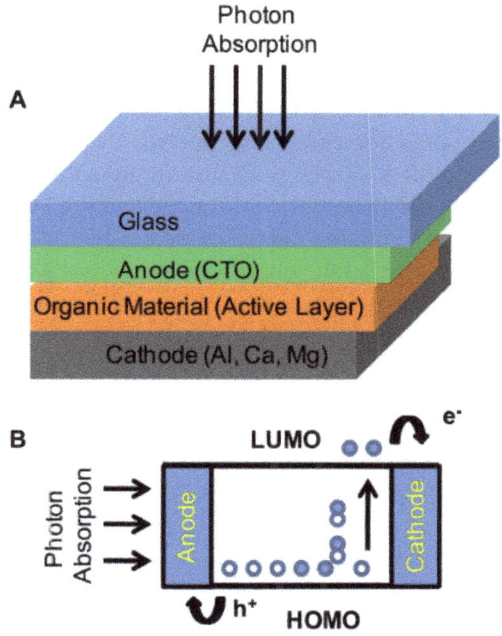

Fig. 2 (**a**) Single organic layer photovoltaic, (**b**) energy level diagram of a single layer OSC [12]

[13]. Another groundbreaking concept for OSCs was bulk heterojunction (BHJ) solar cells. Bulk heterojunctions mix donor and acceptor materials at the nanoscale to create a bicontinuous interpenetrating network that improves charge transport and interfacial area for stimulation dissociation [14]. The combination of organic materials in both layers resulted in a very quick advancement in OSC efficiencies that saw an over 2% efficiency improvement from the beginning and small improvements thereafter with enhancements in material development and device engineering. OSCs are different from their inorganic varieties in that they use small-molecules or π-conjugated polymers as active materials [15]. For these reasons, OSCs tend to have many advantages over inorganic photovoltaics, including mechanical flexibility, low-cost fabrication methods utilizing solution-based processing, and conformance to lightweight substrates [16]. OSCs do however face issues in terms of limited environmental stability, poor exciton diffusion lengths, and low charge-carrier mobilities. There have been significant achievements in OSC development, with improvements made to donor–acceptor materials, improved control over morphology, and interfacial engineering of materials to achieve reported power-conversion efficiencies (PCEs) over 18.69% in laboratory-scale devices [17]. For OSCs interfacial layers are designed to reduce non-radiative recombination, same or better stability of the devices and to assist in energy level alignment and charge selectivity. Organic photovoltaics are also under research and development in hybrid systems by combining both organic materials and inorganic semiconductors or nanocrystals, ultimately benefiting from both material classes. While still not as commercially viable as silicon-based solar cells, OSCs represent a strong future prospect for photovoltaic devices, especially in applications requiring lightweight,

flexible, semi-transparent energy harvesting [14]. In 2024, reliable representation and documentation, OSCs are improving their competitiveness with other thin-film technologies based on recent demonstrated efficiencies in excess of 19%. Compared to other leading PV technologies, for example, perovskites, OSCs are especially appealing for sustainable applications due to their environmental sustainability based off low-energy manufacturing and negligible use of rare or hazardous elements [18]. The real-world applications of OSCs that have improved market appearance are building-integrated photovoltaics (BIPV), wearable electronics, portable power generation, and indoor energy harvesting exploiting their unique flexibility, semi-transparency and variable optical properties.

2.2 *Integration of PVDF in Organic Solar Cells*

Organic solar cells provide key benefits such as mechanical flexibility, low weight, and cost-effective manufacturing. However, their performance tends to decline when exposed to environmental factors like moisture, oxygen, ultraviolet (UV) light, and elevated temperatures. To address this, poly(vinylidene fluoride) (PVDF), a ferroelectric, hydrophobic, piezoelectric, and semi-crystalline polymer, has been widely integrated into OSC architectures. PVDF contributes not only to morphological and interfacial stabilization, but also actively improves charge dynamics, efficiency, and long-term durability.

2.2.1 Charge Dynamics Modulation and Electric Field Enhancement

In OSCs, poor phase separation and interface defects are critical factors that limit exciton dissociation and charge collection efficiency. PVDF addresses these limitations through several synergistic mechanisms. Its fluorine-rich backbone facilitates strong interactions with donor and acceptor molecules, enabling trap passivation and reducing the density of non-radiative recombination centers. The high dielectric constant of PVDF screens the Coulombic interaction between electrons and holes, facilitating exciton dissociation and promoting free charge generation. This function is critical in bilayer and planar heterojunctions, where interfacial charge separation is otherwise insufficient.

The most notable attribute of PVDF is its ferroelectric β-phase, which introduces spontaneous molecular polarization. When incorporated into the active or interfacial layers, this dipole alignment reinforces the built-in electric field (BIF), thus promoting more effective carrier separation and directed charge transport. Wang et al. demonstrated this effect in planar p–n heterojunction (PHJ) OSCs, where the polarization of PVDF-TrFE enabled Wannier–Mott-type free charge generation, significantly improving PCE [19]. This internal polarization reduces the probability of bimolecular recombination and enables field-assisted extraction, even without external poling. Moreover, PVDF contributes to vertical phase separation, favoring vertical orientation of donor and acceptor domains. This spontaneous alignment leads to better charge transport pathways and reduces recombination at disordered interfaces. As an interface layer, PVDF modulates energy-level alignment and smooths interfacial dipoles, ensuring more balanced carrier extraction and minimal leakage currents. The introduction of PVDF is thus not only morphological but electronic in impact, making it a highly functional layer in optimized OSC architectures [20].

2.2.2 Morphological Stabilization and Crystallinity Control

PVDF has a significant influence on the crystallinity, molecular ordering, and morphology of active materials in OSCs. In the commonly used P3HT:PCBM blend, PVDF promotes π–π stacking of donor polymers and enhances nanoscale phase separation. It helps in achieving better ordering and larger domain formation, reducing interfacial area and thereby lowering recombination losses. These properties result in improved carrier mobility and higher current output. In composite systems, such as PANI:PVDF, the polymer blend enhances electrical conductivity and optimizes interfacial interactions. Sardi et al. showed that the β-phase fraction of PVDF exceeded 90% when processed at appropriate temperatures, leading to superior buffer layer conductivity and effective work function alignment. Devices fabricated using ITO/PVDF:PANI layers showed enhanced charge transfer and PCE [21]. Moreover, PVDF is well-suited for solution-based fabrication methods such as spin-coating and inkjet printing, making it ideal for large-scale production and roll-to-roll manufacturing. Its role in dual-interface engineering is also notable. When combined with other interfacial materials such as PS-r-PMMA, PVDF ensures symmetry in energy-level alignment at both electrodes. This balance in interfacial potential reduces hysteresis and enhances fill factor. PVDF is also found to be compatible with non-fullerene acceptors (NFAs) such as ITIC and Y6. These NFAs benefit from PVDF-induced stabilization, as the polymer minimizes voltage losses and improves morphological control for high-efficiency devices [20–22].

2.2.3 Environmental Stability, Moisture Resistance, and Mechanical Durability

Stability is one of the most critical factors determining the commercial viability of OSCs. PVDF's inherent hydrophobic nature, attributed to its fluorinated backbone, makes it a powerful barrier against water vapor and oxygen infiltration. When used as an interlayer or encapsulation layer, PVDF significantly delays the onset of hydrolysis, oxidation, and interfacial degradation. For instance, in a dual-interface configuration with PVDF serving as the cathode interfacial layer, Ge et al. demonstrated strong retention of the initial PCE even after 3600 h of exposure to ambient conditions. Similarly, Li et al. explored an alternative strategy [22], combined PVDF with 3-hydroxyflavone (3-HF) to create a bilayer that not only physically blocked moisture but also chemically scavenged oxygen species. The barrier function of PVDF also limits metal ion migration, preventing electrode-induced parasitic reactions that often lead to irreversible damage. In high-temperature conditions, PVDF maintains structural integrity due to its semi-crystalline matrix, helping reduce stress-induced delamination or interlayer cracking. Furthermore, PVDF's piezoelectric response under mechanical stress introduces the piezophototronic effect, which enables transient internal fields that assist charge separation. This effect is especially valuable in flexible and wearable OSCs, where bending or strain can be converted into an advantage rather than a performance loss. PVDF stabilizes films under cyclic stress, absorbing and redistributing mechanical strain, while also minimizing the formation of microcracks and maintaining long-term device functionality. PVDF stabilizes films under cyclic stress, absorbing and redistributing mechanical strain, while also minimizing the formation of microcracks and maintaining long-term device functionality.

Overall, the incorporation of PVDF has been shown to simultaneously improve device efficiency, environmental resilience, and mechanical endurance, making it a multifunctional material that bridges several bottlenecks in organic solar cell design.

2.3 *Perovskite Solar Cells*

The perovskite solar cell has recently emerged as technology, which has shown great potential and promises to be highly cost effective and more efficient than any other thin-film solar cell technology known so far. Perovskite materials do not incorporate any rare elements that keep their cost very high. These devices can easily be prepared on flexible substrates at room temperature with the spin coating process on a small scale, whereas via the printing process on a large scale [23]. Due to its high efficiency and high-cost effectiveness, this technology made a breakthrough in the field of solar cells. The first perovskite solar cell was prepared in 2009 and had an efficiency of 3.8%. Since then, there have been remarkable improvements in their performance and efficiency has gone beyond 25% and has potential to outperform the Shockley Queisser limit [24, 25]. Their unique properties, along with versatile processing techniques and relatively low fabrication costs, position them at the forefront of next-generation solar technologies. A breakthrough in this domain occurred in 2009 when Kojima et al. first utilized methylammonium lead bromide ($MAPbBr_3$) as a light sensitizer in dye-sensitized solar cells, achieving a PCE of 3.8% [26, 27]. This milestone triggered a wave of research and innovation. Over the years, developments in lead-based, tin-based, germanium-based, polymer-based, lead-free halide double perovskites, and tandem configurations have expanded the landscape of PSCs, making them increasingly adaptable to different application environments and design architectures.

Perovskites refer to materials that adopt the ABX_3 crystal structure as shown in Fig. 3. In this configuration, 'A' typically represents an organic or inorganic cation like methylammonium (MA^+), formamidinium (FA^+), or cesium (Cs^+); 'B' is a

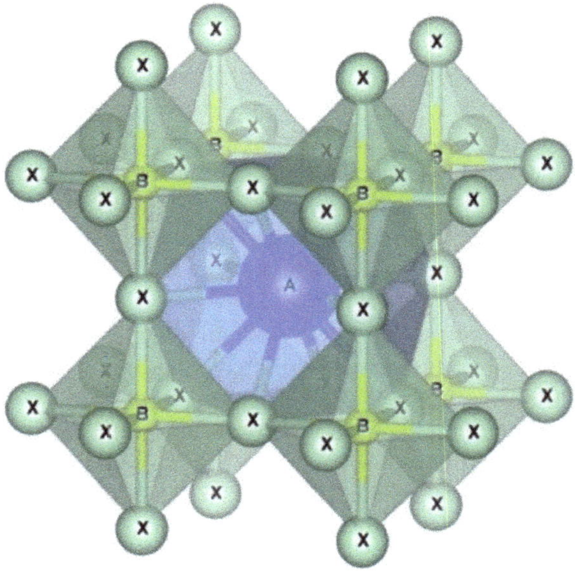

Fig. 3 Crystal structure of perovskite materials [28]

metal cation such as lead (Pb^{2+}) or tin (Sn^{2+}); and 'X' is a halide anion like iodide, bromide, or chloride. This highly tunable structure enables the customization of physical and optoelectronic properties, particularly the bandgap, which can range from 1.3 to 2.2 eV. This tunability allows PSCs to effectively absorb different portions of the solar spectrum, making them ideal candidates for tandem applications with silicon or copper indium gallium selenide (CIGS) solar cells.

The high absorption coefficient of perovskite materials (e.g., 5.7×10^4/cm at 600 nm) means they can harvest solar energy efficiently even in thin-film form. This feature not only contributes to the production of lightweight and flexible solar panels but also minimizes material usage, reducing fabrication costs. In contrast to silicon, which requires thick wafers, perovskites deliver high performance with much thinner layers, opening new possibilities for building-integrated photovoltaics (BIPV), portable electronics, and wearable energy systems.

Another key advantage of PSCs is their excellent charge carrier mobility, typically ranging from 1 to 10 cm^2/V/s. This high mobility facilitates efficient transport of photogenerated charge carriers to the respective electrodes, minimizing recombination losses and boosting overall efficiency. Even under low-light conditions, PSCs can maintain relatively high output, a trait that makes them suitable for indoor applications and regions with variable sunlight exposure. Equally important is the low exciton binding energy of perovskite materials. Excitons bound electron–hole pairs must be dissociated into free carriers to generate current. In PSCs, the low exciton binding energy enables efficient charge separation at room temperature, ensuring effective photoconversion without the need for external thermal assistance. The high dielectric constant of perovskite compounds further aids in reducing recombination by screening electrostatic interactions between oppositely charged carriers.

The success of PSCs also hinges on long charge diffusion lengths, which refer to the average distance a charge carrier can travel before recombining. Longer diffusion lengths enable more efficient charge extraction, especially in devices with relatively thick absorber layers. This property complements the high absorption coefficient and contributes to enhanced open-circuit voltages and short-circuit current densities, both crucial performance indicators for solar cell technologies.

Among the various classes of perovskites, organic–inorganic hybrid perovskites (OHIPs) have garnered particular interest. These materials, which combine an organic A-site cation with an inorganic metal-halide framework, exhibit unique electrical and optical properties not commonly found in purely organic or inorganic semiconductors. With weak exciton binding energies, large Bohr radii, and excellent dielectric properties, OHIPs offer superior light absorption and charge transport characteristics.

In practical PSC device configurations, perovskite materials such as $CH_3NH_3PbX_3$ ($X = I^-$, Br^-, Cl^-) or $CH_3CH_2NH_3SnX_3$ are typically sandwiched between an electron transport layer (ETL) and a hole transport layer (HTL) as illustrated in Fig. 4. When the device is exposed to sunlight, the perovskite layer absorbs photons and generates electron–hole pairs. These charge carriers are then separated and transported to their respective electrodes via ETL and HTL, creating a flow of current through the external circuit [29]. The thickness of the perovskite layer plays a crucial role in determining the efficiency and stability of the device, as it affects both absorption and recombination dynamics. Fabrication of PSCs benefits from several scalable and low-temperature processing techniques, including solution processing,

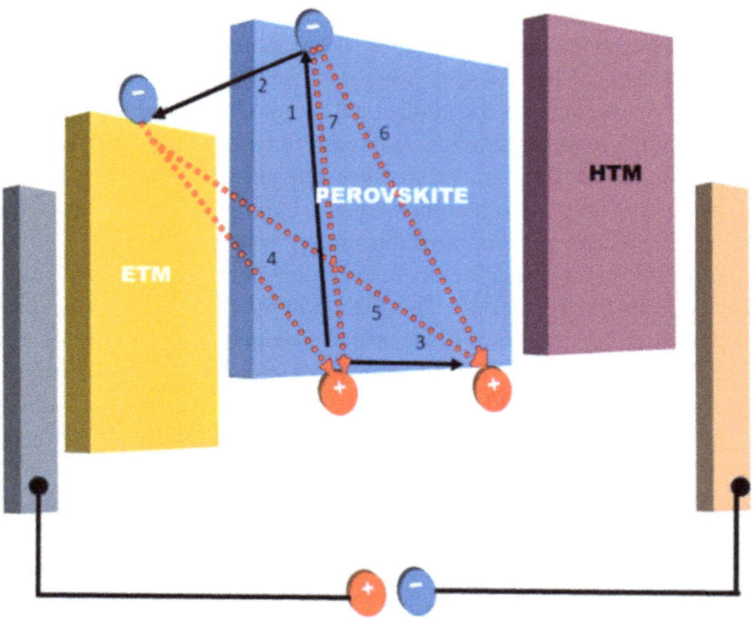

Fig. 4 Schematic representation of PSC device with charge transfer processes [28]

spin coating, slot-die coating, and vacuum deposition. These methods are compatible with flexible substrates and can be adapted for roll-to-roll manufacturing, significantly lowering the production cost and carbon footprint. Notably, many perovskite materials can be synthesized using earth-abundant precursors, adding to the sustainability profile of PSCs.

Despite their exceptional promise, PSCs still face several critical challenges before they can achieve widespread commercial success. Stability under ambient conditions, sensitivity to moisture and heat, and concerns over the toxicity of lead-based compounds remain key issues. Efforts are underway to develop more stable encapsulation methods, replace lead with less toxic metals like tin or bismuth, and improve the intrinsic stability of the perovskite phase through compositional engineering and additive strategies.

Nonetheless, ongoing advancements continue to improve the performance and resilience of PSCs. For example, tandem solar cells that combine perovskites with silicon have already demonstrated efficiencies surpassing 30%, and multi-junction configurations hold the potential for even higher performance. These trends highlight the adaptability and forward momentum of perovskite-based technologies in addressing global energy challenges.

In addition to performance, the future of PSCs will also depend on their ability to integrate with other functional materials and device architectures. One such material is PVDF, a ferroelectric polymer known for its mechanical flexibility, chemical resistance, and excellent dielectric properties. PVDF has recently attracted attention for its potential role in improving the structural integrity and environmental stability of perovskite solar cells, especially when integrated as a protective layer or component of composite films.

2.4 *Integration of PVDF in Perovskite Photovoltaics*

The use of polymer additives in PSCs has emerged as a promising strategy to enhance film quality, passivate defects, and improve long-term operational stability. Among the various polymers explored, PVDF has attracted considerable attention due to its unique combination of electrical, chemical, and mechanical properties. PVDF is not only compatible with solution-processable fabrication techniques but also offers multiple functional roles including defect mitigation, moisture resistance, charge modulation, and structural reinforcement. This section highlights the mechanisms through which PVDF contributes to the improved performance and durability of PSCs, with a focus on its roles in crystallization control, electric field enhancement, environmental protection, and suitability for flexible device applications.

2.4.1 Defect Passivation and Grain Boundary Stabilization

In PSCs, grain boundaries and surface defects are among the primary causes of non-radiative recombination, ion migration, and long-term instability. These regions are typically rich in under-coordinated lead ions, halide vacancies, and organic cation deficiencies, all of which create electronic trap states. The integration of PVDF into the perovskite matrix presents a powerful strategy to suppress these defect-related issues and enhance both performance and stability. PVDF, due to its rich fluorine content and semi-crystalline polymer backbone, offers functional groups capable of forming strong electrostatic interactions and chemical coordination with ions in the perovskite lattice. Fluorine atoms in PVDF can interact with lead ions (Pb^{2+}) and hydrogen from organic cations (such as FA^+ and MA^+), establishing coordinate bonds and hydrogen bonds that effectively passivate under-coordinated sites. These interactions reduce the density of trap states, suppress carrier recombination, and increase carrier lifetimes within the absorber layer. One of the most notable advantages of PVDF is its influence on the crystallization process of perovskites. By modulating ion mobility and solvent evaporation during film formation, PVDF slows down the kinetics of nucleation and grain growth. This controlled crystallization enables the formation of larger and more uniform grains with fewer boundaries. As grain boundaries are a major source of recombination and ionic diffusion, their reduction contributes directly to performance enhancement. PVDF also acts as a structural modulator during the transition from intermediate solvated phases to the final black perovskite phase. Its polymer chains assist in stabilizing the perovskite framework during this conversion, resulting in smoother morphology and more phase-pure films. The presence of PVDF ensures better crystallographic ordering and minimizes the occurrence of secondary phases, which can otherwise interfere with light absorption and charge transport.

Post-crystallization, PVDF molecules tend to localize at grain boundaries and interfaces, where they continue to play a defect passivation role. These regions often exhibit high ionic disorder and serve as pathways for ion migration. PVDF's presence inhibits the accumulation and movement of mobile ionic species such as I^- or MA^+, which are known to cause hysteresis and operational degradation. The polymer matrix effectively blocks these migration channels, improving device stability under illumination and voltage bias. Additionally, PVDF contributes to the mechanical and structural stability of the perovskite layer. Its semi-crystalline nature and flexibility allow it to act as a stress-relieving network within the perovskite film. This is particularly beneficial for flexible PSCs, where bending and thermal stress can lead to cracking or delamination. PVDF helps absorb and distribute mechanical stress, preserving the film's integrity and maintaining long-term performance. Furthermore, PVDF's presence helps create a more uniform dielectric environment within the active layer. This dielectric smoothing minimizes local variations in electric field intensity that can cause spatial inhomogeneity in carrier collection or charge imbalance. The result is more consistent electronic performance, lower leakage currents, and improved fill factor.

2.4.2 Enhancement of Built-in Field via Polarization Effects

PVDF, particularly in its ferroelectric β-phase, demonstrates inherent dipolar characteristics that can be strategically exploited to enhance the internal electrostatics of PSCs. The built-in electric field (BIF) in a PSC plays a vital role in facilitating the separation of photogenerated electron–hole pairs and directing their motion toward the respective electrodes. Efficient charge separation underpins the open-circuit voltage, current density, and ultimately the power conversion efficiency of the device. Integration of PVDF into PSCs offers a passive but robust method to reinforce this internal electric field through permanent molecular polarization.

When incorporated into the active layer or interfacial regions of the PSC, PVDF's aligned dipole moments interact with the existing electric field within the device. The dipoles naturally orient in a direction complementary to the internal field, effectively enhancing the electrostatic driving force across the perovskite layer. This augmented BIF reduces the potential barrier for charge carrier drift and suppresses the probability of recombination by accelerating charge separation. Consequently, photogenerated carriers are extracted more efficiently, contributing to improved Voc, higher charge collection efficiency, and increased fill factor. PVDF's ability to transition from a non-polar α-phase to a highly polar β-phase is central to this phenomenon. The β-phase exhibits an all-trans conformation of its molecular chains, resulting in a strong net dipole moment. This transformation can be induced by thermal processing or through blending with polar molecules such as $dabcoHReO_4$ (DH), which promotes multiaxial polarization. In such PVDF-DH composite systems, the resulting field is not only stronger but also better aligned with the device's inherent field. This synergy generates a supplementary internal polarization that enhances the internal field homogeneity and energy-level alignment across the device layers. The effects of this internal polarization are multifaceted. Beyond promoting carrier separation, PVDF-induced polarization contributes to the suppression of ion migration within the perovskite lattice. Mobile ionic species, particularly iodide and methylammonium ions, are responsible for hysteresis and long-term instability in PSCs. The electrostatic environment created by polarized PVDF dipoles inhibits the mobility of these species by increasing the energy barrier for ion displacement, thus stabilizing the perovskite structure over time.

Furthermore, PVDF's dipolar network contributes to the electrostatic smoothing of interfacial potentials. Fluctuations in surface potential and local field irregularities often lead to the formation of defect states and recombination sites. The uniform dipolar field introduced by PVDF acts as an electrostatic stabilizer, mitigating such potential variations and ensuring a more consistent electronic landscape across the perovskite and adjacent layers. The ferroelectric nature of PVDF also opens possibilities for field-responsive behavior under mechanical or electrical stimuli. Though the polarization effect in this context is spontaneous and does not require external poling, its permanence offers a long-term method for field reinforcement without additional energy input. As a result, PVDF functions not only as an additive that enhances physical morphology but also as an active electronic component that improves the internal operational dynamics of PSCs.

2.4.3 Crystallization Control and Interface Engineering

The crystallization behavior of perovskite films plays a critical role in defining their optoelectronic properties and overall device efficiency. Rapid or uncontrolled crystallization often leads to polycrystalline films with small grains, high defect densities, and non-uniform morphology. These issues increase recombination losses and compromise the performance and stability of PSCs. PVDF, due to its chemical composition and polymeric nature, serves as a highly effective agent for regulating crystallization and enhancing interface properties. During perovskite film formation, PVDF slows down the transition from the solvated intermediate phase to the final black perovskite phase. This delayed conversion is primarily attributed to the interaction of PVDF's fluorine atoms with metal ions and organic cations in the precursor. These interactions control the release and movement of ions, resulting in a more uniform and gradual crystallization process. As a consequence, the final perovskite layer develops larger and more continuous grains with well-defined boundaries and fewer defect states.

This improvement in grain quality directly reduces the number of trap states within the bulk and at grain boundaries. The larger grains minimize the total interfacial area where recombination typically occurs, while the passivation effect of PVDF further suppresses the formation of deep-level traps. This dual benefit ensures more efficient charge carrier transport and extraction, leading to improved Jsc and FF. Moreover, PVDF influences the morphology of the film by enhancing surface coverage and reducing pinhole formation. Its polymeric matrix contributes to a smoother film texture, which facilitates better contact with adjacent transport layers. A uniform perovskite layer is essential for preventing shunting paths and ensuring consistent current flow across the device.

In addition to morphological benefits, PVDF contributes to the alignment of energy levels between the perovskite absorber and charge transport layers. The presence of PVDF at the interface can induce favorable dipole orientations that shift energy bands, thereby improving carrier injection and extraction. This leads to reduced interfacial resistance and better rectifying behavior within the device. Furthermore, PVDF's polymer network acts as a mechanical buffer layer at the interface, absorbing stress during thermal cycling or bending in flexible devices. This reinforces the mechanical integrity of the interface, which is often the most vulnerable point in multilayered photovoltaic architectures. PVDF also aids in stabilizing interfacial chemical reactions. In PSCs, uncontrolled interfacial degradation due to ion migration or moisture can quickly deteriorate performance. PVDF serves as a protective interfacial layer that limits ion diffusion and shields reactive species from environmental exposure. This protective function extends the operational lifetime of PSCs and helps maintain consistent performance over time.

2.4.4 Piezoelectric Functionalities and Application in Flexible PSCs

In addition to its polarization properties, PVDF especially in the form of its copolymer P(VDF-TrFE) exhibits strong piezoelectric behavior, making it highly attractive for use in flexible and self-powered perovskite solar cells (PSCs). Piezoelectric materials generate an internal electric field when subjected to mechanical strain, such as bending, stretching, or pressure. This strain-induced internal field can interact with the charge transport dynamics of PSCs, especially in flexible configurations. When incorporated into flexible PSCs, PVDF-based materials act as mechanically responsive components. Upon mechanical deformation—such as during bending or wearable application use—the piezoelectric effect induces a transient electric field that enhances carrier separation within the perovskite layer. This effect, known as the piezophototronic effect, allows for improved charge extraction by amplifying the internal potential in strained regions. As a result, devices exhibit enhanced current output and reduced recombination under dynamic mechanical conditions.

Unlike conventional devices that may suffer performance drops under bending, PVDF-enabled structures can benefit from deformation. This makes PVDF not only a mechanical stabilizer but also an active enhancer of photovoltaic behavior in flexible systems. Such characteristics are crucial for emerging technologies including wearable electronics, foldable displays, and motion-powered solar panels. In practical applications, the dynamic polarization induced by the piezoelectric response can be harnessed to modulate junction properties, effectively tuning the local energy landscape in real time. This leads to a more adaptive and responsive solar cell architecture, which is especially beneficial in environments where devices are continuously exposed to movement or vibration. Moreover, the incorporation of piezoelectric PVDF layers contributes to mechanical durability. Its semi-flexible nature helps distribute mechanical stress more evenly across the device layers, reducing the risk of microcracks, delamination, or electrode damage during repetitive bending. This enhances both performance consistency and operational lifespan.

2.4.5 Stability and Moisture Resistance

One of the most critical challenges limiting the commercialization of PSCs is their sensitivity to environmental conditions particularly moisture, oxygen, temperature, and prolonged light exposure. These external stressors can degrade the perovskite layer, reduce efficiency, and cause long-term operational instability. PVDF, due to its high hydrophobicity, chemical inertness, and polymeric structure, offers an effective means of enhancing the environmental stability of PSCs. The hydrophobic character of PVDF originates from its fluorine-rich backbone. Fluorine atoms repel polar molecules like water, forming a moisture-resistant barrier when PVDF is incorporated into or coated over perovskite films. This barrier prevents water vapor from reaching and reacting with the perovskite's ionic lattice, thereby avoiding hydrolysis and phase degradation. Even under ambient conditions, the presence of

PVDF helps maintain the integrity of the perovskite crystal structure. In addition to resisting water ingress, PVDF also plays a role in blocking oxygen and metal ion diffusion, which are known to trigger unwanted redox reactions at the perovskite–electrode interface. For instance, metal electrodes like silver or gold can slowly diffuse into the perovskite layer over time, forming metallic halide compounds that reduce device performance. A PVDF layer at the interface acts as a physical and chemical shield, preventing such interfacial degradation. Furthermore, PVDF contributes to thermal and photostability. Its semi-crystalline polymer matrix can tolerate a broad range of temperatures without decomposition or structural distortion. This thermal resilience helps reduce the formation of temperature-induced defects or layer delamination, especially under fluctuating operating conditions. Under sunlight exposure, PVDF's chemical robustness prevents photo-oxidation and suppresses the formation of unstable intermediate species in the perovskite film. The barrier properties of PVDF also help in suppressing ion migration under electric bias, an issue that contributes to hysteresis and performance drift. By hindering the movement of mobile species like iodide ions, PVDF stabilizes the internal ion distribution and supports consistent electrical behavior over time. Taking together, these mechanisms contribute to significantly enhanced operational lifetime for PSCs. Devices with PVDF integration have been shown to retain a large percentage of their initial efficiency even after extended exposure to humidity, light, and heat, without requiring additional encapsulation layers. This simplifies the device architecture and supports scalable fabrication for real-world applications.

2.4.6 Comparative Advantage Over Other Polymers

Various polymers have been explored as additives in perovskite solar cells (PSCs) to improve film quality, passivate defects, and enhance stability. Common examples include poly(methyl methacrylate) (PMMA), polyvinylpyrrolidone (PVP), polyvinyl alcohol (PVA), and polyethylene glycol (PEG). While these materials contribute to specific aspects of device performance, poly(vinylidene fluoride) (PVDF) offers a broader and more multifunctional impact, making it distinctively advantageous among polymeric additives.

Most conventional polymers used in PSCs serve primarily as morphology regulators or surface passivators. For instance, PVP and PEG are effective in modifying grain boundaries and reducing trap states, while PMMA and PVA help enhance surface uniformity. However, these polymers often suffer from issues like poor environmental stability, limited long-term compatibility with perovskite chemistry, and lack of functional versatility. Their stabilizing effects are generally passive and often rely on weak physical interactions such as van der Waals forces or hydrogen bonding. In contrast, PVDF exhibits strong chemical affinity toward perovskite constituents due to the presence of highly electronegative fluorine atoms, which can form coordinate bonds and strong hydrogen bonds with metal ions and organic cations. This enables PVDF to engage in effective defect passivation, not only at grain boundaries but also throughout the bulk of the film. Unlike other polymers, PVDF also actively modulates crystallization kinetics, allowing for the growth of large, uniform grains and suppression of non-photoactive phases.

Another significant distinction lies in PVDF's hydrophobic nature. Unlike hydrophilic polymers such as PVA or PEG—which can attract moisture and lead to faster degradation—PVDF acts as a moisture barrier. It resists water penetration and prevents the degradation of the perovskite lattice in humid environments. This property makes PVDF especially valuable for long-term outdoor applications, where humidity is a major degradation factor. Furthermore, PVDF offers ferroelectric and piezoelectric functionalities, which are absent in most conventional polymers. These properties enable PVDF to contribute to the internal electric field of the device, enhance charge separation, and improve carrier transport. Its dipolar nature also aids in suppressing ion migration, a feature critical for minimizing hysteresis and ensuring stable operation. The combination of electronic modulation and mechanical responsiveness makes PVDF uniquely suited for flexible and self-powered devices, unlike traditional polymers that are limited to static roles. Additionally, PVDF's thermal and chemical stability is superior to many commonly used polymers. It maintains structural integrity under thermal cycling and UV exposure, whereas other polymers tend to degrade, yellow, or delaminate over time. This chemical resilience adds another layer of long-term protection to perovskite devices.

3 Flexible Solar Cells

Flexible solar cells revolutionize photovoltaic technology by enabling lightweight, conformable, and versatile power generation, moving beyond the limitations of rigid panels to integrate seamlessly into diverse applications. This transformative approach relies on depositing thin-film light-absorbing materials onto flexible substrates, allowing devices to maintain electrical performance even when bent, folded, or stretched. The substrate, therefore, is not merely structural support but a critical determinant of the solar cell's mechanical durability, optical efficiency, thermal stability, and overall manufacturability, opening new frontiers for solar energy deployment in previously inaccessible or impractical scenarios.

3.1 PVDF: A Flexible Substrate

PVDF serves as an excellent flexible substrate due to its unique combination of inherent properties. Its semi-crystalline molecular structure provides a crucial balance: the amorphous regions allow for significant bending without fracture, while the crystalline areas impart necessary mechanical strength and thermal stability. This results in a material with high tensile strength and elongation at break, capable of withstanding considerable pulling forces and stretching, coupled with a relatively low Young's modulus that ensures its conformability. Beyond its mechanical prowess, PVDF boasts robust chemical resistance to a wide array of substances, good thermal stability allowing it to endure fabrication and operational temperatures, and advantageous electrical properties including a low dielectric constant and high breakdown strength. Its resilience extends to UV and weathering resistance, making it suitable for outdoor applications, and its low water absorption ensures reliable performance in humid conditions.

These combined attributes make PVDF a highly versatile and reliable flexible substrate across numerous advanced applications. It is particularly valuable in flexible photovoltaics for creating conformable solar cells, wearable electronics where durability and flexibility are paramount, and flexible sensors and actuators that leverage its inherent piezoelectric properties. Furthermore, PVDF finds utility in flexible printed circuit boards and various energy storage devices due to its electrochemical stability. Its ability to maintain structural and functional integrity under mechanical stress and environmental exposure positions PVDF as a critical enabler for the next generation of flexible electronic devices.

3.2 Deposition Techniques on PVDF Substrates

As PVDF stands out as a highly attractive flexible substrate for the burgeoning fields of OSCs and PSCs. Its exceptional mechanical flexibility, robust chemical resistance, and moderate thermal stability make it an ideal candidate for wearable electronics and flexible power sources. However, successful thin-film deposition onto PVDF necessitates careful consideration of its inherent properties, particularly its relatively low melting point (typically around 160–170 °C) and inert surface chemistry. Consequently, deposition strategies must prioritize low-temperature processing and often incorporate surface pre-treatment to ensure optimal film quality and device performance.

3.2.1 Roll-to-Roll (R2R) Processing

Roll-to-roll (R2R) processing [30] refers to high-output manufacturing processing technology that employs a flexible substrate, transported in a continuous loop between rollers while subjecting it to different processing sequences to complete fabrication. The technology is commonly employed for the manufacture of flexible solar cells (FSCs), displays, batteries and many other printed electronics. R2R processing may contain multi methodologies including sputtering, chemical vapor deposition (CVD), inkjet printing, and plasma treatment, as a means to build thin-film layers on flexible substrates (plastic, metal foil, paper). The process typically moves raw materials through a series of rollers, for thin film growth and selective deposition steps, followed by lamination, and then an output as finished components. The primary challenge when performing R2R processing is to maintain proper web tension to get the registration and alignment of the layers right, which is

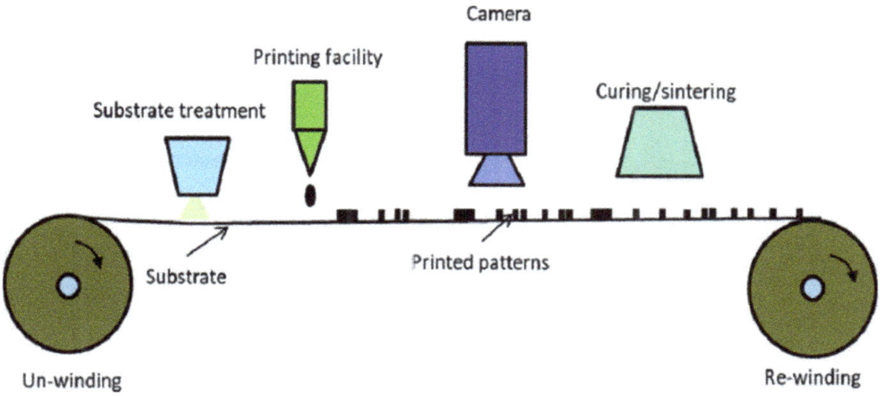

Fig. 5 Schematic illustration of typical roll-to-roll system [31]

critical to producing quality, repeatable product. Tension control systems are used to monitor and control the substrate movement to prevent defects like misalignment and wrinkling (Fig. 5).

Although R2R processing involves some complexity and is a large capital investment in the beginning, it has many advantages including high throughput, low per-part price, material waste, and scalability. This is especially true for manufacturing products where a large area needs to be utilized, and both additive and subtractive patterning can be used. R2R processing can be used in many sectors. In photovoltaics, R2R is used to fabricate flexible PV modules including CIGS-based devices. In energy storage, R2R is employed for printing thin-film batteries, such as laminar lithium-ion devices. R2R processing also plays an important role in the large-scale manufacturing of flexible electronics products such as RFID tags, OLED displays, sensors, and multilayer printed circuits. Beyond this, R2R has applications in the manufacturing of thick film sensor materials, multilayer capacitors, and components in solid oxide fuel cells. Because R2R processing is fast and scalable, it is a foundational technology for the development of new lightweight, flexible, and portable electronic devices.

3.2.2 Chemical Bath Deposition (CBD)

Chemical Bath Deposition (CBD) [33] is a widely recognized method for fabricating thin films, particularly effective for large-area manufacturing. It is commonly employed in the formation of buffer layers in solar cell structures. Also known as the chemical solution method or solution growth method, CBD involves immersing a substrate into a precursor solution containing metal ions, chalcogenide ions, and a base to facilitate controlled film growth on the substrate surface. The ions are released and react in a controlled manner aided by complexing agents and buffered solutions, ultimately depositing a thin film as the process progresses. The reaction is typically performed at temperatures not exceeding 100 °C, useful if some

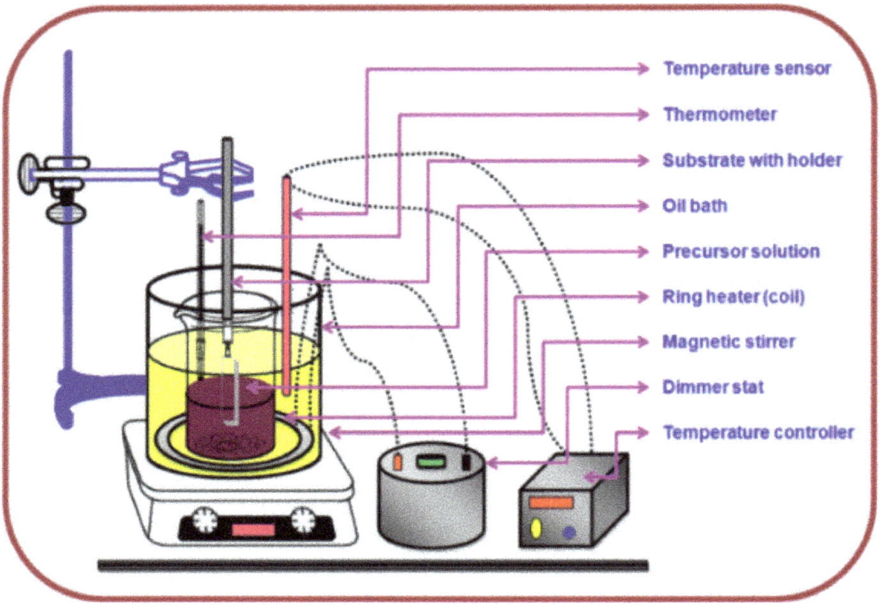

Fig. 6 Schematic representation of chemical bath deposition [32]

temperature-sensitive materials such as polymers are used. To achieve greater uniformity of the film, you may make use of a rotating substrate holder. CBD is preferred for its simple processing and low-temperature capabilities allowing for the deposition of sulfide, selenide, and oxide films without further thermal treatments. The method produces reproducibly uniform and adherent thin films. One of the primary limitations of CBD is the waste disposal of the solution after each deposition cycle, which involves a large volume of liquids. Nevertheless, CBD is entirely applicable to layers that are intended to be functional such as transparent conductive oxides (TCOs), catalysts, and superhydrophobic coatings. CBD has also been especially important for flexible solar cell (FSC) applications where it has been used to deposit materials such as ZnO on glass substrates, where it was an affordable and scalable way to develop high-performing thin films (Fig. 6).

3.2.3 Chemical Vapor Deposition (CVD)

Chemical vapor deposition (CVD) is a widely used thin film fabrication technique known for its precise deposition rates, high reproducibility, and excellent uniformity [35]. This versatile process finds extensive applications in microelectronics, photovoltaics, optoelectronics, and protective coatings. In a CVD system, volatile precursors react on or near a heated substrate, depositing a solid thin film while gaseous by-products are carried away. CVD can deposit a wide array of materials, including metals, oxides, nitrides, carbides, and various ceramics, with common

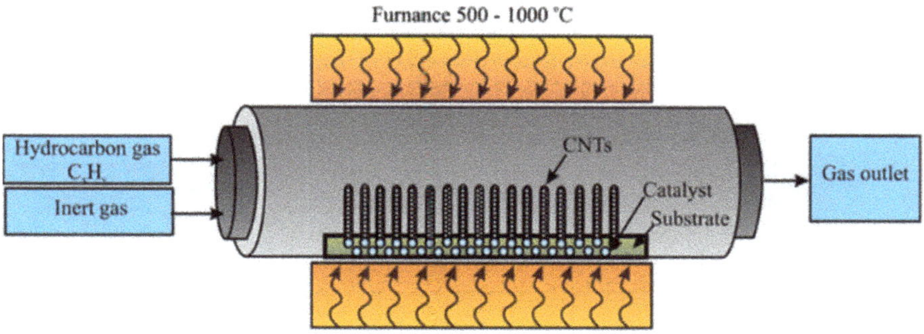

Fig. 7 Schematic representation of chemical vapor deposition (CVD) process [34]

examples being silicon dioxide, silicon nitride, and carbon-based materials like graphene. Its key advantages, such as producing ultra-high-purity films, achieving excellent conformity on complex surfaces, and ensuring strong adhesion, make CVD indispensable for semiconductor device fabrication requiring insulation, barrier, or gate dielectric layers [35] (Fig. 7).

3.2.4 Spin Coating

Spin coating [37] is a widely used technique for producing uniform thin films on flat substrates. It is commonly applied in fields such as microelectronics, solar cell fabrication, and optical coatings. The process is straightforward: a liquid solution is dispensed onto the center of a substrate, which is then rapidly rotated typically at speeds ranging from 1000 to 10,000 revolutions per minute (RPM). The centrifugal force spreads the solution evenly across the surface, resulting in a smooth, uniform film. When the substrate rotates, the liquid will spread radially outward due to centrifugal force, creating a thin layer of liquid. In the meantime, the liquid will evaporate out of the solution due to the spinning and contact with ambient air or from heat, and a solid thin film will be created. The thickness of the final film will depend on many factors including solution viscosity, final spin speed, spin acceleration, and time spun. Higher speed spinning and lower viscosity will end up with a thinner film, while with low speeds and high viscosity solutions, end up with thicker films. Spin coating is popular due to its reproducibility, it is easy to use, with limited processing steps, and very uniform thin films can be produced on planar surfaces. Spin coating is generally limited to small area, flat substrates due to the centrifugal spreading. Nevertheless, spin coating remains an important technique for the deposition of photoresists, dielectric layers, organic semiconductors and in photovoltaic applications (Fig. 8).

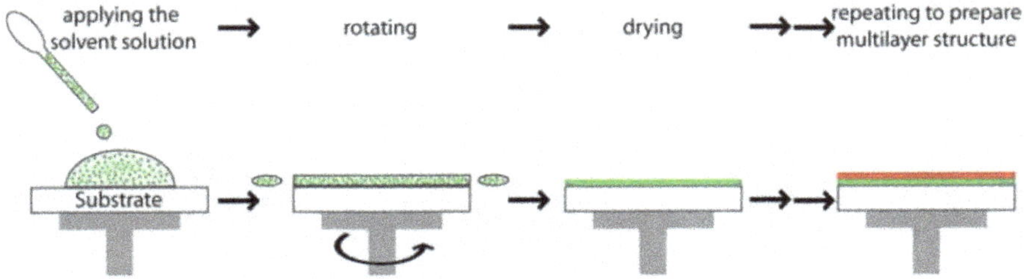

Fig. 8 Schematic of the spin coating process [36]

3.2.5 Dip Coating

Dip coating [39] is a straightforward deposition technique in which a substrate is immersed into a solution containing the desired coating material and then withdrawn vertically at a controlled speed. As the substrate is pulled out, a thin liquid film adheres to its surface. Upon solvent evaporation, this liquid film solidifies into a uniform thin layer. The final thickness and uniformity of the coating are influenced by several factors, including the solution's viscosity, surface tension, withdrawal speed, concentration, and ambient temperature. The substrate after withdrawal is frequently annealed to improve uniformity and adhesion of the thin film to the substrate. The dip method is highly compatible with a variety of substrate materials, and the dip method can deposit coatings on both sides of a substrate in one immersion. The dip method allows the use of coatings on high accuracy, coatings with high precision, or production to a larger scale. Also, it is relatively easy to modify the dip method for lower concentrations simply by varying the withdrawal speed. While the dip coating method has certain advantages, such as ease of conducting, compactness of the equipment, or applicability to R2R, it also has its disadvantages, including longer drying times, and the risk of contamination of evaporating wet films. Despite its advantages and disadvantages, dip coating is still a very efficient method for making functional films, particularly in research or small-scale industrial applications (Fig. 9).

3.2.6 Langmuir–Blodgett (LB) Deposition

Langmuir–Blodgett (LB) deposition [41] is a thin-film deposition technology that builds ultra-thin, highly ordered films through the transfer of monolayers or multilayers of ordered molecules from a liquid surface to a solid substrate, usually water, to containerize the liquid (or in some cases, even directly from the atmosphere) or aqueous substrates. LB deposition is especially useful for producing ultra-thin (<100 nm), highly ordered films. The process begins by spreading amphiphilic liquid molecules (molecules with hydrophilic and hydrophobic) in a monolayer on top of the water layer. At the water-air interface, amphiphilic molecules organize themselves and form a stable film of ordered molecules called a Langmuir film. Once a

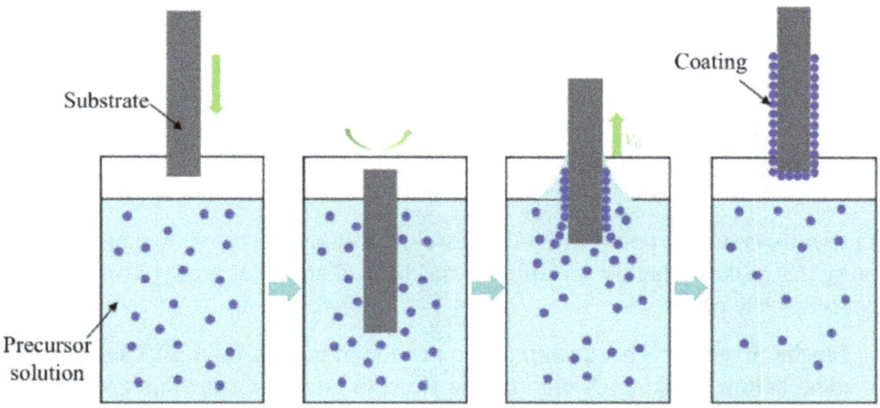

Fig. 9 Schematic of dip coating process [38]

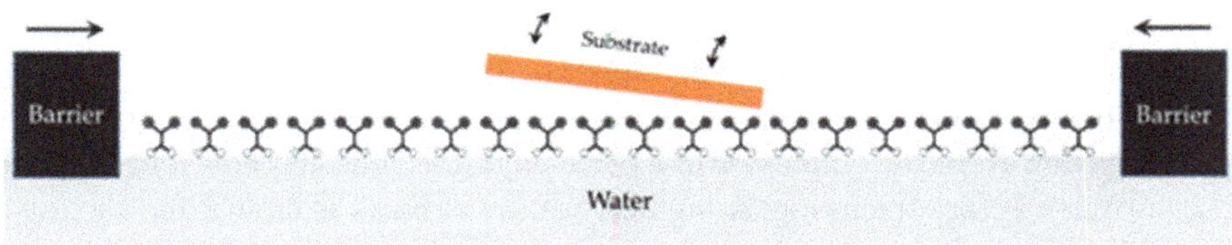

Fig. 10 Illustration of transferring monolayers to a solid substrate in the LB method [40]

Langmuir film is established, the packing density of the film can be manipulated by compressing the monolayer. After the desired surface pressure and arrangement of the molecules achieved, a solid substrate can then be vertically dipped into or pulled from the subphase, which allows for the transfer of the organized monolayer onto the substrate material. The process is repeatable depending on how many multilayer films results from reuse of the solid substrate, and have accurate control of the desired thickness and structure of the multilayers. LB is very useful for studying chemical molecular self-assembly, producing functional coatings, and thin film fabrication for sensors and organic electronics and various bio surface fabrication. While LB offers excellent film uniformity and quality control, the limitations of the method is that it is limited to lab-scale use and the deposition is a relatively slow processing method to use (Fig. 10).

4 Mechanical Characterization of PVDF Flexible Substrate

Understanding the mechanical behavior of PVDF substrates is crucial for designing durable, efficient, and long-lasting flexible PV devices. This section will detail the various mechanical characterization techniques employed to assess PVDF's suitability, differentiating between the intrinsic material properties and the specialized considerations for flexible substrates.

4.1 Intrinsic Mechanical Properties of PVDF

As a semi-crystalline polymer, PVDF possesses a unique blend of strength and flexibility that makes it highly versatile. Its mechanical attributes are typically defined by several key parameters:

- **Tensile Strength**: This quantifies the maximum pulling force PVDF can withstand before breaking. A high tensile strength is vital for flexible PV applications, ensuring the material can endure the stresses of handling, installation, and general operation without fracturing.
- **Young's Modulus (Elastic Modulus)**: This measures the material's stiffness or rigidity. For flexible PV, a lower Young's modulus signifies greater flexibility, allowing the material to bend without permanent deformation. PVDF strikes a good balance, offering medium stiffness that is often beneficial.
- **Elongation at Break**: Expressed as a percentage, this indicates how much PVDF can stretch before fracturing. A high elongation at break is crucial for ductility, enabling the substrate to accommodate significant bending and stretching without catastrophic failure.
- **Yield Strength**: This vital property indicates the stress level at which PVDF begins to deform permanently. Materials with high yield strength like PVDF maintain their original shape even after being subjected to considerable mechanical stress.
- **Impact Strength**: Represented in joules per meter (J/m), this measures PVDF's capacity to absorb and dissipate energy from a sudden force or shock. Its excellent impact resistance is valuable for protecting sensitive PV components from accidental drops or impacts.
- **Fatigue Resistance**: PVDF exhibits strong resistance to fatigue, meaning it can endure repeated cycles of mechanical stress without significant degradation. This is particularly important for flexible components that undergo constant bending or cyclic loading.
- **Creep Resistance**: This property describes PVDF's ability to resist permanent deformation under a sustained mechanical load over time. Good creep resistance ensures the long-term dimensional stability of the substrate, crucial for device reliability.

These properties are commonly evaluated using standardized tests like those from ASTM (e.g., ASTM D638 for tensile properties, ASTM D256 for impact strength). Furthermore, PVDF's mechanical properties can often be enhanced through specific processing methods such as molding, stretching, and annealing, which influence its molecular alignment and crystallinity, leading to improvements in strength and ductility.

4.2 Specialized Characterization for Flexible Substrates

The unique nature of flexible substrates necessitates additional mechanical characterization beyond the fundamental material properties. These tests specifically address the challenges associated with thin, bendable materials in electronic applications:

- **Residual Stress Measurement**: Unlike rigid semiconductors, flexible substrates can accumulate internal stress during fabrication. Residual stress analysis is critical because these inherent stresses can lead to warping, cracking, or delamination over time, compromising device integrity. Techniques are employed to quantify these internal stresses to optimize manufacturing processes.
- **Fracture Toughness Measurement**: This evaluates a material's ability to resist the propagation of cracks in the presence of existing flaws. For flexible solar cells, understanding fracture toughness is vital for ensuring the substrate won't fail prematurely from minor defects. Methods such as load-versus-depth indentation curves and controlled buckling tests are used to assess this critical property.
- **Adhesion Measurement**: The reliability of flexible electronic devices heavily relies on strong interfaces between different layers. Adhesion refers to the interfacial forces holding two surfaces together. Poor adhesion can lead to spontaneous delamination, especially when combined with residual stress. Characterizing adhesion quality is crucial before fabricating reliable FSCs. While conventional adhesion tests for rigid substrates often aren't suitable, specialized methods like adhesion strength measurement and interfacial toughness measurements are employed, often focusing on initiating and modeling the propagation of stable cracks at interfaces. It's challenging to find a universal technique, as failure mechanisms can vary greatly.
- **Hardness and Elastic Modulus Measurement (Surface Specific)**: While general hardness and elastic modulus are characterized, specific tests like indentation, surface wrinkling, and the bubble method are used to assess these properties at the surface or thin-film level. These methods often analyze loading/unloading curves during indentation and are sensitive enough to measure nanometer-level deformations, which are particularly relevant for flexible electronics with thin functional layers.

4.3 Encapsulation Characterization

The encapsulation layer, which protects the sensitive PV components from environmental degradation, also undergoes rigorous mechanical and thermal characterization. Techniques include:

- **X-ray Topography (XRT)**: This non-destructive method is used to identify process-induced defects and growth striations in thin films within the encapsulation, ensuring the quality of the protective layers without damaging the solar module.

- **Thermally Stimulated Techniques (TSCs)**: TSCs assess the thermal properties of encapsulating materials, typically over a broad temperature range (e.g., −150 to +70 °C). Analyzing the peak maximum current and the area under the TSC current peak helps determine key parameters like glass transition temperature, activation energy, and relaxation frequency, which are crucial for predicting the long-term stability and performance of the encapsulation under varying thermal conditions.
- **Adhesion and Peel Tests**: These are vital for evaluating the bond strength between the encapsulation material and the underlying PV components, preventing delamination and ensuring a robust protective seal.

By thoroughly characterizing PVDF flexible substrates using these diverse mechanical tests, researchers and manufacturers can ensure the development of highly reliable, durable, and performant flexible photovoltaic devices capable of withstanding the rigors of various application environments.

5 Enhancement in Power Conversion Efficiency

5.1 Performance Enhancement of Organic Solar Cells Through PVDF Integration

Enhancing power conversion efficiency is central to the evolution of organic solar cells into commercially viable technologies. In recent years, the introduction of functional polymer additives, especially PVDF and its ferroelectric copolymers like PVDF-TrFE has opened a new dimension in active layer and interfacial engineering. These materials not only play passive roles in morphological tuning but also actively contribute to the internal charge dynamics and device electrostatics, enabling significant performance gains. This was presented by Wang et al., who demonstrated the potential of PVDF-TrFE in planar p–n heterojunction (PHJ) OSCs. Traditionally, bulk heterojunctions (BHJs) have dominated OSC research due to their extensive donor–acceptor intermixing, which facilitates exciton dissociation. However, PHJ architectures offer simpler processing, better structural control, and superior reproducibility, though their efficiency has lagged due to limited interfacial charge generation. The ferroelectric polarization induced by PVDF-TrFE introduces a strong built-in electric field that enhances exciton dissociation at the donor–acceptor interface. This internal polarization promotes a Wannier–Mott-type free charge generation, shifting the operational paradigm from tightly bound Frenkel excitons to more delocalized, easily dissociated carriers thus bridging the efficiency gap between PHJ and BHJ devices. This integration led to an impressive power conversion efficiency (PCE) of 19.3%, accompanied by a short-circuit current density (Jsc) of 24.71 mA/cm^2, an open-circuit voltage (Voc) of 0.923 V, and a fill factor (FF) of 84.2% [19].

In addition, PVDF-based layers contribute to vertical phase separation, driving the preferential orientation of donor and acceptor materials in the vertical direction. This alignment ensures more efficient charge transport to respective electrodes and reduces the bimolecular recombination typically observed in disordered blends. The ferroelectric nature of PVDF also allows for dipole alignment without external poling, meaning that high internal polarization can be achieved spontaneously during film formation. This results in improved energy-level alignment and more efficient charge collection at the electrodes. Additionally, PVDF contributes to enhanced molecular ordering and increased crystallinity in donor polymers. For example, in the P3HT:PCBM system a well-known BHJ blend PVDF has been found to enhance the degree of crystallinity and π–π stacking of the P3HT domains when incorporated at optimal concentrations. This ordering reduces energetic disorder, facilitates intra-chain charge hopping, and enhances carrier mobility, thereby increasing both the short-circuit current and fill factor. Moreover, PVDF can guide the formation of larger grain domains, decreasing grain boundaries that act as recombination centers.

In dual-interface engineering strategies, PVDF serves not only as a performance enhancer but also as a functional gatekeeper at the cathode side. In combination with materials like PS-r-PMMA at the anode, PVDF helps preserve the energetic landscape at both interfaces, reducing interfacial dipoles and stabilizing charge transport pathways. This symmetry in interface design contributes to balanced charge extraction, a key determinant of high fill factor and suppressed hysteresis. For instance, Ge et al. reported a PCE of 13.4%, with a Jsc of 21.1 mA/cm^2, an Voc of 0.87 V, and FF of 73.2% in devices utilizing PVDF/PS-r-PMMA dual-interface protection. Remarkably, these devices maintained 81.2% of their initial efficiency (10.9%) after 3600 h of exposure to humid air, demonstrating that the performance improvements did not come at the cost of long-term stability [20] (Fig. 11).

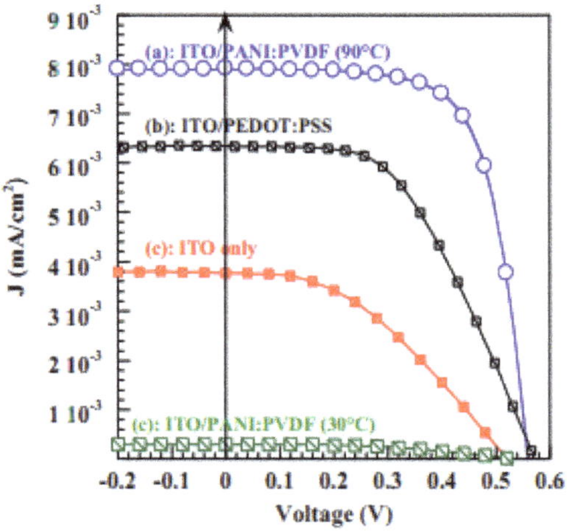

Fig. 11 J–V characteristics of (**a**) ITO/PANI:PVDF/P3HT:PCBM/Al, (**b**) ITO/PEDOT:PSS/P3HT:PCBM/Al, (**c**) ITO/P3HT:PCBM/Al and (**d**) ITO/PANI:PVDF/P3HT:PCBM/Al

Another important approach involves the use of PVDF composite buffer layers, such as PANI:PVDF thin films [21]. These structures were shown by Sardi et al. to exhibit high β-phase content in PVDF (Fβ > 90%) when processed at optimized temperatures. The β-phase is critical for enabling dipolar activity, which translates to enhanced carrier mobility and interfacial conductivity (Table 1).

Devices fabricated using ITO/PVDF:PANI buffer layers outperformed those with traditional PEDOT:PSS, owing to better charge transfer characteristics and a more favorable work function alignment with the active layer. Their study reported a PCE of 3.06%, with a Jsc of 7.54 mA/cm^2, Voc of 0.71 V, and FF of 57.2%. Importantly, PVDF-based buffer or interfacial layers can be seamlessly integrated into low-temperature, solution-processable fabrication methods, including blade coating, spin-coating, and inkjet printing. This scalability makes them highly attractive for roll-to-roll manufacturing of large-area OSC modules (Fig. 12).

Lastly, PVDF incorporation into OSCs enables greater adaptability to newer active layer systems, including non-fullerene acceptors (NFAs) like ITIC, Y6, and their derivatives. These low-bandgap molecules offer superior absorption profiles and tunable energy levels. When paired with PVDF-based layers, they show reduced voltage losses and improved morphological stability, supporting high-efficiency operation without the need for highly complex processing additives. In one such configuration, a bilayer PVDF with 3-hydroxyflavone (3-HF) achieved a PCE of 11.83%, Jsc of 20.1 mA/cm^2, Voc of 0.88 V, and FF of 66.2%, with retention of 8.17% efficiency even after 60 days under 77% relative humidity [22].

Table 1 Summary of the photovoltaic devices performance using different buffer layer

Photo-anode	J_{sc} (mA/cm^2)	V_{oc} (V)	FF	η (%)	R_{shunt} (Ω cm^2)	R_s (Ω cm^2)
ITO	3.81	0.54	0.48	0.96	2631.6	76.92
ITO/PEDOT-PSS	6.34	0.57	0.53	1.78	4566.2	38.46
ITO/PVDF-PANI (30 °C)	0.29	0.54	0.45	0.11	700.99	833
ITO/PVDF-PANI (90 °C)	7.96	0.55	0.70	3.06	4651.2	4.16

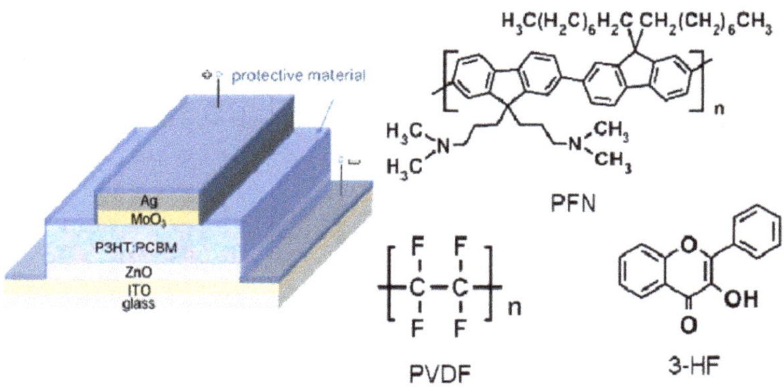

Fig. 12 Schematic representation of fabricated inverted solar cells covered with a protective layer along with chemical structures of used protective materials, PFN, 3-HF, and PVDF

5.2 *Performance Enhancement of Perovskite Solar Cells Through PVDF Integration*

The power conversion efficiency of PSCs has witnessed rapid improvement over the past decade. However, performance bottlenecks persist due to non-radiative recombination, poor crystallinity, and defect states within the perovskite films. To overcome these limitations, the integration of functional polymer additives such as PVDF and its derivatives has emerged as a promising strategy. These fluorinated polymers facilitate controlled crystal growth, improve film morphology, and introduce beneficial internal electric fields that enhance charge separation and carrier extraction. This section reviews various approaches utilizing PVDF, PVDF-HFP, and ferroelectric PVDF-based composites to optimize the structural and electronic quality of the perovskite layer, thereby boosting overall device efficiency.

Sun et al. demonstrated substantial improvement in the photovoltaic performance of PSCs by incorporating PVDF into the perovskite layer. The fluorine atoms in PVDF form strong hydrogen bonds with organic cations and coordinate with Pb^{2+} ions, resulting in slower and more controlled crystal growth and effective defect passivation, particularly at grain boundaries. These effects significantly reduce non-radiative recombination losses and minimize energy loss in the Voc. As a result, PVDF-based PSCs achieved a champion efficiency of 24.21% with an impressively high Voc of 1.22 V, one of the highest reported for FAMAPb $(I/Br)_3$-based devices [42].

Zheng et al. proposed a similar approach to enhance the quality of perovskite films by introducing PVDF, which contains abundant hydrophobic fluorine atoms. Acting as a growth template, PVDF promotes the crystallization process, improves film morphology, reduces defect density, and suppresses carrier recombination as shown in Fig. 13. The optimized device achieved a PCE of 21.42%, reflecting a 10.87% improvement compared to the control [43].

Zhang et al. focused on the use of poly(vinylidene fluoride-*co*-hexafluoropropylene) (PVDF-HFP) as an additive in $CH_3NH_3PbI_3$-based PSCs, particularly targeting visible-light-semitransparent applications with thinner perovskite films (~150 nm). The addition of PVDF-HFP significantly improved film coverage and surface smoothness, while reducing voids and grain boundaries especially important for thin-film applications. Devices fabricated using a 12 mg/mL PVDF-HFP solution achieved over 30% higher efficiency than control devices, with PCEs of 10.6% (80 nm Au) and 8.8% (20 nm Au). Supporting characterizations such as time-resolved photoluminescence and electrochemical impedance spectroscopy revealed enhanced carrier lifetimes and reduced charge transfer resistance, confirming the improved optoelectronic quality of the devices [44].

Similarly, Ahn et al. explored the enhancement of perovskite solar cell stability and performance for space photovoltaic applications by incorporating PVDF-HFP

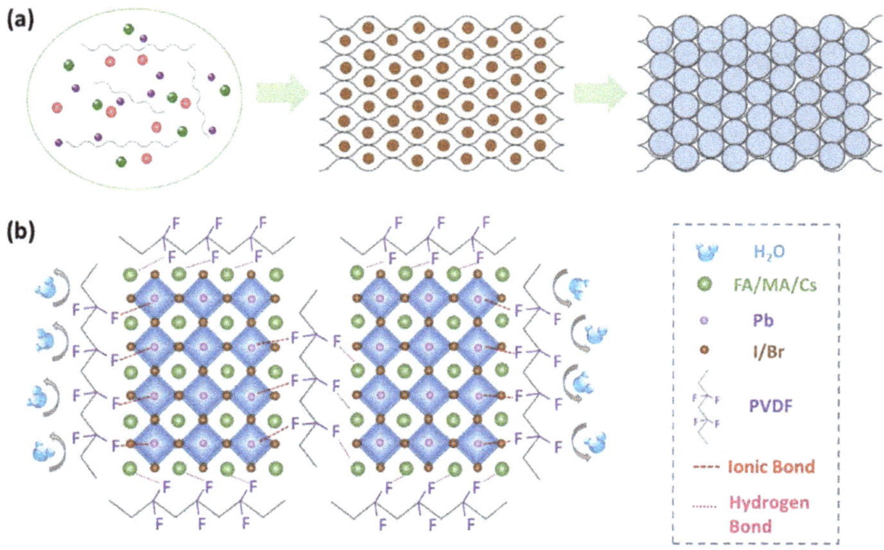

Fig. 13 (**a**) Schematic representation of the PVDF-based perovskite crystal growth process, and (**b**) the proposed mechanism for the observed improvement in stability [43]

as an additive into the perovskite composition. The high fluorine content in PVDF-HFP forms hydrogen bonds with organic cations and coordinate bonds with Pb^{2+}, effectively mitigating defects within the perovskite film. The PVDF-HFP-modified PSCs exhibited a significant increase in power conversion efficiency, achieving 22.54% compared to 19.85% in pristine devices. Moreover, these devices maintained strong performance under extreme conditions, retaining 70% of initial efficiency after 600 h in high vacuum (2×10^{-7} torr), and demonstrated high resilience to gamma-ray irradiation (as in Fig. 14), confirming PVDF-HFP's suitability for harsh environments [45].

Narendhiran et al. focused on improving hole-transport-layer-free, carbon-based PSCs by using PVDF-HFP as a polymer additive to passivate perovskite grains and promote the growth of large, well-oriented crystals. The fluoro-polymer systematically altered perovskite morphology and established H–F bonding with organic cations, leading to reduced trap density and an increase in Voc by ~14%. The final device achieved a PCE of 11.11%. These findings underscore how PVDF-HFP not only aids crystallization but also modulates film structure for enhanced performance in low-cost, HTL-free device configurations [46].

Chen et al. introduced a novel approach to performance enhancement by doping the perovskite layer with ferroelectric PVDF:DH (polyvinylidene fluoride:dabcoHReO$_4$). This material possesses high polarizability and introduces a permanent internal electric field aligned with the built-in field (BEF) of the perovskite solar cell, thereby boosting charge-carrier transport and extraction while suppressing non-radiative recombination. Additionally, PVDF:DH facilitates the formation of mesoporous PbI_2, supporting high-crystallinity perovskite growth with

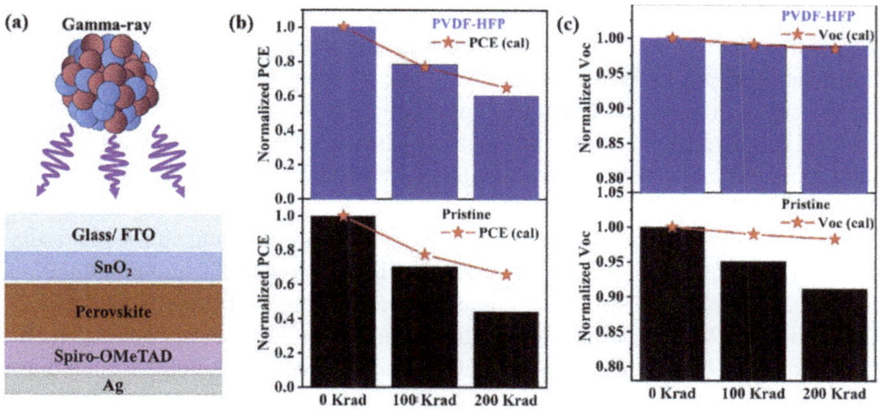

Fig. 14 (**a**) Schematic showing a PSC being irradiated with gamma rays. (**b**) and (**c**) depict the changes in PCE and Voc, respectively, for PSCs subjected to gamma radiation [45]

minimal defects. The resulting device achieved a PCE of 24.23% for a 0.062 cm^2 device (certified 23.45%) and 22.69% for a 1 cm^2 cell, along with improvements in moisture resistance and operational stability [47].

Nie et al. applied the principle of the piezophototronic effect to enhance the performance of PSCs by doping them with polarized P(VDF-TrFE) (polyvinylidene fluoride-*co*-trifluoroethylene). The composite film was polarized via external potential, introducing internal polarization without compromising optical absorption. This polarization enhanced the built-in potential and reduced interface barriers, improving carrier separation and extraction. Their approach resulted in a PCE of 22.1%, with a Jsc of 24.2 mA/cm^2 and a Voc of 1.18 V, demonstrating a simple and effective polarization method for boosting performance [48].

6 Stability and Durability Improvements

PVDF significantly enhances the long-term stability and durability of OSCs and PSCs, which are typically vulnerable to environmental degradation from moisture, oxygen, heat, and UV exposure. Its flucrine-rich, semi-crystalline structure provides exceptional hydrophobicity, chemical resistance, and thermal stability. As an interfacial layer or encapsulant, PVDF effectively blocks moisture and oxygen pressure; for instance, OSCs with a PVDF cathode interlayer retained over 81% efficiency after 3600 h in humid air [20], while dual-layer systems with antioxidants maintained up to 77% efficiency after 60 days in high humidity [22]. PVDF also suppresses ion migration, enhances thermal/photo stability, protects against photooxidation, and prevents phase separation under prolonged exposure. Beyond protection, PVDF improves interfacial electronic stability by reducing electrode corrosion and metal diffusion, preserving charge transport, and minimizing recombination losses through energy-level smoothing. In flexible and wearable OSCs, PVDF (especially PVDF-TrFE) offers mechanical reinforcement, preventing microcracks and delamination during bending, and its piezo-phototronic effect further boosts charge separation under strain, benefiting motion-responsive and self-powered devices. Its chemical inertness and electrical insulation ensure seamless integration and support scalable fabrication, ultimately boosting both performance and environmental resilience.

For PSCs, PVDF and its derivatives address moisture, thermal stress, and radiation degradation through their inherent hydrophobicity, crystallization modulation, and defect passivation. The hydrophobic fluorine backbone effectively blocks moisture, allowing PVDF-modified devices to maintain over 90% power conversion efficiency for 2500 h under ambient humidity [42]. PVDF acts as a nucleation template, improving crystal growth and film uniformity, leading to enhanced thermal and humidity resilience, with unencapsulated devices retaining over 90% efficiency after 2400 h at 45% RH and 300 h at 85 °C [43]. The incorporation of PVDF-HFP copolymer into the perovskite matrix enhances both moisture resistance and structural robustness, yielding smoother, void-free layers [44]. These devices show remarkable stability under extreme conditions, retaining 70% efficiency after 600 h in high vacuum and exhibiting strong gamma-ray tolerance for aerospace applications [45]. PVDF-HFP also facilitates larger, well-oriented grains and reduced trap densities in carbon-based, hole-transport-layer-free devices, improving shelf life (86% efficiency after 30 days at 60–70% RH) [46]. Advanced ferroelectric composites like PVDF:DH and P(VDF-TrFE) further boost stability by introducing a permanent internal polarization field that aids charge extraction, suppresses recombination, improves crystallinity, and reduces defects [47]. P(VDF-TrFE) also generates a piezophototronic effect, reinforcing the internal electric field and stabilizing interface energetics for sustained high photovoltaic output [48]. Overall, PVDF-based materials provide crucial multi-functional advantages moisture protection, thermal stability, improved film morphology, and internal field reinforcement for robust PSC performance in diverse environments, from terrestrial to extreme.

7 Conclusion

Polyvinylidene fluoride has established itself as a multifunctional and indispensable material in the domain of flexible photovoltaic technologies. Its unique combination of ferroelectricity, piezoelectricity, chemical inertness, and hydrophobicity enables PVDF to serve as more than just a structural support it functions as an active layer material that directly influences device performance, longevity, and reliability. By promoting enhanced charge separation, defect passivation, crystallinity control, and environmental stability, PVDF significantly addresses the limitations of both organic and perovskite solar cells. The material's compatibility with solution-based and roll-to-roll processing techniques aligns well with the goals of low-cost, large-area, and scalable photovoltaic manufacturing. Furthermore, its mechanical durability and flexibility make it especially attractive for emerging applications such as wearable electronics, portable power generation, and building-integrated photovoltaics.

As discussed throughout this chapter, the multifunctional role of PVDF spans structural, electronic, and environmental domains, marking it as a critical component in the evolution of next-generation solar technologies. Moving forward, the integration of PVDF with other functional materials and device engineering strategies is expected to unlock new performance thresholds. Continued interdisciplinary research will be essential to fully realize PVDF's potential in shaping a sustainable and flexible energy future.

References

1. Neville, R.C.: Solar Energy Conversion: The Solar Cell, 2nd edn. Elsevier Science, Amsterdam (1995)
2. Panwar, N.L., Kaushik, S.C., Kothari, S.: Role of renewable energy sources in environmental protection: a review. Renew. Sustain. Energy Rev. **15**, 1513–1524 (2011)
3. Owusu, P.A., Asumadu-Sarkodie, S.: A review of renewable energy sources, sustainability issues and climate change mitigation. Cogent Eng. **3**, 1167990 (2016). https://doi.org/10.1080/23311916.2016.1167990
4. Kabir, E., Kumar, P., Kumar, S., et al.: Solar energy: potential and future prospects. Renew. Sustain. Energy Rev. **82**, 894–900 (2018). https://doi.org/10.1016/J.RSER.2017.09.094
5. Gong, J., Li, C., Wasielewski, M.R.: Advances in solar energy conversion. Chem. Soc. Rev. **48**, 1862–1864 (2019). https://doi.org/10.1039/C9CS90020A
6. Cardon, F.: Photovoltaic and Photoelectrochemical Solar Energy Conversion. Springer US, New York (2012)
7. Ibn-Mohammed, T., Koh, S.C.L., Reaney, I.M., et al.: Perovskite solar cells: an integrated hybrid lifecycle assessment and review in comparison with other photovoltaic technologies. Renew. Sustain. Energy Rev. **80**, 1321–1344 (2017). https://doi.org/10.1016/j.rser.2017.05.095
8. Conibeer, G.: Third-generation photovoltaics. Mater. Today. **10**, 42–50 (2007). https://doi.org/10.1016/S1369-7021(07)70278-X
9. Sall, M.L., Diaw, A.K.D., Gningue-Sall, D., et al.: Toxic heavy metals: impact on the environment and human health, and treatment with conducting organic polymers, a review. Environ. Sci. Pollut. Res. **27**, 29927–29942 (2020). https://doi.org/10.1007/s11356-020-09354-3
10. Dong, H., Ran, C., Gao, W., et al.: Metal halide perovskite for next-generation optoelectronics: progresses and prospects. eLight. **3**, 3 (2023). https://doi.org/10.1186/s43593-022-00033-z
11. Spanggaard, H., Krebs, F.C.: A brief history of the development of organic and polymeric photovoltaics. Sol. Energy Mater. Sol. Cells. **83**, 125–146 (2004). https://doi.org/10.1016/j.solmat.2004.02.021
12. Dash, B.P., Kumar Beriha, S., Naik, B., Kumar Sahoo, P.: Organic materials based solar cells. Mater. Today Proc. **67**, 1057–1063 (2022). https://doi.org/10.1016/J.MATPR.2022.07.002
13. Solak, E.K., Irmak, E.: Advances in organic photovoltaic cells: a comprehensive review of materials, technologies, and performance. RSC Adv. **13**, 12244–12269 (2023). https://doi.org/10.1039/D3RA01454A
14. Xu, X., Yu, L., Meng, H., et al.: Polymer solar cells with 18.74% efficiency: from bulk heterojunction to interdigitated bulk heterojunction. Adv. Funct. Mater. **32**, 2108797 (2022). https://doi.org/10.1002/adfm.202108797

15. Li, Y., Huang, W., Zhao, D., et al.: Recent progress in organic solar cells: a review on materials from acceptor to donor. Molecules. **27**, 1800 (2022). https://doi.org/10.3390/molecules27061800
16. Wei, W.-R., Tsai, M.-L., Ho, S.-T., et al.: Above-11%-efficiency organic–inorganic hybrid solar cells with omnidirectional harvesting characteristics by employing hierarchical photon-trapping structures. Nano Lett. **13**, 3658–3663 (2013). https://doi.org/10.1021/nl401540h
17. Jin, K., Xiao, Z., Ding, L.: 18.69% PCE from organic solar cells. J. Semicond. **42**, 060502 (2021). https://doi.org/10.1088/1674-4926/42/6/060502
18. Antolin, E., Urieta-Mora, J., Molina-Ontoria, A., Martín, N.: Organic solar cells: principles, materials, and working mechanism. Curr. Opin. Colloid Interface Sci. **76**, 101893 (2025). https://doi.org/10.1016/j.cocis.2024.101893
19. Wang, J., Li, Y., Bi, F., et al.: Revealing intrinsic free charge generation: promoting the construction of over 19% efficient planar p–n heterojunction organic solar cells. Angew. Chem. Int. Ed. **64**, e202417143 (2025). https://doi.org/10.1002/anie.202417143
20. Ge, Y., Xiao, X., Yao, G., et al.: Dual interface protection for high performance and excellent long-term stability of organic solar cells. ACS Appl. Mater. Interfaces. **13**, 57664–57672 (2021). https://doi.org/10.1021/acsami.1c15792
21. Saïdi, S., Mannaï, A., Derouiche, H., Belhadj Mohamed, A.: Effect of drying temperature on structural and electrical properties of PANI:PVDF composite thin films and their application as buffer layer for organic solar cells. Mater. Sci. Semicond. Process. **19**, 130–135 (2014). https://doi.org/10.1016/j.mssp.2013.12.002
22. Li, C.-P., Yang, C.-S., Chen, P.-C., Hsu, F.-C.: Stability improvement of inverted organic solar cells with thin organic protective layer. Org. Electron. **108**, 106602 (2022). https://doi.org/10.1016/j.orgel.2022.106602
23. Subudhi, P., Punetha, D.: Progress, challenges, and perspectives on polymer substrates for emerging flexible solar cells: a holistic panoramic review. Prog. Photovolt. Res. Appl. **31**, 753–789 (2023). https://doi.org/10.1002/pip.3703
24. Nozik, A., Conibeer, G., Beard, M.C.: Advanced Concepts in Photovoltaics. The Royal Society of Chemistry (2014)
25. Cho, K.T., Paek, S., Grancini, G., et al.: Highly efficient perovskite solar cells with a compositionally engineered perovskite/hole transporting material interface. Energy Environ. Sci. **10**, 621–627 (2017). https://doi.org/10.1039/C6EE03182J
26. Kojima, A., Teshima, K., Shirai, Y., Miyasaka, T.: Organometal halide perovskites as visible-light sensitizers for photovoltaic cells. J. Am. Chem. Soc. **131**, 6050–6051 (2009). https://doi.org/10.1021/ja809598r
27. Ubani, C.A., Ibrahim, M.A., Teridi, M.A.M.: Moving into the domain of perovskite sensitized solar cell. Renew. Sustain. Energy Rev. **72**, 907–915 (2017). https://doi.org/10.1016/j.rser.2017.01.025
28. Noman, M., Khan, Z., Jan, S.T.: A comprehensive review on the advancements and challenges in perovskite solar cell technology. RSC Adv. **14**, 5085–5131 (2024). https://doi.org/10.1039/D3RA07518D
29. Subudhi, P., Punetha, D.: Pivotal avenue for hybrid electron transport layer-based perovskite solar cells with improved efficiency. Sci. Rep. **13**, 19485 (2023). https://doi.org/10.1038/s41598-023-33419-1
30. Krebs, F.C., Gevorgyan, S.A., Alstrup, J.: A roll-to-roll process to flexible polymer solar cells: model studies, manufacture and operational stability studies. J. Mater. Chem. **19**, 5442 (2009). https://doi.org/10.1039/b823001c
31. Khan, S., Ali, S., Bermak, A.: Smart manufacturing technologies for printed electronics. In: Hybrid Nanomaterials—Flexible Electronics Materials. IntechOpen (2020)
32. Dhandayuthapani, T., Girish, M., Sivakumar, R., et al.: Tuning the morphology of metastable MnS films by simple chemical bath deposition technique. Appl. Surf. Sci. **353**, 449–458 (2015). https://doi.org/10.1016/j.apsusc.2015.06.154

33. Nair, P.K., Nair, M.T.S., García, V.M., et al.: Semiconductor thin films by chemical bath deposition for solar energy related applications. Sol. Energy Mater. Sol. Cells. **52**, 313–344 (1998). https://doi.org/10.1016/S0927-0248(97)00237-7
34. Zaytseva, O., Neumann, G.: Carbon nanomaterials: production, impact on plant development, agricultural and environmental applications. Chem. Biol. Technol. Agric. **3**, 17 (2016). https://doi.org/10.1186/s40538-016-0070-8
35. Sun, L., Yuan, G., Gao, L., et al.: Chemical vapour deposition. Nat. Rev. Methods Prim. **1**, 5 (2021). https://doi.org/10.1038/s43586-020-00005-y
36. Kandjani, S.A., Mirershadi, S., Nikniaz, A.: Inorganic–organic perovskite solar cells. In: Solar Cells—New Approaches and Reviews. IntechOpen (2015)
37. Krebs, F.C.: Fabrication and processing of polymer solar cells: a review of printing and coating techniques. Sol. Energy Mater. Sol. Cells. **93**, 394–412 (2009). https://doi.org/10.1016/J.SOLMAT.2008.10.004
38. Ma, Y.-F., Wang, Y.-M., Wen, J., et al.: Review of roll-to-roll fabrication techniques for colloidal quantum dot solar cells. J. Electron. Sci. Technol. **21**, 100189 (2023). https://doi.org/10.1016/j.jnlest.2023.100189
39. Hu, Z., Zhang, J., Xiong, S., Zhao, Y.: Performance of polymer solar cells fabricated by dip coating process. Sol. Energy Mater. Sol. Cells. **99**, 221–225 (2012). https://doi.org/10.1016/J.SOLMAT.2011.12.002
40. Yin, Z., Tian, B., Zhu, Q., Duan, C.: Characterization and application of PVDF and its copolymer films prepared by spin-coating and Langmuir–Blodgett method. Polymers (Basel). **11**, 2033 (2019). https://doi.org/10.3390/polym11122033
41. De Oliveira, R.F., De Barros, A., Ferreira, M.: Nanostructured films: Langmuir–Blodgett (LB) and layer-by-layer (LbL) techniques. In: Nanostructures, pp. 105–123. William Andrew Publishing (2017). https://doi.org/10.1016/B978-0-323-49782-4.00004-8
42. Sun, R., Tian, Q., Li, M., et al.: Over 24% efficient poly(vinylidene fluoride) (PVDF)-coordinated perovskite solar cells with a photovoltage up to 1.22 V. Adv. Funct. Mater. **33**, 2210071 (2023). https://doi.org/10.1002/adfm.202210071
43. Zheng, H., Liu, G., Wu, W., et al.: Highly efficient and stable perovskite solar cells with strong hydrophobic barrier via introducing poly(vinylidene fluoride) additive. J. Energy Chem. **57**, 593–600 (2021). https://doi.org/10.1016/j.jechem.2020.09.026
44. Zhang, S., Lu, Y., Lin, B., et al.: PVDF-HFP additive for visible-light-semitransparent perovskite films yielding enhanced photovoltaic performance. Sol. Energy Mater. Sol. Cells. **170**, 178–186 (2017). https://doi.org/10.1016/j.solmat.2017.05.057
45. Ahn, S., Chiu, W.-H., Chu, W.-C., et al.: A systematic investigation of PVDF-HFP in perovskite solar cells for improved space mission reliability. Chem. Eng. J. **496**, 153974 (2024). https://doi.org/10.1016/j.cej.2024.153974
46. Narendhiran, S., Kunka Ravindran, A., Rajan Thomas, I.D., et al.: Poly(vinylidene fluoride-*co*-hexafluoropropylene) additive in perovskite for stable performance of carbon-based perovskite solar cells. Int. J. Energy Res. **46**, 1565–1574 (2022). https://doi.org/10.1002/er.7270
47. Chen, W., Liu, S., Li, Q., et al.: High-polarizability organic ferroelectric materials doping for enhancing the built-in electric field of perovskite solar cells realizing efficiency over 24%. Adv. Mater. **34**, 2110482 (2022). https://doi.org/10.1002/adma.202110482
48. Nie, J., Zhang, Y., Dan, M., et al.: Piezophototronic effect enhanced perovskite solar cell based on P(VDF-TrFE). Sol. RRL. **5**, 35 (2021). https://doi.org/10.1002/solr.202170124

PVDF for Supercapacitors and Energy Storage Devices

A. Bharathi Sankar Ammaiyappan, Manavalan Vijayakumar, and Mani Karthik

Abstract Piezo-electric energy harvesting (PEH) is a technology that converts mechanical energy into electrical energy using piezoelectric materials. These materials can generate an electrical charge when subjected to mechanical stress or pressure. This process is known as the piezoelectric effect. Piezoelectric materials can be made from a variety of materials such as ceramics, crystals, and polymers. Some common examples of piezoelectric materials include lead zirconate titanate (PZT), lithium niobate, and quartz. Vibration-based energy harvesting, which captures energy from ambient vibrations, such as those generated by traffic or machinery. Impact-based energy harvesting, which captures energy from impact loads, such as those generated by foot traffic or falling objects. Strain-based energy harvesting, which captures energy from deformation of the piezoelectric material, such as that generated by bending or twisting. Piezoelectric energy harvesting systems are being used in a wide range of applications, including wireless sensors, IoT devices, wearables, and other portable electronic devices. They are also being used in various industrial and infrastructure applications, such as in bridges and buildings to convert vibrations generated by wind and traffic into electricity. When piezoelectric materials are subjected to mechanical stress, an electrical potential will be generated. There are many products that contain this direct piezoelectric effect, such as strain sensors, gas lighters, microphones, and ultrasonic detectors.

Keywords Supercapacitors · Energy storage and conversion · Energy harvesting · IoT · Flexible electronics

A. B. S. Ammaiyappan (✉)
School of Electronics Engineering, Vellore Institute of Technology (VIT), Chennai, India
e-mail: bharathisankar.a@vit.ac.in

M. Vijayakumar
Chemical Engineering and Applied Chemistry, School of Infrastucture and Sustainable Engineering, College of Engineering and Physical Sciences, Birmingham, UK

M. Karthik
Centre for Solar Energy Materials, International Advanced Research Centre for Powder Metallurgy and New Materials (ARCI), Hyderabad, India

D. Punetha (ed.), *Polyvinylidene Fluoride (PVDF)*, Engineering Materials,
https://doi.org/10.1007/978-981-95-6718-8_4

1 Introduction

Wearable technology, such as fitness trackers, smartwatches, and Internet of Things sensors, has advanced dramatically in recent years. Batteries, the primary power source for these devices, require regular recharging and contribute to the production of e-waste. This project explores an alternate energy source by using piezoelectric transducers to convert mechanical energy from human motions, such as walking, into electrical energy. The stored energy is used to power wearable electronics, reducing the need for traditional batteries.

The proposed system consists of piezoelectric strips integrated with the LTC3588 energy harvester and a capacitor to store the converted energy. A simulation model was made in LTSpice using the LTC3331 System Model in order to evaluate the system's efficiency. In order to demonstrate energy harvesting capabilities, the hardware prototype used piezoelectric transducers to generate an AC voltage that was then rectified and regulated for storage. The experimental results showed that the piezo sensor had a peak voltage of 38–50 V and the regulated output had a peak voltage of 4.8 V. This work provides a sustainable energy solution for low-power wearables, which paves the way for self-sustaining electronics.

With capabilities like health monitoring and real-time communication, wearable electronics have become a crucial element of modern life. As the requirement for these devices increases, rechargeable batteries which have a short lifespan, frequent charging cycles, along with significant environmental issues associated with battery disposal become increasingly necessary [1–4]. One of the most promising alternative energy options being investigated in response to the demand for a reliable and sustainable power source is energy harvesting technology. Energy harvesting refers to the process of capturing ambient energy from sources such as kinetic motion, thermal gradients, solar radiation, or radiofrequency signals and converting it into usable electrical energy [5]. Among these, kinetic energy harvesting using piezoelectric materials is particularly well-suited for wearable applications due to its ability to generate electricity from human motion. Piezoelectric materials possess unique properties that allow them to generate an electric charge when subjected to mechanical stress, such as footsteps or body movements [6]. This makes them an excellent candidate for integration into wearable devices, enabling continuous and sustainable power generation without the need for frequent recharging.

The process of capturing ambient energy from sources including kinetic motion, thermal gradients, exposure to sunlight, or radioactive signals and developing it into usable electrical energy is called energy harvesting [7]. Among these, kinetic energy i.e., energy developed from motion of any object harvesting using piezoelectric materials seems particularly well-suited for wearable applications because to its capability for generating power from human movement. A combination of unusual qualities, piezoelectric materials create an electric charge under mechanical stress, including body movement. This qualifies them to be included in wearable devices, therefore allowing constant and sustainable power production without regular recharging.

The goal of this project is to use MEAS piezoelectric strips in conjunction with an LTC3588 energy harvester to design and implement an effective piezoelectric energy harvesting system [8, 9]. IoT sensors, communication modules, and fitness trackers are examples of low-power electronic devices that can be powered by the captured energy, which is stored in a capacitor. Prior to hardware implementation, the system's electrical performance was examined using LTSpice simulations with LTC3331 System model, guaranteeing the best possible energy conversion efficiency [10].

A critical aspect of this study is the evaluation of different configurations of piezoelectric transducers to determine the most effective energy harvesting approach. Additionally, the system's response to varying vibration frequencies was examined to understand its behaviour under real-world conditions [9]. By conducting these experiments, this project aims to provide valuable insights into the optimization of piezoelectric energy harvesting systems for practical applications. Another major challenge addressed in this work is the efficient management of harvested energy. Since piezoelectric transducers generate alternating current (AC) voltage, a rectification and regulation stage are required to ensure a stable direct current (DC) output suitable for electronic devices. The LTC3588 energy harvester plays a crucial role in this process, as it efficiently converts and regulates the harvested energy, ensuring reliable power delivery. The energy is then stored in a capacitor, allowing for intermittent power usage and enabling continuous operation of the connected device.

The practical implementation of this energy harvesting system involved fabricating a hardware prototype and conducting real-time voltage and power measurements. The system was tested under different mechanical input conditions, simulating real-life scenarios such as walking, running, and applying controlled pressure [11]. Experimental results demonstrated the feasibility of the proposed system, with peak AC voltages ranging from 38 to 50 V generated by the piezoelectric transducers and a stable 4.8 V DC output after rectification and regulation.

In conclusion, the increasing prevalence of wearable electronics necessitates the development of efficient and sustainable energy solutions. Piezoelectric energy harvesting presents a viable approach to achieving self-powered wearable devices, eliminating the need for frequent battery recharges, and promoting environmental sustainability. Through a systematic study of different configurations, energy management techniques, and real-world performance analysis, this project provides a comprehensive framework for implementing piezoelectric-based energy harvesting in next-generation wearable electronics.

2 Methodology

Energy harvesting technologies have gained significant attention in recent years due to the growing demand for sustainable power sources. Various energy harvesting methods have been explored, including solar, thermal, electromagnetic, and kinetic energy harvesting [12]. Among these, kinetic energy harvesting using piezoelectric materials has emerged as a viable option for powering wearable electronics due to its ability to generate energy from body movements.

Piezoelectric materials are widely recognized for their ability to transform mechanical stress into electrical energy, making them particularly useful in scenarios where mechanical vibrations or human movement can be converted into a reliable power source. Extensive research has highlighted the efficiency of various piezoelectric materials in generating adequate energy for low-power electronic devices [13–15]. Among the most studied materials, lead zirconate titanate (PZT) and polyvinylidene fluoride (PVDF) have demonstrated high energy conversion efficiency and durability, making them suitable for wearable and portable energy-harvesting applications [16]. These materials have been rigorously analysed in numerous studies to assess their feasibility in powering small-scale electronics, ensuring their compatibility with emerging wearable technologies. The efficiency of energy harvesting systems is greatly influenced by the energy storage and power management strategies employed. Studies have shown that integrating energy harvesting circuits with efficient power converters, such as the LTC3588, can significantly enhance the energy conversion efficiency of the system [8]. Capacitors and rechargeable thin-film batteries are commonly used for energy storage in wearable applications.

Extensive research has been conducted to evaluate the feasibility of integrating energy harvesting systems into wearable electronics. These systems have been successfully applied in various fields, including self-powered health monitoring devices, GPS trackers, and wireless communication systems [17]. Studies suggest that strategic placement of piezoelectric transducers in regions of maximum mechanical stress, such as shoe soles and joint areas, can significantly improve energy conversion efficiency [18].

Additionally, research has explored the integration of flexible and stretchable piezoelectric materials into textiles, enhancing the adaptability of these systems for different applications. The combination of smart materials and efficient power management circuits has further optimized the potential of energy harvesting technologies for wearables as shown in Fig. 1.

Despite the promising potential of energy harvesting systems, challenges such as low energy output, efficiency limitations, and material durability remain areas of active research. Ongoing advancements in nanotechnology and flexible electronics are expected to address these limitations and improve the feasibility of self-powered wearable devices [19]. Future studies are focusing on hybrid energy harvesting systems that combine multiple sources to maximize energy output.

The circuit represents an energy harvesting system using the LTC3331 energy harvester [20]. This system is designed to collect energy from two primary sources: a solar panel and a piezoelectric transducer. Figure 2 pictorially explains the LTC architecture.

The solar panel provides a variable DC voltage ranging from 3 to 19 V. The piezoelectric transducer converts mechanical energy into electrical energy, generating an AC signal that requires rectification. The LTC3331 integrates a rectifier, a

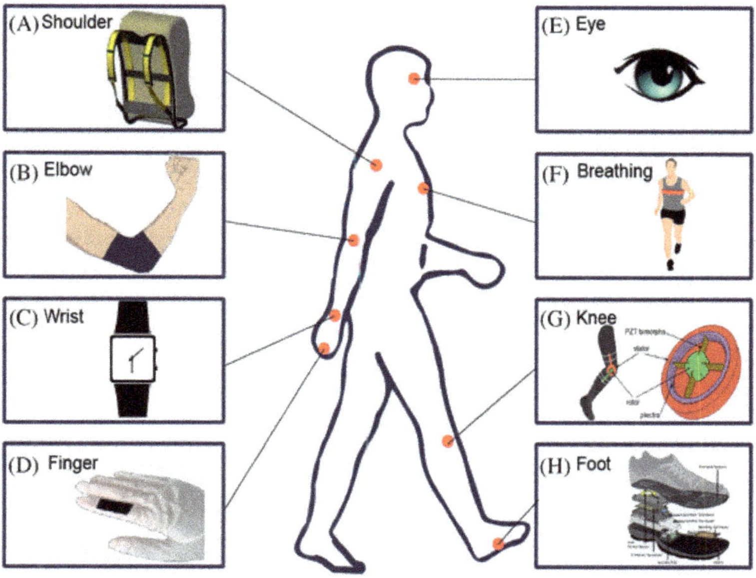

Fig. 1 Applications of wearable technology

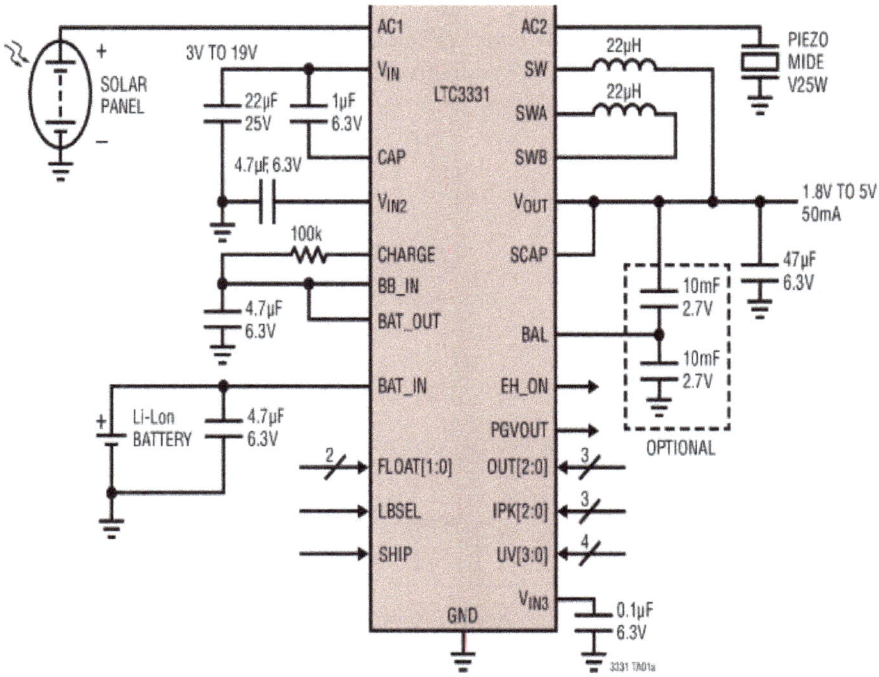

Fig. 2 LTC3331 architecture

buck-boost converter, and an energy storage management circuit. It is responsible for regulating input power from both solar and piezo sources, providing a stable DC output in the range of 1.8 to 5 V, which is suitable for wearable electronics, managing battery charging, allowing energy to be stored in a lithium-ion battery, the use of capacitors (4.7 μF, 22 μF, and 47 μF) ensures voltage stabilization and noise filtering. This model is optimized for low-power applications, such as IoT-based wearable devices, where efficient energy harvesting and storage are critical.

To evaluate the performance of the system, data collection involved real-time voltage and power measurements under varying mechanical inputs [21]. The energy harvesting performance was analysed based on:

Voltage and Power Output: The AC voltage generated by the piezoelectric transducers was measured and rectified. The output power was calculated to determine the efficiency of the energy conversion process.

Load Testing: The system was tested under different load conditions to analyze its ability to power wearable electronic devices.

LTSpice Simulations: The electrical characteristics of the system were simulated in LTSpice to predict its performance before hardware implementation.

Mechanical Input Variations: Different levels of pressure, impact frequencies, and durations of applied force were tested to understand how mechanical motion affects energy harvesting efficiency.

Solar Integration Performance: The contribution of the solar panel as a secondary energy source was evaluated under different lighting conditions. This helped determine the hybrid system's overall effectiveness in supplementing piezoelectric harvesting (Fig. 3).

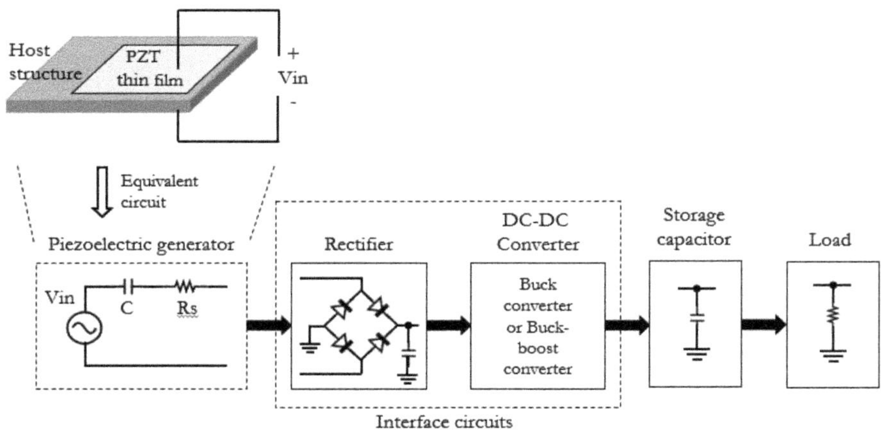

Fig. 3 Block diagram of piezo energy harvester

3 Piezo-electric Energy Harvesting Converters

In addition to converting vibration energy to regulated voltages, power management products that convert the piezoelectric energy to batteries and super capacitors provide high efficiency conversions. By analog devices there are four converters for energy harvesting the converters used are LTC3588-2, LTC3588-1, LTC3330, LTC3331 each of them has different application a buck converter, also known as a step-down converter, is a type of DC–DC converter that is used to reduce the voltage of a DC power source. It is widely used in electronic devices that require a lower voltage than the available power source. In a buck converter, the input voltage is applied to a switching circuit that rapidly turns on and off. This creates a series of voltage pulses that are then smoothed out by a filter circuit to produce a steady output voltage that is lower than the input voltage. The basic components of a buck converter include an inductor, a diode, a capacitor, a switching transistor, and a control circuit. The switching transistor is used to control the on/off cycles of the converter, while the inductor and capacitor are used to filter the output voltage and reduce any ripple. The efficiency of a buck converter is dependent on t"e du'y cycle of the switching transistor, which is controlled by the control circuit. When the duty cycle is increased, the output voltage is decreased, and the efficiency of the converter increases. Buck converters have many advantages, including their high efficiency, low cost, and small size. They are used in a wide range of applications, including mobile devices, computer power supplies, LED lighting, and electric vehicles. Energy harvesting buck converters are a type of DC–DC converter that are designed to harvest and convert energy from various sources such as solar panels, thermal generators, and piezoelectric transducers into usable electrical energy for low-power devices. These converters use the same basic principle as a traditional buck converter, but they have specific design considerations to optimize efficiency and output voltage for the specific energy source being harvested [10, 22, 23].

For example, a solar-powered buck converter will have a different design than a thermal energy harvesting buck converter due to the differences in the characteristics of the energy sources. A solar-powered buck converter will need to be designed to handle the variable input voltage and current from the solar panel, while a thermal energy harvesting buck converter will need to be designed to operate with low input voltages and low input currents. Energy harvesting buck converters typically include additional circuitry to manage the power source, such as maximum power point tracking (MPPT) to optimize energy extraction from the source, and voltage regulators to maintain a steady output voltage. Energy harvesting buck converters are used in a wide range of applications, including wireless sensors, IoT devices, wearable electronics, and remote monitoring systems. They allow these devices to operate without the need for a traditional power source, making them ideal for remote or difficult to access locations [24–26].

4 LTC3588 Energy Harvester Converter

The LTC3588 is a highly integrated energy harvesting power management IC (PMIC) designed for low-power applications. It is manufactured by Analog Devices, Inc. and is widely used in energy harvesting applications such as wireless sensor networks, environmental monitoring systems, and remote sensing systems. The LTC3588 provides a complete power management solution for energy harvesting applications by combining a full-bridge energy harvesting power supply, a high-efficiency synchronous buck-boost converter, a programmable maximum power point tracking (MPPT) circuit, and a high-accuracy LDO voltage regulator. Here are some key details about the LTC3588 chip: Input Voltage: 2.7–20 V Output Voltage: 1.8–5.25 V Output Current: up to 100 mA Maximum Power Point Tracking (MPPT): programmable with an external resistor Full-Bridge Rectifier: for AC energy harvesting Synchronous Buck-Boost Converter: high-efficiency converter that can step up or step down the input voltage LDO Regulator: high-accuracy output voltage regulation for low-power loads Low Power Consumption: 950 nA quiescent current Shutdown Mode: 10 nA quiescent current Package: 24-lead QFN (4 mm × 5 mm). The LTC3588 is a highly efficient and versatile chip that provides a complete power management solution for energy harvesting applications. Its low power consumption and small size make it an ideal choice for low-power, battery-free devices that require a stable and efficient power source shown in Figs. 4 and 5.

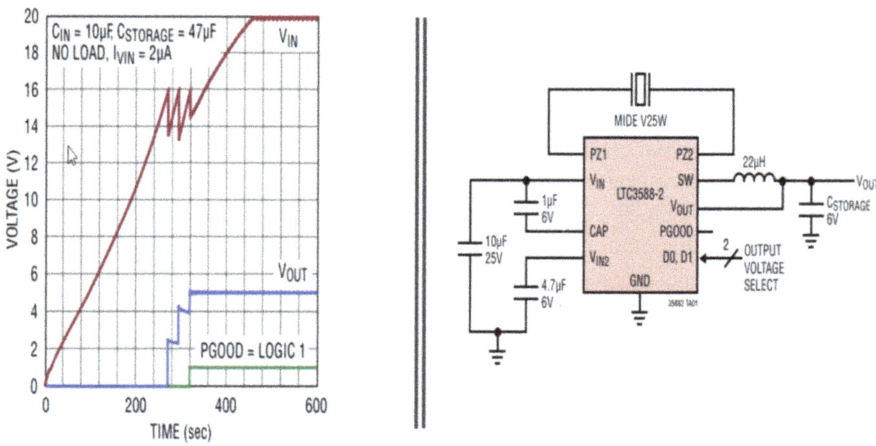

Fig. 4 Nano energy harvester converter pin configuration and circuit diagram

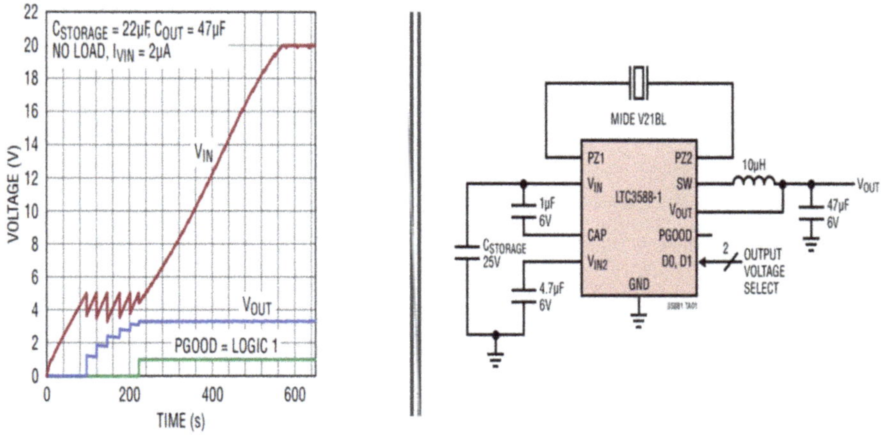

Fig. 5 Piezoelectric energy harvesting converter pin configuration and circuit diagram

5 LTC3331 Energy Harvester Converter

The LTC3331 is a highly integrated energy harvesting power management IC (PMIC) designed for low-power applications. It is manufactured by Analog Devices, Inc. and is widely used in energy harvesting applications such as wireless sensor networks, environmental monitoring systems, and remote sensing systems. The LTC3331 provides a complete power management solution for energy harvesting applications by combining a full-bridge energy harvesting power supply, a high-efficiency synchronous buck-boost converter, and a low-noise LDO voltage regulator. It is designed to work with a wide range of energy harvesting sources, including solar cells, thermoelectric generators, and piezoelectric harvesters. Here are some key details about the LTC3331 chip: Input Voltage: 3–19 V Output Voltage: 1.8–5 V Output Current: up to 50 mA Full-Bridge Rectifier: for AC energy harvesting Synchronous Buck-Boost Converter: high-efficiency converter that can step up or step down the input voltage LDO Regulator: low-noise output voltage regulation for low-power loads MPPT: programmable with an external resistor Battery Charger: for rechargeable batteries Power Good Indicator: for output voltage monitoring Low Power Consumption: 3 µA quiescent current Package: 28-lead QFN (4 mm × 5 mm) Compared to the LTC3330, the LTC3331 has a lower quiescent current, which makes it ideal for applications where the power source is intermittent and energy harvesting efficiency is critical. The LTC3331 also features an adjustable input voltage window, which allows the chip to start up and operate with lower input voltages. This feature enables operation with energy sources that might otherwise be unusable. Overall, the LTC3331 is a highly efficient and versatile chip that provides a complete power management solution for energy harvesting applications. Its low power consumption and small size make it an ideal choice for low-power, battery-free devices that require a stable and efficient power source. Additionally, its

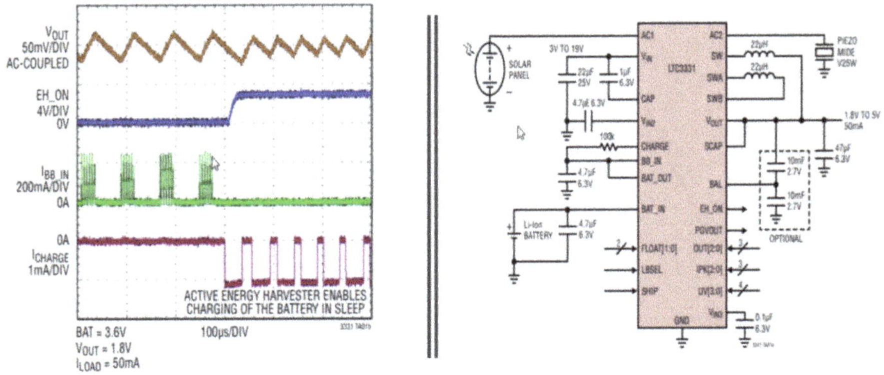

Fig. 6 Nanopower buck-boost DC/DC with energy harvesting battery charger pin and circuit diagram

integrated battery charger feature allows for energy storage, increasing the operational lifetime of the device shown in Fig. 6.

6 LTC3330 Energy Harvester Converter

The LTC3330 is a highly integrated energy harvesting power management IC (PMIC) designed for low-power applications. It is manufactured by Analog Devices, Inc. and is widely used in energy harvesting applications such as wireless sensor networks, environmental monitoring systems, and remote sensing systems. The LTC3330 provides a complete power management solution for energy harvesting applications by combining a full-bridge energy harvesting power supply, a high-efficiency synchronous buck-boost converter, and a low-noise LDO voltage regulator. It is designed to work with a wide range of energy harvesting sources, including solar cells, thermoelectric generators, and piezoelectric harvesters. Here are some key details about the LTC3330 chip: Input Voltage: 3–19 V Output Voltage: 1.8–5 V Output Current: up to 50 mA Full-Bridge Rectifier: for AC energy harvesting Synchronous Buck-Boost Converter: high-efficiency converter that can step up or step down the input voltage LDO Regulator: low-noise output voltage regulation for low-power loads MPPT: programmable with an external resistor Battery Charger: for rechargeable batteries Power Good Indicator: for output voltage monitoring Low Power Consumption: 6 μA quiescent current Package: 28-lead QFN (4 mm × 5 mm). The LTC3330 is a highly efficient and versatile chip that provides a complete power management solution for energy harvesting applications. Its low power consumption and small size make it an ideal choice for low-power, battery-free devices that require a stable and efficient power source. Additionally, its integrated battery charger feature allows for energy storage, increasing the operational lifetime of the device shown in Fig. 7.

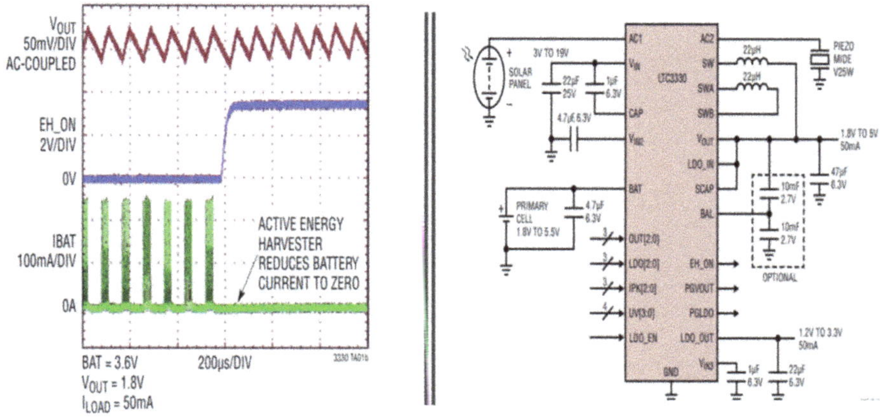

Fig. 7 Energy harvester converter pin and schematic circuit diagram

7 Results and Discussions

This chapter aims to thoroughly evaluate and interpret the findings obtained from both experimental setups and simulations. The primary focus is to analyze the performance of the proposed piezoelectric energy harvesting system and its feasibility for powering wearable electronics. The system utilizes MEAS piezoelectric strips to generate electrical energy from mechanical motion, which is then processed through the LTC3588 energy harvester for rectification and regulation. This section will provide insights into the voltage and power generation capabilities of the system under different mechanical inputs, comparing both experimental and simulation results.

8 Simulation of the Energy Harvester

The circuit shown in Fig. 8 is a simulation of the LTC3331, an energy-harvesting power management unit (PMU) designed to efficiently extract power from low-energy sources like solar panels or thermoelectric generators. The primary function of the LTC3331 is to manage energy flow between a harvester, a rechargeable battery, and a load. It does this by using an integrated buck-boost DC–DC converter, which ensures a stable output voltage even when the input voltage varies. The circuit also incorporates a battery backup system, which takes over when the energy harvester cannot supply sufficient power.

The input section of the circuit consists of a 4 mA current source (I1), simulating an energy harvester such as a solar cell. A diode (D1) prevents backflow of current, ensuring that energy only moves toward the PMU. The resistor R1 (10 Ω) acts as a current limiter to protect the circuit, while R2 (5 kΩ) serves as part of a voltage

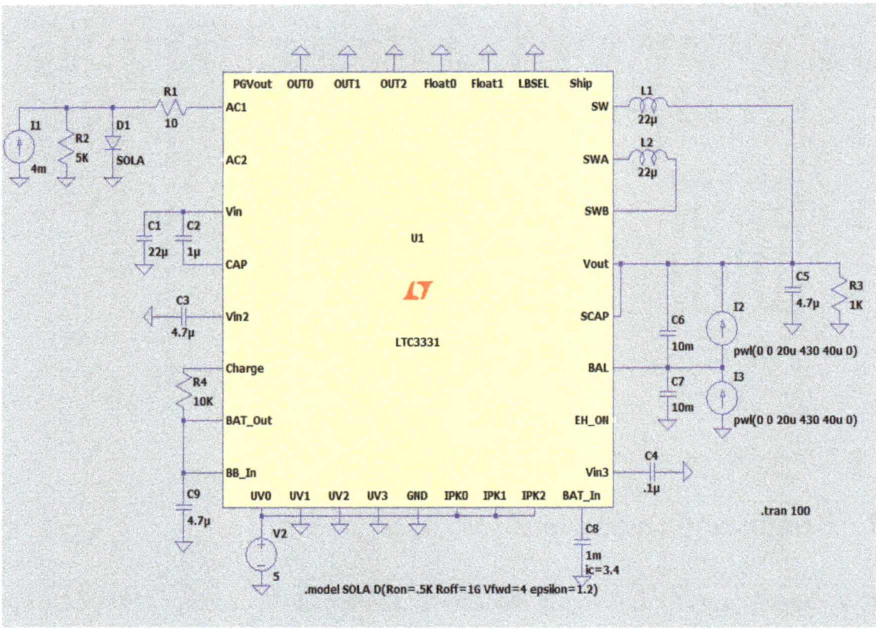

Fig. 8 LTC3331 energy harvester simulation model

sensing network. Multiple capacitors, including C1 (22 µF) and C2 (1 µF), are placed across the input to smooth voltage fluctuations, providing a stable input to the LTC3331.

Inside the LTC3331, the Vin and Vin2 pins receive the input power from the harvester, which is then processed through the internal buck-boost converter. The SW, SWA, and SWB pins, connected to inductors L1 and L2 (each 22 µH), play a crucial role in switching and energy conversion. These inductors store and release energy to maintain a stable Vout, even when the input voltage fluctuates. The Charge and BAT_Out pins are used for managing a rechargeable battery, ensuring the system has a secondary power source when harvesting conditions are poor. The BAT_In pin allows a backup battery to be integrated into the system, and the BB_In pin serves as a logic input for enabling battery backup.

Overall, this circuit ensures continuous and reliable power delivery by intelligently switching between the energy harvester and battery as needed. The LTC3331 optimizes energy extraction and provides a stable output voltage, making it ideal for ultra-low-power applications such as IoT devices, remote sensors, and wearable electronics.

9 Conclusion and Future Scope

The chapter successfully demonstrates the feasibility of piezoelectric energy harvesting for low-power wearable applications. By utilizing MEAS piezoelectric strips in combination with the LTC3588 energy harvester, mechanical energy from vibrations was efficiently converted into usable electrical energy. The harvested energy was stored in a 25 V, 1000 μF capacitor and later used to power an LED, proving the system's capability to support small-scale electronic devices.

The simulation results confirmed the expected performance of the LTC3588-based energy harvester, demonstrating proper voltage regulation and energy storage dynamics. The voltage across the capacitor exhibited a charge–discharge cycle, indicating efficient energy transfer from the piezoelectric harvester to the storage unit. Furthermore, the current analysis in the simulation verified that the harvester operates in an intermittent pulsed manner, aligning with the real-world behavior observed in the oscilloscope waveforms. These simulations provided valuable insights into optimizing component selection and circuit efficiency before physical implementation, reducing design errors and improving overall system reliability.

The experimental results, analyzed using oscilloscope waveforms, confirmed the proper functioning of the system, showing the AC voltage output from the piezoelectric strip and its subsequent rectification and regulation by the LTC3588 module. The voltage and current characteristics observed throughout the study validate the system's ability to provide a steady power supply, making it a viable alternative to conventional battery-powered solutions.

Despite the success of both experimental and simulated results, some challenges were identified. One key limitation is the presence of feedback current, which can negatively impact the efficiency of the energy storage system. This issue needs to be addressed to prevent any unintended discharge of the stored energy, especially when using supercapacitors. Although supercapacitors offer high energy storage capabilities, their fast discharge rate makes them less suitable for long-term power supply applications in the current setup.

This chapter lays the base for further advancements in piezoelectric energy harvesting, particularly for wearable electronics and IoT applications. The project can be extended in several ways to improve energy efficiency, storage capacity, and practical usability. Future work should explore the use of supercapacitors or hybrid energy storage systems that combine lithium-ion batteries with capacitors. This would allow for efficient energy buffering, ensuring that energy harvested from mechanical motion is stored effectively and used as needed without rapid discharge losses.

To improve the reliability of energy harvesting, hybrid approaches combining piezoelectric, thermoelectric, and solar energy harvesting could be investigated. A multi-source energy harvesting system would ensure continuous power supply even when one energy source is insufficient. For instance, integrating a flexible solar panel with the piezoelectric system could provide additional energy harvesting capability during daylight hours.

The successful implementation of this project suggests promising applications In smart wearables, particularly for health monitoring devices, GPS tracking units, and military communication systems. Wearable electronics that rely on self-sustaining energy sources would eliminate frequent battery replacements and reduce electronic waste.

References

1. Han, D.-Y., Song, C.K., Lee, G., Song, W.-J., Park, S.: A comprehensive review of battery-integrated energy harvesting systems. Adv. Mater. Technol. **9**(21), 2302236 (2024)
2. Liang, Y., Zhao, C.-Z., Yuan, H., Chen, Y., Zhang, W., Huang, J.-Q., Yu, D., et al.: A review of rechargeable batteries for portable electronic devices. InfoMat. **1**(1), 6–32 (2019)
3. Li, X., Zeng, X., Li, J., Li, B., Chen, Y., Zhang, X.: Biomechanical energy harvesting technologies for wearable electronics: theories and devices. Friction. **12**(8), 1655–1679 (2024)
4. Kim, C.Y., Lee, J., Jeong, E.Y., Jang, Y., Kim, H., Choi, B., Han, D., Oh, Y., Jeong, J.-W.: Wireless technologies for wearable electronics: a review. Adv. Electron. Mater. **11**, 2400884 (2025)
5. Covaci, C., Gontean, A.: Energy harvesting with piezoelectric materials for IoT—review. In: ITM Web of Conferences, vol. 29, p. 03010. EDP Sciences (2019)
6. Munawar, T., Akram, N., Rashid, K.T., Mansha, A., Ali, A.: Piezoelectric materials for biomedical and energy harvesting applications. Adv. Funct. Piezoelectr. Mater. Appl. **131**, 165–185 (2022)
7. Kazmierski, T.J., Beeby, S.: Energy harvesting systems. In: Principles, Modeling and Applications. Springer (2011)
8. Patil, A., Jadhav, M., Joshi, S., Britto, E., Vasaikar, A.: Energy harvesting using piezoelectricity. In: 2015 International Conference on Energy Systems and Applications, pp. 517–521. IEEE (2015)
9. Mishra, R., Jain, S.: Energy Harvesting Using Piezoelectric Elements to Power Microelectronic Devices. Jaypee University of Information Technology, Solan (2016)
10. Tsenempis, I., Filios, G., Katsidimas, I., Nikoletseas, S.: Energy harvesting and smart management platform for low power IoT systems. In: 2019 15th International Conference on Distributed Computing in Sensor Systems (DCOSS), pp. 339–346. IEEE (2019)
11. Atout, Mohammed. "Exploration of Piezoelectric Sensors for Signal Processing and Health Monitoring." (2024)
12. Ali, A., Shaukat, H., Bibi, S., Altabey, W.A., Noori, M., Kouritem, S.A.: Recent progress in energy harvesting systems for wearable technology. Energy Strategy Rev. **49**, 101124 (2023)
13. Erturk, A., Inman, D.J.: Piezoelectric Energy Harvesting. Wiley (2011)
14. Sezer, N., Koç, M.: A comprehensive review on the state-of-the-art of piezoelectric energy harvesting. Nano Energy. **80**, 105567 (2021)
15. Covaci, C., Gontean, A.: Piezoelectric energy harvesting solutions: a review. Sensors. **20**(12), 3512 (2020)
16. Fahad, M., Waqar, A., Kim, B.: Effective-performance of inorganic and organic (barium zirconium titanate/polyvinylidene fluoride) piezoelectric composite for energy harvesting and self-powered smart IoT-based electronics. J. Alloys Compd. **985**, 174033 (2024)
17. Kang, M., Yeo, W.-H.: Advances in energy harvesting technologies for wearable devices. Micromachines. **15**(7), 884 (2024)
18. Turkmen, A.C., Celik, C.: Energy harvesting with the piezoelectric material integrated shoe. Energy. **150**, 556–564 (2018)
19. Shafer, M.W., Garcia, E.: The power and efficiency limits of piezoelectric energy harvesting. J. Vib. Acoust. **136**(2), 021007 (2014)

20. Analog Devices.: LTC3331 datasheet and product information. https://www.analog.com/en/products/ltc3331.html (2022). Accessed 22 Mar 2022
21. Mohamad, T.N., Tengku, J.S., Berhanuddin, D.D.: Architecture of micro energy harvesting using hybrid input of RF, thermal and vibration for semi-active RFID tag. Eng. J. **21**(2), 183–197 (2017)
22. Hasan, M.K., Mahmud, M., Habib, A.K.M.A., Motakabber, S.M.A., Islam, S.: Review of electric vehicle energy storage and management system: standards, issues, and challenges. J. Energy Storage. **41**, 102940 (2021)
23. Skavnes, S.: Design and Implementation of a Magnetic Energy Harvesting System for Low Primary Current Applications. NTNU (2021)
24. Gaponenko, S.O., Kondratiev, A.E.: Device for calibration of piezoelectric sensors. Proced. Eng. **206**, 146–150 (2017)
25. Dalzell, P., Bonello, P.: Analysis of an energy harvesting piezoelectric beam with energy storage circuit. Smart Mater. Struct. **21**(10), 105029 (2012)
26. Mateu, L., Moll, F.: Appropriate charge control of the storage capacitor in a piezoelectric energy harvesting device for discontinuous load operation. Sens. Actuators A Phys. **132**(1), 302–310 (2006)

PVDF-Based Antennas for Wireless Communication

Ankit Mayur and Anand Sharma

Abstract The need for flexible and high-performing antenna materials has increased as wireless communication systems develop into more integrated, small, and flexible platforms—particularly with the global rollout of 5G infrastructure and the explosive growth of IoT devices. Due to its inherent ferroelectric characteristics and superior dielectric qualities, polyvinylidene fluoride (PVDF), a semi-crystalline fluoropolymer, has become a popular substrate for flexible antennas. The dielectric stability of PVDF, which has a relative permittivity ranging from 8 to 12 and a consistently low loss tangent (<0.02) over a broad frequency spectrum relevant to RF applications, is examined in this chapter. Furthermore, ferroelectric modulation of antenna performance metrics such as beam steering, bandwidth, and impedance is made easier by its reversible spontaneous polarization in the β-phase. A thorough analysis is conducted of the design of PVDF-based conformal antennas, their construction on curved and deformable substrates, and their integration with 5G and IoT modules. Applications include biomedical telemetry, wearable electronics, smart textiles, and mobile 5G devices—fields where mechanical flexibility and dielectric integrity are critical. PVDF's role in adaptive frequency-selective antennas and reconfigurable intelligent surfaces (RIS), supported by its potent ferroelectric switching capabilities, are examples of emerging applications. PVDF offers a ground-breaking method for flexible, conformal, and reconfigurable antenna designs in upcoming wireless networks by combining ferroelectric tunability and dielectric efficiency. In recent years, advancements in additive manufacturing and printable electronics have further accelerated the deployment of PVDF-based antennas across diverse platforms. Techniques such as inkjet printing, screen printing, and fused filament fabrication (FFF) have enabled high-resolution patterning of conductive elements on PVDF films, allowing for scalable and cost-effective antenna prototyping. Moreover, hybrid integration of PVDF with nanofillers like $BaTiO_3$, graphene, and MXenes has opened new pathways for tunable permittivity and enhanced RF

A. Mayur · A. Sharma (✉)
Department of Electronics and Communication Engineering, Motilal Nehru National Institute of Technology, Allahabad, Prayagraj, India
e-mail: anandsharma@mnnit.ac.in

D. Punetha (ed.), *Polyvinylidene Fluoride (PVDF)*, Engineering Materials,
https://doi.org/10.1007/978-981-95-6718-8_5

functionality. These composite formulations can be engineered to optimize performance across multiband and ultrawideband (UWB) regimes, making them ideal candidates for cognitive radio, vehicular networks, and next gen mm Wave communications. The convergence of material science, smart manufacturing, and reconfigurable RF design solidifies PVDF's position as a cornerstone material for future wireless antenna systems. Additionally, PVDF's application in conformal antenna design also promotes the creation of body-worn and implantable antennas, where biocompatibility and mechanical compliance are vital. Unlike rigid substrates, which often fall short of performance, surface-mounted PVDF-based antennas can preserve consistent radiation properties despite stretching, bending, or folding. Moreover, the anisotropic permittivity and polarization behavior of PVDF enable designers to dynamically control the propagation of electromagnetic waves, therefore providing possibilities for real-time beam reconfigurability and directional radiation patterns in MIMO and beamforming arrays. Research has looked at field-induced tunability in PVDF-based metasurfaces and frequency-selective surfaces (FSS) to enable smart RF environments responding to electrical bias or environmental stimuli, therefore maximizing its ferroelectric behaviour. All vital in crowded spectrum situations like urban 5G cells and dense IoT deployments, this allows dynamic impedance matching, programmable frequency hopping, and environment-adaptive antenna performance. Improvements in 3D-printed dielectric structures using PVDF blends also indicate potential in designing volumetric antennas with maximum gain and reduced side lobes. Finally, PVDF's use in ultra-conformal platforms like epidermal electronics and implantable devices is enhanced by its combination with stretchable interconnects and soft encapsulation layers. PVDF naturally supports these new systems' need for high RF efficiency with low signal distortion even under biomechanical stress. The adaptability of PVDF improved by chemical changes, composite engineering, and microstructural alignment will help to shape the next frontier of flexible, smart, and high-frequency antenna systems as wireless standards evolve toward 6G, terahertz (THz) applications, and intelligent reflecting surfaces (IRS).

Keywords PVDF antennas · Dielectric properties · Ferroelectric polymers · Flexible antennas · Conformal antenna design · IoT devices

1 Introduction

Driven by the unrelenting need for quicker data transfer rates, universal connectivity, and seamless integration of electronics into daily life, wireless communication systems have undergone a profound transformation in the last few decades. The very fabric of communication technologies is being rewoven as society moves toward more digital and connected lifestyles: away from rigid, centralized, and heavy infrastructures to compact, flexible, and decentralized systems. These comprise wearable electronics, smart medical implants, flexible phones, and embedded sensor networks inside textiles or the human body. Once regarded as secondary and sometimes large items, conventional antenna systems are now essential for the performance, appearance, and dependability of new platforms. A key turning point in this development is the worldwide rollout of fifth generation (5G) communication networks. 5G is the backbone for applications like autonomous cars, telemedicine, remote industrial monitoring, and augmented reality since it not only improved mobile broadband (eMBB) but also ultra-reliable low-latency communication (URLLC) and large machine-type communication (mMTC). Expectations are even more demanding with sixth generation (6G) networks already on the horizon—pushing toward terahertz communication, AI-native wireless protocols, and ultra-high spectral and energy efficiencies. The antenna is a smart, active subsystem rather than a passive appendage in this changing scene. Modern antennas must be conformal, tiny, and integrated into unusual form factors all while supporting several functions: beam steering, polarization agility, environmental adaptability, and dynamic reconfiguration. Particularly at the substrate level, these paradigm changes call for a comparable revolution in antenna material science. Though mature and commonly used, traditional antenna substrates such as FR4 (epoxy-based glass), PTFE (e.g., Rogers's materials), and ceramics have many drawbacks [1–8]. For example, FR4 is inefficient for small RF or millimeter-wave systems since it has a low dielectric constant (usually about 4.4) and high dielectric losses at high frequencies. It is also stiff, weighty, and inappropriate for integration into flexible or deformable platforms. Though pricey, less mechanically strong, and challenging to manufacture in flexible forms, PTFE-based materials provide superior dielectric performance. These constraints highlight the pressing need for substitute materials able to satisfy the sophisticated needs of the next communication systems. One such material, polyvinylidene fluoride (PVDF), promises a radical change in the design and use of antennas. A semi-crystalline fluoropolymer, PVDF offers great ferroelectric and piezoelectric qualities, low Dielectric loss tan $\delta < 0.02$ and high dielectric constant $\varepsilon r \approx 8$–12. It distinguishes PVDF is not only its electromagnetic compatibility but also its mechanical flexibility, chemical inertness, and simplicity of manufacture. These qualities taken together allow the creation of electrically tunable, low-profile, conformable, lightweight antennas—qualities absolutely required for soft robotics, unmanned systems, implantable, and wearables. Apart from its natural qualities, PVDF works well with a great variety of sophisticated manufacturing techniques. Inkjet printing, screen printing, electrospinning, and fused filament fabrication (FFF) are among the methods that allow the deposition of antenna structures on flexible PVDF films with great resolution and scalable throughput [9]. For example, inkjet printing lets silver nanoparticle inks or graphene-based formulations digitally, mask-less pattern conductive traces on PVDF. For low-cost, large-area antennas, screen printing offers quick deposition of thick conductive layers. Electrospinning makes it possible to create PVDF nanofibers for use in micro-antennas and flexible sensors. Often compatible with roll-to-roll manufacturing, these techniques let low-cost, flexible communication devices be produced at high volume. PVDF's ferroelectric character also allows dynamic tuning of antenna performance. Polarization of PVDF molecules under an outside electric field changes the dielectric environment. Tunable resonance frequencies, beam steering, and adaptive impedance matching can all be achieved by using this reconfigurability. Cognitive radio systems, which must dynamically access underused spectral bands and reconfigure in real-time, need such capabilities. In large MIMO arrays as well, dynamic beamforming is vital to improve coverage and capacity in congested metropolitan areas.

Furthermore, PVDF's mechanical durability is a major benefit. Its performance is not compromised by repeated bending, stretching, and torsional stresses, so making it appropriate for tough uses include wearable antennas for soldiers, aerospace skins, and automotive body integration [10]. Its resistance to UV radiation, high and low temperatures, and environmental pollutants (moisture, chemicals, etc.) makes it perfect for outdoor deployments and severe conditions, including satellite communication modules and space probes. PVDF is also suitable for composite material engineering. To increase its dielectric constant or modify its frequency response, researchers have doped PVDF with ceramic particles such $BaTiO_3$, ZnO, and $SrTiO_3$ [11]. PVDF matrices have been embedded with conductive fillers including carbon nanotubes (CNTs), graphene, and MXenes to produce flexible, high-conductivity patterns, enabling multifunctional devices combining sensing, energy harvesting, and wireless communication in a single module. These nanocomposites have adjustable qualities that let antenna designers maximize size, bandwidth, and efficiency depend on application requirements. PVDF-based antennas have many applications and are expanding. In the healthcare industry, PVDF antennas are being investigated for wireless delivery systems, neural interfaces, and implantable biosensors [12]. These antennas must be tiny, biocompatible, and able to operate steadily inside the conductive environment of the body. PVDF has shown good results in vivo studies and fits these requirements. Essential for autonomous driving systems, PVDF antennas integrated inside windshields, bumpers, or dashboards support vehicle-to-vehicle (V2V) and vehicle-to-infrastructure (V2I) communication. PVDF can monitor strain, vibration, and structural health in smart infrastructure while also allowing wireless data transmission, therefore lowering the need for wired sensing networks. PVDF-based antennas are attracting interest in defense and aerospace for their low weight, conformability, and stealth qualities. These antennas can be included into the surface of soldier uniforms, drones, or aircraft without compromising aerodynamics or mobility. Their low profile and electromagnetic transparency also make them appropriate for integration into frequency-selective surfaces (FSS) and metamaterial absorbers, where customized EM response is needed [13]. Another field where PVDF antennas are set to have major influence is the growth of reconfigurable intelligent surfaces (RIS). Surface-mounted antenna arrays that reflect or refract signals in controlled ways are used in RIS technology to manipulate the propagation environment. PVDF can allow dynamic RIS components responding to electrical, thermal, or mechanical stimuli—offering real-time control over EM wavefronts without needing large RF chains by means of its ferroelectric tunability. From a system integration standpoint, PVDF's simplicity of lamination onto textiles, plastics, and biodegradable polymers lets for conformal electronics that are ergonomically unobtrusive and visually appealing [14]. PVDF antennas, for instance, can be woven straight into garments or applied to bandages, so enabling wireless body area networks (WBANs) for continuous health monitoring [15]. Operating in concert with body energy harvesters and low-power communication protocols such as BLE or LoRa, these systems create a strong and autonomous sensing environment. The study scene around PVDF antennas is growing fast. Advanced multiphysics simulation tools are required for modeling and simulation of PVDF-based antennas since they include complex coupling between electromagnetic solvers and ferroelectric hysteresis models [16]. Fabrication studies are more concentrating on optimizing ink formulations, curing techniques, and mechanical adhesion on soft substrates. At the same time, empirical research looks at fatigue behavior, signal stability, biocompatibility, and long-term environmental dependability.

Historically speaking, the evolution of antenna technology has mirrored developments in system-level miniaturization and materials engineering. Early wireless systems depended on large dipole or Yagi-Uda structures built from metal conductors fastened on stiff boards. Planar patch antennas constructed on substrates like Rogers RT/Duroid were adopted by the industry as microstrip technology developed in the 1970s since they supported high-frequency operation with fair efficiency and repeatability. These materials, though, were frequently too expensive for scalable consumer-grade uses and lacked mechanical flexibility [17]. Material scientists started looking at substitutes that could equal or surpass conventional RF performance criteria with the arrival of flexible electronics in the 2000s and the push toward wearable and implantable devices, so opening the investigation of polymers like PVDF. Its functional behavior is largely determined by PVDF's molecular structure.

As shown in the Fig. 1 each monomer consists of two hydrogen atoms and two fluorine atoms attached to a carbon-carbon backbone. Fluorine and hydrogen atom uneven distribution along the polymer chain produces a net dipole moment opposite to the chain axis. PVDF shows significant ferroelectric and dielectric behaviour when these dipoles are aligned—most especially in the β-phase configuration. Its use in flexible RF devices, sensors, and energy harvesting systems is supported by this unusual molecular structure, which is especially important for dielectric performance and polarization tunability. Composed of repeating units of - (CH2-CF2)-, it exists in various crystalline phases: alpha, beta, gamma, and delta. Among these, the beta-phase shows the greatest ferroelectric and piezoelectric reactions because of its all-trans planar zigzag conformation, which produces net dipole alignment along the polymer chain as shown in Fig. 2. Mechanical stretching, electrical poling, or annealing treatments can cause the non-polar alpha-phase to transition to the electroactive beta-phase. This phase change not only improves the dielectric

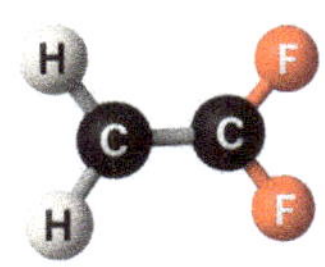

Fig. 1 The chemical structure of the PVDF monomer

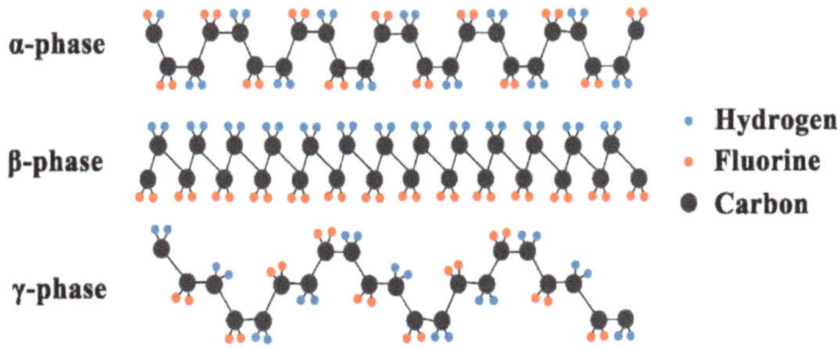

Fig. 2 Structures of α, β, and γα, β, and γ phase polyvinylidene fluoride (PVDF) [18]

characteristics of the material but also adds dynamic tunability, therefore allowing the creation of antennas directionally steerable in real-time, frequency-reconfigurable, and polarization-agile.

The ferroelectric character of PVDF also lets it store and release energy depending on polarization reversal, which has been applied in memory-based antenna tuning and programmable RF devices. This effect offers a useful memory effect in which the antenna's state may be maintained even after external stimulus removal, therefore offering a low-power solution for reconfigurable wireless modules. Moreover, the coupling between mechanical deformation and electrical response makes it possible to create strain-sensitive antennas that can perform sensing as well as communication functions concurrently. In wearable and prosthetic applications, where compactness and multifunctionality are top priorities, such dual-mode systems are becoming more relevant. PVDF has demonstrated notable promise in the domain of frequency-selective structures and metasurfaces [19]. Metasurfaces manipulate incoming electromagnetic waves by means of subwavelength resonators and customized surface impedances. Including PVDF as a dielectric host or active component inside metasurface unit cells produces electrically tunable phase profiles, enabling dynamic control over absorption, transmission, and reflection properties. Beam shaping, stealth technologies, and electromagnetic cloaking all benefit greatly from this since it allows for on-demand spatial response reconfiguration. Recent studies have also looked at stretchable and serpentine PVDF antenna geometries created using fractal patterns, kirigami cuts, or mesh-like topologies. Ideal for uses in soft robotics, smart textiles, and epidermal communication devices, these designs preserve electrical continuity while allowing significant strains [20]. Antennas built on serpentine dipoles printed on PVDF-elastomer composites, outperforming traditional copper-based stretchable interconnects. Moreover, particularly in thin-film or nanofiber setups, PVDF's frequency response reaches far into the millimeter-wave and terahertz (THz) ranges. On-chip wireless connections, ultra-fast communication, and non-invasive imaging all require low-profile antennas in these domains. Compact THz antenna arrays are very compatible with PVDF's high dielectric constant and low dispersion properties. Research teams have shown PVDF-based slot and ring resonator antennas running beyond 100 GHz, with possible uses in terahertz spectroscopy, sub-mm wave radar, and high-capacity backhaul links.

Though it has benefits, using PVDF-based antennas presents some technical issues as well. Long-term stability of dielectric characteristics under cyclic mechanical loading and environmental exposure is one such difficulty. Moisture absorption, UV-induced depolarization, or thermal relaxation of PVDF's ferroelectric domains can cause its response to deteriorate. Dealing with these problems calls for hybrid material integration with protective overlays, surface passivation, or encapsulation techniques [21]. Fabrication repeatability and substrate adhesion are also issues, especially in multilayer antenna systems where dielectric interfaces are essential for impedance matching and radiation efficiency [22]. Standardization and commercialization are also significant factors. Although PVDF films come in different thicknesses and shapes—sheets, spun fibers, membranes—there is little standardization in RF-grade quality, dielectric uniformity, or processing techniques for antenna uses. PVDF-based antennas' integration into mainstream products has been somewhat delayed by the absence of defined commercial manufacturing channels [23]. But developing printable PVDF composites, roll-to-roll printing lines, and hybrid integration platforms co-fabricating sensors, energy harvesters, and antennas on the same substrate is closing this gap for emerging industrial partnerships and academic spin-offs [24].

From a systems engineering viewpoint, the integration of PVDF-based antennas must also consider electromagnetic interference (EMI), biocompatibility, legal compliance, and packaging limitations [25]. These factors are especially important in medical implants and aerospace applications since interference with sensitive electronics or external fields can have serious repercussions. Researchers are looking at composite shielding layers, frequency band allocation plans, and on-demand encapsulation methods to offset such hazards in reaction [26].

The field is also seeing PVDF-based antennas converge with design optimization driven by machine learning. Dramatically cutting development time and enhancing performance predictability, AI-based electromagnetic solvers and genetic algorithms are now used to optimize PVDF antenna geometries, matching networks, and reconfiguration protocols. Key characteristics of PVDF-based devices—non-linear hysteresis, multiphysics interactions, and environmental adaptation—such computational tools can model. PVDF is set to be quite important in the next generation 6G and beyond communication systems where universal connectivity, high data rates, and energy neutrality are the main goals. Realizing these visions will depend on PVDF's ability to support self-powered, shape-adaptive, multi-band antennas. Moreover, hybrid PVDF-based platforms could act as interfaces between actuation, sensing, and communication systems—ushering in an age of "smart surfaces" and "intelligent materials" that not only transmit data but also interpret and react to their environment. Ultimately, PVDF's adaptability goes well beyond that of a traditional antenna substrate. Its unusual combination of electrical, mechanical, and chemical qualities lets it work as a multifunctional enabler in flexible, conformal, and adaptive wireless systems. PVDF not only allows miniaturization and integration but also reconfigurability, multifunctionality, and environmental compatibility as discussed in this chapter [27]. The following sections will offer a thorough analysis of the dielectric and ferroelectric mechanisms of PVDF, the design concepts and manufacturing techniques for PVDF-based antennas, and their deployment across important applications including 5G, IoT, medical telemetry, and smart surfaces. The chapter intends to express PVDF's strategic role in rethinking antenna technology for the future of wireless communication by means of this multidisciplinary perspective.

The following chapters of this book will offer a thorough examination of the dielectric and ferroelectric mechanisms supporting PVDF's RF behavior, novel applications across consumer, industrial, medical, and military sectors, and advanced methods for conformal antenna design. Grounded in present research, the conversation will be enhanced with actual implementation case studies and backed by high-impact experimental studies [28]. This work intends to show PVDF's transforming potential in next-generation wireless communication networks by closing the gap between material invention and antenna system integration [29].

2 Dielectric and Ferroelectric Properties for RF Applications

Radio frequency (RF) applications depend heavily on the dielectric and ferroelectric characteristics of materials, particularly in contemporary systems that require mechanical flexibility, tunability, and miniaturization. Materials like Polyvinylidene Fluoride (PVDF) and its composites are perfect for reconfigurable radio frequency components because they provide a special combination of switchable polarization, moderate to high relative permittivity, and low dielectric loss. In its dielectric function, PVDF ensures low loss tangent ($\tan\delta < 0.02$) and stable permittivity ($\varepsilon_r \approx 8$–12), which minimizes signal attenuation and facilitates efficient impedance matching for transmission lines, antennas, and resonators. Voltage-controlled modulation of permittivity is made possible by PVDF's ferroelectric behavior, which is characterized by spontaneous and reversible dipole alignment when processed into its β-phase. When dynamic frequency or radiation pattern control is needed, this property is especially useful in tunable filters, phase shifters, varactors, and antennas. Compact, adaptable systems appropriate for 5G, IoT, and wearable electronics are made possible by the incorporation of ferroelectric materials into RF devices, which enable field-driven tunability without adding extra mechanical complexity or size. The next generation of intelligent communication and sensing technologies is supported by ferroelectric polymers' compatibility with flexible substrates, which also increases their applicability to conformal and body-worn RF systems.

2.1 Dielectric Properties

Particularly in high-frequency wireless communication systems, the design and performance of antennas are significantly influenced by the dielectric characteristics of a substrate material [30]. These qualities have a direct impact on antenna impedance, bandwidth, efficiency, and radiation pattern. Polyvinylidene fluoride (PVDF) shows exceptional dielectric qualities that make it a desirable choice for radio frequency (RF) and microwave applications since its unique molecular structure and phase-dependent behaviour. Being a semi-crystalline polymer, PVDF has a mix of amorphous and crystalline areas that produces a special interaction of dielectric behaviour and mechanical flexibility. Depending on the degree of crystallinity, the processing technique, and the phase (α, β, γ, or δ) present in the polymer, its dielectric constant (εr) usually falls between 8 and 12. Among these, the β-phase of PVDF is especially significant for dielectric applications because of its very polar structure and improved dipole alignment under an electric field. Unlike many traditional substrates whose dielectric properties degrade at higher frequencies, PVDF keeps a relatively consistent εr value up to several gigahertz, making it appropriate for 5G sub-6 GHz bands, Wi-Fi 6, Bluetooth, and developing 6G applications. Even with rising operating frequencies, this consistency guarantees consistent performance, low insertion losses, and dependable impedance matching. Another important factor influencing PVDF's performance as an RF substrate is its dielectric loss tangent (tan δ). Tan δ is the percentage of electromagnetic energy consumed as heat inside the material (Fig. 3).

To reduce energy loss, high-performance antenna systems benefit from a low loss tangent. Depending on its molecular structure, processing conditions, and environmental elements including temperature and humidity, PVDF usually shows a loss tangent in the range of 0.01 to 0.02. For battery-powered and energy-constrained wireless devices, this low loss factor guarantees great radiation efficiency, which is

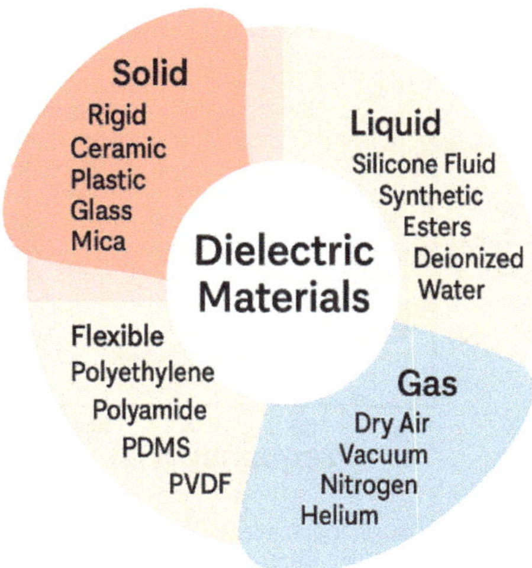

Fig. 3 Various dielectric materials

quite important. PVDF's wide operational temperature range is a further advantage. It keeps its dielectric qualities over temperatures from −40 °C to +150 °C, therefore providing consistent performance under various environmental conditions. Compared to its composites, pure PVDF has superior thermal and dielectric stability. For high-frequency and high-temperature electronic applications, it is still a more reliable choice.

The Fig. 4 shows how the dielectric loss tangent (tan δ) varies at ambient temperature, 50 °C, 75 °C, and 100 °C for pure PVDF and its composites (CdZ-P and CaZ-P) across a wide frequency range (1 kHz to 10 MHz). At any temperature, pure PVDF continuously shows the lowest tan δ, suggesting excellent dielectric insulation and negligible energy loss. Due to improved dipolar relaxation, all samples exhibit larger tan δ as the temperature rises; nonetheless, pure PVDF retains superior thermal stability in comparison to the composites. Because of interfacial polarization and enhanced conductivity, the inclusion of CdZ and CaZ fillers raises tan δ values, particularly at higher filler concentrations and higher temperatures. The influence of fillers on dielectric dissipation at high thermal loads is shown by the composites' noticeable increase in loss tangent at 100 °C. PVDF is especially appropriate for outdoor antennas, aerospace communication systems, and automotive radar applications where temperature changes can be severe as a result.

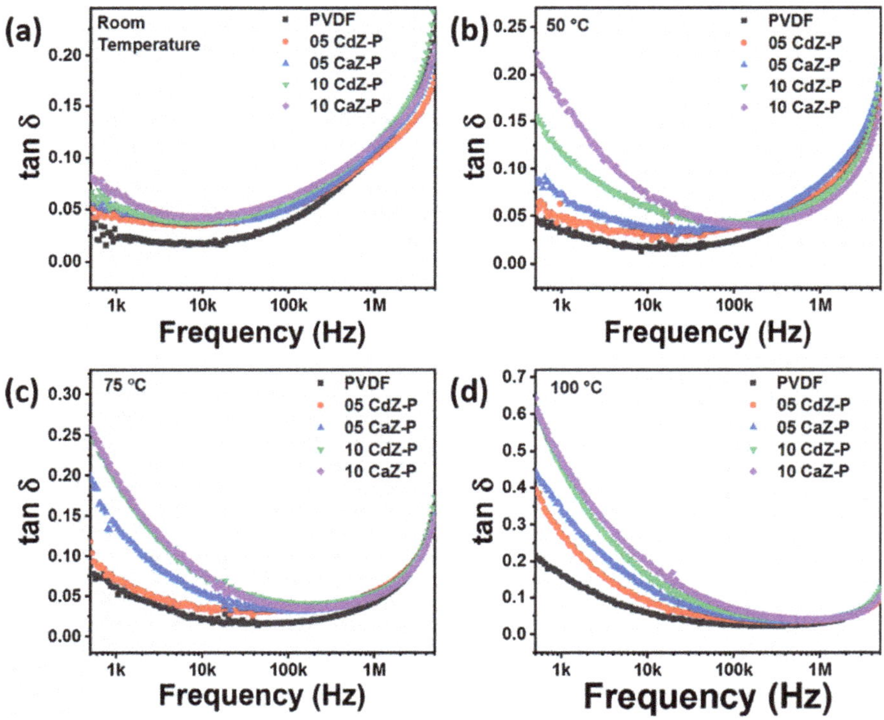

Fig. 4 Dielectric loss tangent of pure PVDF and PVDF composite films at (**a**) room temperature, (**b**) 50 °C. (**c**) 75 °C, and (**d**) 100 °C [31]

Moreover, PVDF shows great chemical resistance, guaranteeing that its dielectric qualities stay consistent even under severe chemicals, moisture, and ultraviolet (UV) radiation. Antennae deployed in corrosive or high-humidity settings, such as coastal areas, chemical plants, or wearable health-monitoring devices, need this environmental stability. Changing PVDF's processing parameters helps to further optimize its dielectric behaviour. Mechanical stretching of the polymer or electric field application during manufacture can, for instance, cause a greater proportion of β-phase crystallites, so improving the dielectric constant and lowering the loss tangent. Annealing PVDF films at regulated temperatures can likewise increase crystalline alignment and lower flaws, therefore enhancing dielectric performance.

From 10^4 to 10^6 Hz, the graph depicts the frequency-dependent dielectric behaviour of PVDF-HFP and BCZT-based composites (BCZT10 and BCZT15) shows in Fig. 5. Indicating its appropriateness for low-loss RF applications, PVDF-HFP shows a consistent but low dielectric constant (~7–8) with little dielectric loss (<0.05). By contrast, BCZT10 and BCZT15 show much greater dielectric constants—up to about 17 for BCZT15—due to the ceramic filler, although with rising dielectric loss at higher frequencies, particularly for BCZT15 (>0.12). This emphasizes a trade-off between energy dissipation and high permittivity, which is relevant for adjusting material selection in high-frequency electronics. While BCZT composites sacrifice greater dielectric loss for higher permittivity, PVDF-HFP provides low-loss stability perfect for RF use. Researchers have investigated PVDF-based composites in recent years by including nanofillers including barium titanate (BaTiO3), zinc oxide (ZnO), Barium Calcium Zirconate Titanate (BCZT) and graphene [33].

Without sacrificing flexibility, these fillers greatly raise the dielectric constant. PVDF-BaTiO3 composites, for example, have shown εr values over 40 with low tan δ, providing possibilities for ultra-miniaturized antenna designs. Composite

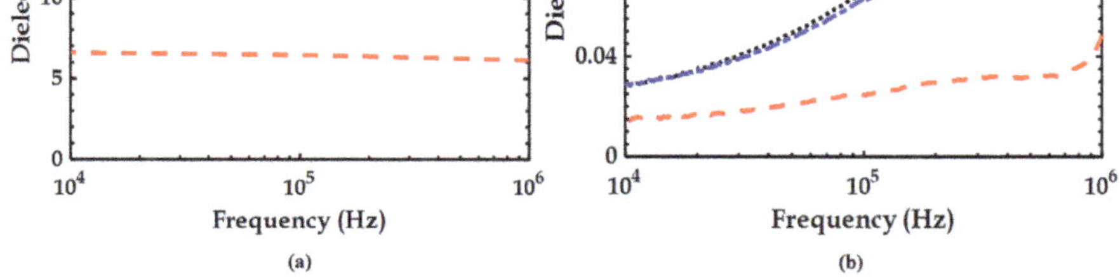

Fig. 5 Frequency dependence of PVDF-HFP/BCZT composites: (**a**) Dielectric constant and (**b**) loss tangent of pure PVDF-HFP compared with PVDF-HFP/BCZT composites at different BCZT concentrations [32]

engineering gives PVDF-based systems even more adaptability by allowing them to customize the dielectric characteristics. Without sacrificing flexibility, these fillers greatly raise the dielectric constant. PVDF-BaTiO3 composites, for example, have shown εr values over 40 with low tan δ, providing possibilities for ultra-miniaturized antenna designs. Composite engineering gives PVDF-based systems even more adaptability by allowing them to customize the dielectric characteristics. Table 1 different PVDF materials and dielectric constant with loss tangent of that material.

Although hardly carefully managed, increasing the dielectric constant can impact bandwidth and radiation patterns even if it can help to lower antenna size. Therefore, a balance must be found between attaining high εr for miniaturization and preserving reasonable bandwidth and efficiency for communication uses. PVDF's

Table 1 Typical dielectric properties of PVDF and PVDF composites

Material	Dielectric Constant (ε_r)	Loss Tangent (tan δ)
Pure PVDF (β-phase rich) [34]	8–12	0.010–0.020
PVDF + $BaTiO_3$ composite [35]	25–45	0.015–0.030
PVDF + ZnO composite [36]	15–25	0.012–0.025
PVDF + graphene composite [37]	10–18	0.010–0.018
PVDF + CNTs composite [38]	12–22	0.012–0.023
PVDF + TiO_2 composite [39, 40]	18–28	0.013–0.024
PVDF + Al_2O_3 composite [40]	10–16	0.010–0.019
PVDF + MWCNTs [41]	14–20	0.011–0.020
PVDF + SiO_2 nanocomposite [40]	11–17	0.009–0.016
PVDF + PZT composite [42]	30–60	0.020–0.035
PVDF + $SrTiO_3$ [43]	22–36	0.014–0.026
PVDF + Fe_3O_4 nanoparticles [44]	14–21	0.013–0.022
PVDF + Ag nanoparticles [45]	13–18	0.011–0.020
PVDF + rGO [46]	9–15	0.010–0.016
PVDF + MoS_2 [47]	11–19	0.012–0.021
PVDF + NiO [48]	17–25	0.015–0.027
PVDF + Co_3O_4 [49]	19–26	0.016–0.028
PVDF + WO_3 [50]	20–30	0.017–0.030
PVDF + MgO [51]	13–19	0.010–0.018
PVDF + $CaCu_3Ti_4O_{12}$ [52]	45–80	0.025–0.045
PVDF + SnO_2 [53]	15–22	0.012–0.020
PVDF + HfO_2 [54, 55]	18–27	0.014–0.024
PVDF + Y_2O_3 [56]	16–23	0.013–0.022
PVDF + Ti_3C_2 Mxene [57]	28–38	0.018–0.030
PVDF + $ZnFe_2O_4$ [58]	12–20	0.011–0.019
PVDF + $BiFeO_3$ [59]	35–55	0.020–0.034
PVDF + ZrO_2 [60]	14–21	0.013–0.022
PVDF + Graphene Oxide [61]	10–16	0.011–0.018
PVDF+Metal–Organic Frameworks [62]	20–32	0.015–0.026

dielectric constant's frequency dependence is rather modest; yet, as the frequency rises into the microwave range, a small drop is seen. Polymer chains cannot reorient fast enough at extremely high frequencies, which causes this phenomenon and a lower polarization response. Though, PVDF shows better frequency stability than other polymers including polyimide or polyethylene terephthalate (PET), so guaranteeing uniform antenna behavior across multiband wireless systems. Often found in flexible and wearable antennas, thin PVDF films can show small changes in εr and tan δ caused by confinement effects and surface roughness. Thus, precise antenna performance forecasting during the design phase depends on thorough characterization and modeling.

The moderate εr of PVDF offers a great compromise in practical antenna designs, particularly for compact devices running in 2.4 GHz (Bluetooth, Wi-Fi) or 5.8 GHz bands. It maintains effective radiation while allowing enough miniaturization (Table 2).

The mechanical flexibility of PVDF allows for integration onto non-planar surfaces without notable loss of dielectric performance, a quality much sought for in conformal antennas [63]. Its dielectric behavior is also influenced by the surface morphology of PVDF. To reduce scattering losses and parasitic capacitance effects, smooth, homogeneous films are preferred. High-quality PVDF films with controlled thickness and surface uniformity have been created using techniques including spin coating, doctor blade casting, and solvent vapor annealing. Particularly if the polymer is not properly treated or protected, environmental elements like humidity can affect the dielectric performance of PVDF somewhat [64]. Moisture ingress can raise dielectric losses by strengthening ionic conduction routes. Therefore, methods of encapsulation and the addition of hydrophobic coatings are frequently employed to maintain the dielectric integrity of PVDF antennas in outdoor or biomedical uses. The dielectric properties of PVDF—high and stable dielectric constant, low loss tangent, environmental resilience, and mechanical flexibility—set it as an ideal material for the new era of wireless communication. Whether used in standalone form or as part of a nanocomposite, PVDF has the capacity to satisfy the exacting standards of next-generation wireless networks, including 5G, 6G, IoT, and wearable electronics. Using the whole potential of PVDF in future antenna designs depends on knowing and optimizing these dielectric qualities (Table 3).

Table 2 Material property comparison table

Properties	Polymer	Textile	Paper	Fluidic
Dielectric loss/loss tangent	Low loss	Low loss	Medium loss	High loss
Tensile strength	High	Low	Low	Low
Deformability	Low	High	High	High
Thermal stability	High	Low	Low	Low
Fabrication complexity	Simple	Complex	Simple	Complex
Robustness to wetness	High	Medium	Low	Low
Cost of fabrication	Medium	Low	Low	High
Weight	Medium	Low	Low	Medium
Overall size	Small	Large	Large	Small

Table 3 Dielectric materials for metamaterial design

Dielectric material	Relative permittivity	Thickness [mm]	Size of Unit cell	Operational Frequency [GHz]	Design Approach	Applications
FR-4	2.4	0.160	$1000 \times 1000 \mu m^2$	38.165, 155.48, 231.71, 275.27	Material perfect absorber (MPA) has two resonators and four resonances	THz absorption applications [65]
Silicon rubber	0.6	0.035	$20 \times 20\ mm^2$	5.26–17.5	MPA has concave up and down stages along with a hemisphere full of graphene ink	Telecommunication filters, as a sensor, and energy harvesting [66]
Polyimide	3.5	240	$142 \times 142 \mu m^2$	15–20, 29–32	Dual-band metamaterial absorber with T-shaped metallic structure	THz imaging for biomedical applications [67]
PDMS	3	0.22	$36 \times 36\ mm^2$	8–12	Reconfigurable metamaterial by loading microfluidic channel layer	Utilized for micro electromechanical system [68]
PVDF	8–12	0.1	$100 \times 100 \mu m^2$	2.4–5.8	Graphene-based antennas and dual-band energy harvesting using β-phase PVDF	Flexible, conformal antenna, wearable antennas and energy harvesting applications [69, 70]

2.2 Ferroelectric Properties

Exhibiting a spontaneous and reversible electric polarization, ferroelectric materials provide unusual electrical characteristics that can be regulated by outside electric fields. This feature lets them be absolutely important in creating tunable RF parts. Among the polymer-based ferroelectrics, polyvinylidene fluoride (PVDF), a semi-crystalline polymer, stands out because of its good electrical properties, process-ability, and mechanical flexibility. Unlike ceramic ferroelectrics, PVDF can be combined with flexible and conformal devices. Particularly in compact, wearable, and miniaturized electronics—which are core to applications like 5G, IoT, and aero-space communication systems—increased study on PVDF for antennas, filters, and other RF subsystems has followed. Ferroelectric materials are defined by spontaneous electric polarization that an outside electric field can reverse. Their tunable dielectric properties, nonlinearity, and possibility for miniaturized reconfigurable electronics have drawn great interest for radio frequency (RF) and microwave applications in recent years. Among polymer-based ferroelectrics, polyvinylidene fluoride (PVDF) and its copolymers stand out as flexible, adaptable, and integrable materials that combine the mechanical benefits of polymers with the functional qualities of ferroelectrics. Particularly the β-phase of PVDF shows significant ferroelectricity and offers great benefits for frequency-agile RF components. Understanding and using the ferroelectric character of PVDF is essential as demand for low-profile, reconfigurable, and conformal antennas in 5G, IoT, and satellite systems grows (Fig. 6).

According to their composition, ferroelectric materials can be classified into three main categories: ceramic-based ($BaTiO_3$, $BiFeO_3$, and $Pb(ZrxTi1–x)O_3$),

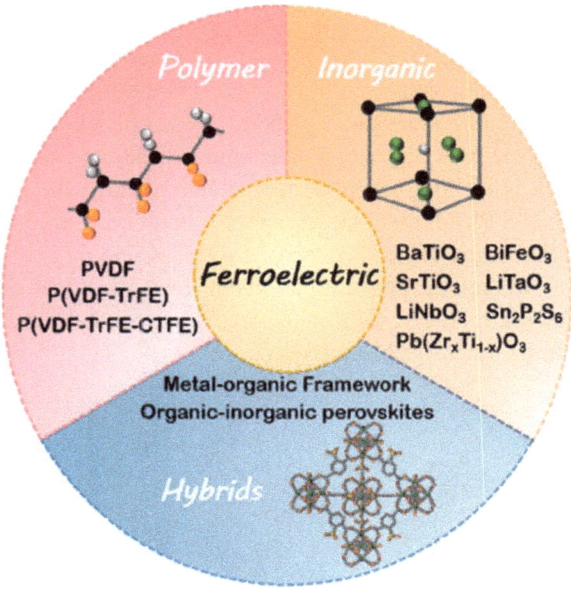

Fig. 6 Classification of ferroelectric materials [71]

polymer-based (polyvinylidene fluoride (PVDF)) and hybridized (compounds containing both inorganic and organic components chalcogenides and compounds with metal atoms bonded to organic backbones) materials.

2.2.1 Polarization Mechanism

The dipole arrangement inside PVDF's crystalline structure gives rise to its ferroelectric behaviour. Aligned along the polymer chain, the molecular dipoles in the β-phase configuration point in the same direction, therefore producing macroscopic polarization. This structural change from the nonpolar alpha-phase to the polar beta-phase is aided by mechanical stretching and electrical poling. This change is important since only the β-phase shows the ferroelectric and dielectric tunability required for RF uses. PVDF can be a dynamically tunable dielectric medium in active RF devices by means of an external field controlling and reversing the direction of dipoles. Processed into its all-trans β.-phase configuration, PVDF becomes ferroelectric. This phase produces net polarization by parallel alignment of molecular dipoles. Applying an outside electric field lets dipoles be reversibly reoriented, therefore allowing hysteresis behaviour vital for tunability.

2.2.2 Polarization–Electric Field (P–E) Hysteresis

A major sign of a material's ferroelectric character is the polarization-electric field (P-E) hysteresis loop. In PVDF, this loop shows the stability of remanent polarization, and the energy needed to change dipole orientation. Typical coercive field values, which range from 50 to 100 MV/m, indicate the strength of the field required to reorient the dipoles. Remanent polarization values between 7 and 10 $\mu C/cm^2$ reveal the material's capacity to keep polarization once the external field is removed. Maintaining consistent performance in RF devices like phase shifters and frequency tunable filters depends on these characteristics. The loop shape also offers knowledge on material fatigue, switching speed, and energy loss, all of which are vital factors for long-term RF operation. Typically, the polarization-electric field (P-E) loop in PVDF films shows coercive fields between 50 and 100 MV/m and remanent polarization values between 7 and 10 $\mu C/cm^2$ [72]. PVDF is especially appropriate for RF tuning applications given these electrical qualities since they allow for desired moderate field-induced changes in dielectric permittivity without causing energy-intensive biasing needs.

2.3 Role of Ferroelectricity in RF Components

RF systems employ ferroelectric PVDF materials to create components including tunable capacitors, ferroelectric phase shifters, and frequency-reconfigurable antennas. PVDF-based varactors use field-dependent permittivity for capacitance modulation, which is crucial in phase shifters, voltage-controlled oscillators (VCOs), and frequency-tunable filters. Under low bias, the dielectric nonlinearity resulting from ferroelectric domain switching allows efficient permittivity modulation. PVDF-based ferroelectric phase shifters use its low acoustic velocity and high electrical tunability to enable effective phase shifting in microstrip or coplanar waveguide (CPW) lines. With their small size, low-loss operation across gigahertz bands, and voltage-controllable phase delay, these devices are appropriate for adaptive RF front ends. Modulating substrate permittivity lets PVDF substrates with embedded electrodes also dynamically change the resonance frequency of antennas. Under DC bias, such antennas can show a resonance shift of 5–10%, which helps them to be used in multi-band and reconfigurable systems like cognitive radios and adaptive wireless communication modules.

2.4 Material Processing and β-Phase Optimization

Realizing its ferroelectric behaviour depends on appropriate PVDF processing. While electrical poling lines the molecular dipoles under a high-voltage field, mechanical stretching at controlled temperatures reorients the polymer chains into the β-phase. A commonly used technique to stabilize the β-phase naturally is copolymerization with trifluoroethylene (TrFE), which lessens the need for high-energy post-processing [73]. While inkjet printing provides maskless patterning for fast prototyping of adjustable devices, electrospinning allows the production of aligned nanofibers. Commonly used to relieve internal stresses and stabilize the dipolar alignment after poling, thermal annealing improves dielectric stability and endurance under cyclic loading [74]. Achieving consistent and effective RF device performance depends on the optimization of these parameters.

Mechanical stretching, electrical poling with fields near 100 MV/m, and copolymerization with trifluoroethylene (TrFE) all help to convert the polymer to its β-phase, thereby inducing ferroelectricity in PVDF. Controlled alignment of thin films is also produced by electrospinning and inkjet deposition among other techniques. Post-treatment annealing increases dipole orientation stability and linearity in field response. One extremely promising processing method is electrospinning. A voltage of 10 to 20 kV is typically added between the nozzle and the substrate after the needle tube containing the PVDF solution sprays the PVDF solution onto a drum via a nozzle and the substrate wraps around the drum [75]. A low-density PVDF film is created by randomly depositing nano- to micro-scale PVDF fibers onto the substrate. The PVDF polarization and deposition processes are finished in a single step by the electrospinning technique. The formation of the β-phase state of PVDF is facilitated by the increased stress between PVDF and the substrate caused by the drum's rotation. Figure 8a displays a schematic diagram of the electrospinning process. To create sensors [77] electrospun a PVDF nanofiber membrane with PET/ITO electrodes and PDMS-Ag NWS. By electrospinning PVDF and a silver mixed solution, Lu et al. [78]. created a flexible bend sensor to track human respiration (Fig. 7b). By electrospinning (BTO)/P(VDF-TrFE) composite nanofibers,

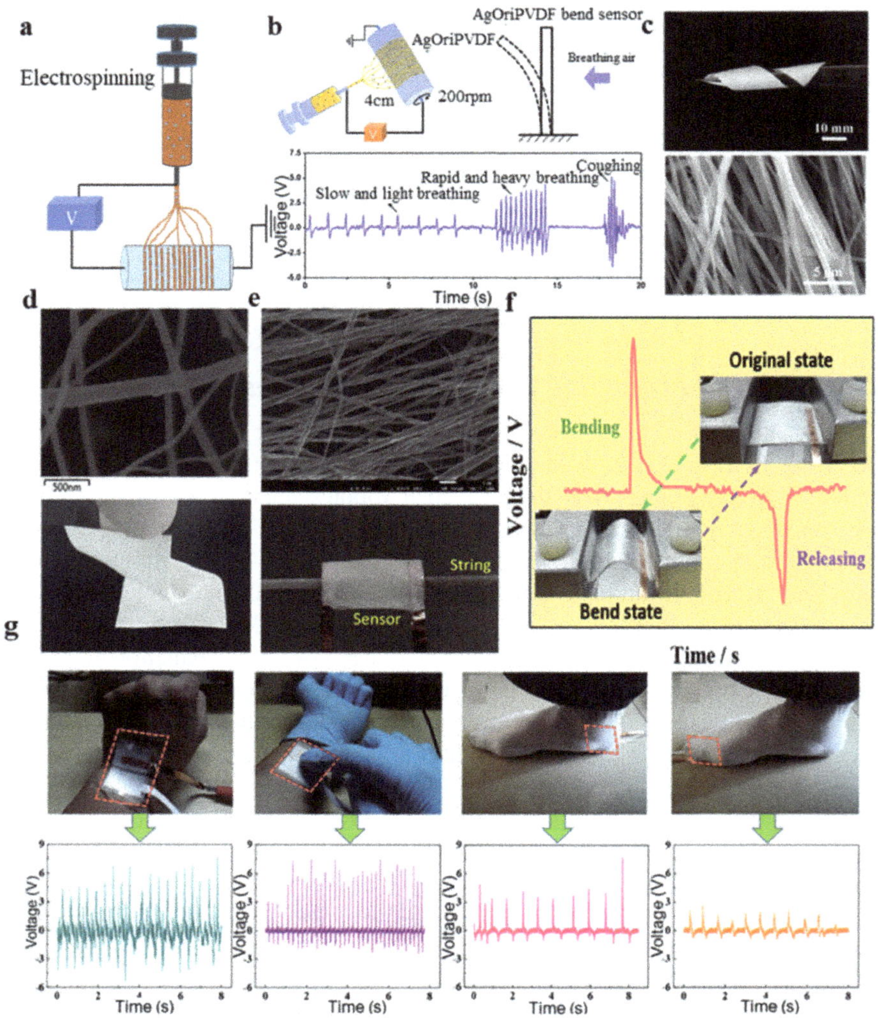

Fig. 7 (**a**) Electrospinning technique schematic diagram; (**b**) flexible bend sensor for human respiration monitoring [78]; (**c**) optical photo and SEM of (BTO)/P(VDF-TrFE) composite nanofibers [78]; (**d**) optical photo and SEM of BNT-ST/PVDF nanofiber composite [79]; (**e**) PVDF strain sensor for string force measurement [80]; (**f**) mechanical movement in both the primitive and bending phases [81]; and (**g**) optical image and output voltage generated by human movement [81]

Xiaohe et al. [79] were able to detect and distinguish walking ants and the energy of a free-falling ball as low as 0.6µJ (Fig. 7c). Sang et al. [76] used electrospinning to create flexible, lead-free piezoelectric nanofibers made of PVDF and BNT-ST ceramic. The BNT-ST/PVDF nanofiber composite module's output voltage measurement system is a frequency function (Fig. 7d). Rahul et al. [80] created a PVDF strain sensor to measure the string force by fabricating PVDF nanofibers with a high β-phase state (Fig. 7e). By electrospinning graphene nanosheets, barium titanate, and PVDF, Kunming et al. [81] created nanocomposite fiber mats with an open-circuit voltage of up to 11 V (Fig. 7f, g).

2.5 *Techniques of Measurement*

RF performance is assessed using methods including time-domain reflectometry (TDR), vector network analyzer (VNA) measurements, and impedance spectroscopy (from 1 MHz to 20 GHz). Typically, between 10 and 20%, dielectric tunability ratios—calculated as ε(Emax)/ε(0)—with S and X bands insertion losses below 0.5 dB [82]. These measures show PVDF's appropriateness for adjustable and high-efficiency RF components.

In Fig. 8 A VNA was used to measure the reflection coefficient (|S11|) directly on the human body. This configuration made it possible to compare the experimental and simulated results directly, providing important information about how well the antenna performs in real-world situations. Figure 8b displays the measured |S11| values in comparison to the simulated data, demonstrating the high degree of agreement between the two datasets. The antenna demonstrated in figure, an impedance bandwidth of 36.6% (4.60–6.65 GHz) for |S11| < −10 dB when it was simulated on

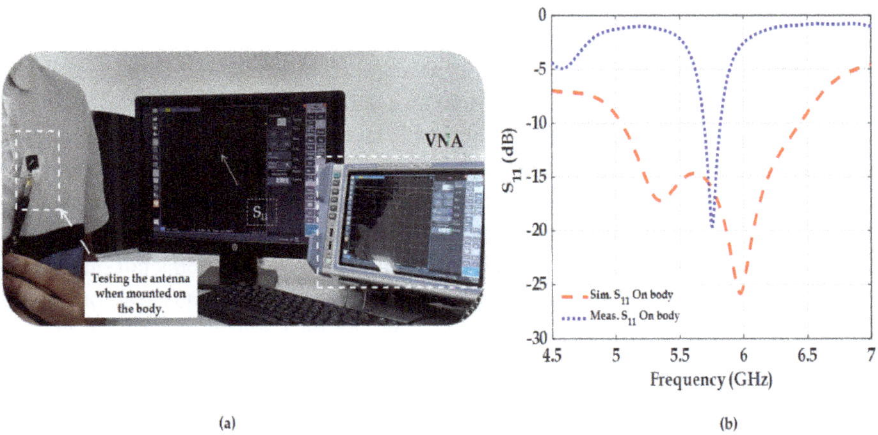

Fig. 8 (**a**) Experimental configuration for utilizing a vector network analyzer to measure antenna performance on a human volunteer. (**b**) The proposed antenna's simulated and measured reflection coefficients (|S11|) in free space and on the human body are compared [76]

the human body phantom. However, due to the higher dielectric loading and absorption effects of human tissues, measurements made directly on the human body showed a reduced bandwidth of 22.0% (4.84–6.04 GHz). This decrease in bandwidth is in line with what is anticipated for wearable antennas and emphasizes how crucial it is to take body loading into consideration when assessing performance [83].

2.6 Future Research Directions and Prospects

Ongoing innovations meant to improve the ferroelectric and dielectric performance of PVDF-based materials for RF and wireless communication systems will therefore drive their future development. Copolymizing PVDF with trifluoroethylene (TrFE) produces poly (vinylidene fluoride-trifluoroethylene), or P(VDF-TrFE), which is one promising solution. The molecular configuration TrFE units introduce causes this copolymer to have a more stable and noticeable ferroelectric response. These units' presence encourages a greater proportion of the β-phase and results in better remanent polarization and lower coercive field requirements, which makes P(VDF-TrFE) systems especially appealing for high-frequency RF uses at low voltage [84]. Reinforcing PVDF with functional nanomaterials to produce nanocomposite structures is another important area of study. Adding materials like carbon nanotubes (CNTs), graphene, and different ceramic nanoparticles has been found to greatly improve the polarization behaviour and mechanical strength of PVDF films. By acting as nucleating agents for beta-phase development and causing interfacial polarization effects, these nanofillers help to enhance dielectric tunability and long-term structural integrity [85]. Such nanocomposites provide the two benefits of electrical performance and mechanical flexibility, both of which are crucial for flexible and wearable electronic platforms. Apart from material changes, increasing attention is being paid to create multifunctional layered architectures combining sensing, actuation, and communication features inside a single PVDF-based structure. Smart systems that can detect environmental stimuli, respond with mechanical or electrical feedback, and wirelessly transmit data can be designed by stacking or co-processing PVDF with other functional layers [86]. These multifunctional designs are especially pertinent for uses in soft robotics, structural health diagnostics, and biomedical monitoring, where mechanical compliance and space constraints are especially important. Together, these developments are propelling the realization of intelligent, adaptive, and autonomous wireless systems not only lightweight and flexible but also able of independent operation in various and sometimes demanding settings [87]. The convergence of materials engineering, device integration, and RF design is set to make PVDF-based ferroelectric systems a central technology in the next generation of smart communication infrastructures. PVDF's ferroelectric characteristics provide an interesting route for creating tunable, flexible, frequency-agile RF components. PVDF's mix of electrical responsiveness and mechanical adaptability fits the increasing needs of next-generation wireless communication systems. PVDF's ferroelectric features are predicted to be transformational in allowing sophisticated, multifunctional RF architectures across 5G, IoT, and future 6G platforms as RF devices get more miniaturized and reconfigurable.

3 Design of Flexible Antennas Using PVDF

The design of flexible antennas has become a key emphasis in the evolution of contemporary wireless communication technologies. The demand for antennas that preserve great performance while changing to non-planar shapes has grown significantly as devices become more portable, wearable, and embedded into irregular surfaces. Given its unusual mix of flexibility, dielectric qualities, piezoelectric activity, and mechanical durability, polyvinylidene fluoride (PVDF) stands out as an outstanding substrate material in this regard. Setting the stage for their widespread application in 5G, IoT, biomedical, aerospace, and wearable systems, this part describes the important design factors, manufacturing techniques, and significant developments in the creation of PVDF-based flexible antennas [88]. Among the technological developments accompanying the fifth generation is flexible industry. Lightweight, affordable, environmentally friendly, and flexible devices are adaptable. Such flexible devices require an antenna to function in certain frequency bands, therefore providing a good wireless connection. Figure 3 shows the typical features of a flexible antenna (Fig. 9).

Fig. 9 Characteristics of a flexible antenna

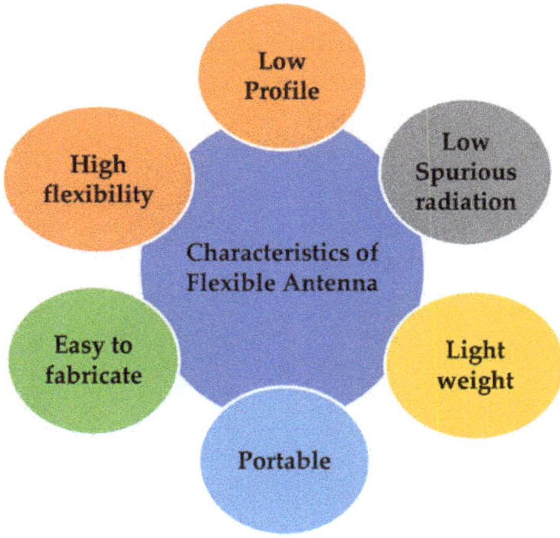

3.1 Design Considerations for PVDF-Based Flexible Antennas

PVDF-based antennas are validated by electrical and mechanical characterization. Of electrical interest are S-parameters (notably S11 and S21), gain (G), radiation efficiency (η), bandwidth (BW), and radiation pattern integrity. Usually, a vector network analyzer (VNA) used with calibrated test jigs and anechoic measurement chambers measures these. Beam directionality, polarization properties, and side lobe levels are revealed by near-field and far-field pattern measurements [89]. The performance of the antenna is evaluated under many deformation cycles to mimic actual use. Performance retention is assessed by bending radius studies, fatigue tests, and torsional loading evaluations [90]. Dynamic mechanical analyzers (DMA) and universal testing machines measure the material's modulus, tensile strength, and strain recovery. Moreover, X-ray diffraction (XRD) and Fourier-transform infrared spectroscopy (FTIR) are used to assess phase purity and crystallinity of the PVDF layers; optical profilometry and atomic force microscopy (AFM) are used to investigate surface morphology. Strain-dependent impedance variation, electromechanical coupling efficiency, and environmental stability under thermal cycling and humidity exposure are among new metrics for next-generation applications as well. These measurements offer thorough understanding of the operational viability of PVDF antennas across several application domains.

The antenna is affixed to a PVDF piezoelectric film that has a thickness of 0.1 mm, a dielectric constant of 9.5, and a loss tangent of 0.05. Figure 10 displays the manufactured prototype. A lower working frequency band is provided by the rectangular copper patch. The L-shaped patches on either side act as a parasitic unit to alter the single-frequency antenna's resonant loop by adjacent coupling, extending a new working frequency band in addition to the metal ground that serves as the coplanar waveguide [91]. It should be noted that PVDF is a type of polymer smart material with a very low melting point (172 °C), making soldering SMA vibrators together impossible (Fig. 11).

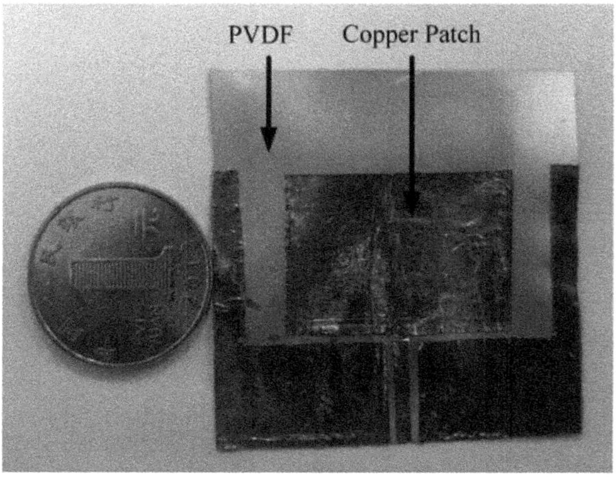

Fig. 10 Fabricated prototype of dual band antenna [91]

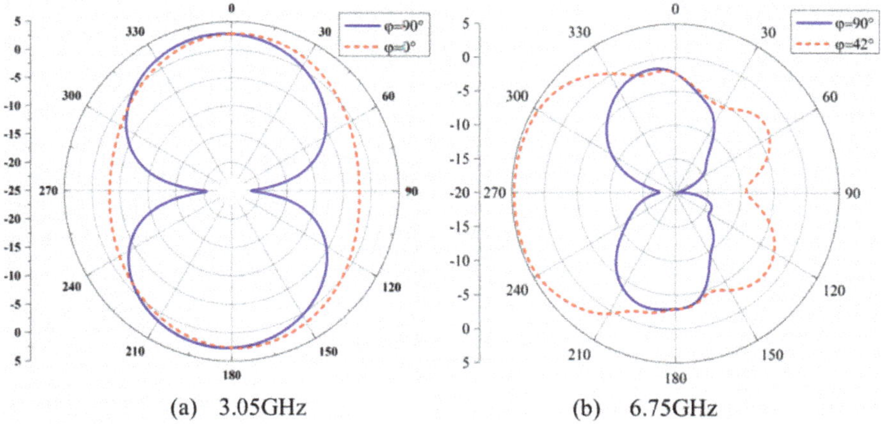

Fig. 11 Radiation Pattern of dual band antenna [91]

The figure clearly shows that the antenna has good omnidirectional radiation characteristics in the low frequency band with a maximum gain of 2.05 dB; it shows great directionality at high frequency, and this characteristic is especially noticeable when φ = 42∘ and maximum gain is 4.88 dB.

3.1.1 Thickness of Substrate

The permittivity and the loss tangent define the characteristics of any substrate. Equation (1) provides the substrate's permittivity [92].

$$\varepsilon = \varepsilon_0 \varepsilon_r \tag{1}$$

where εo is the permittivity of the free space and εr is the relative permittivity of the substrate. Equation (2) can be used to represent the substrate's relative permittivity [92].

$$\varepsilon = \varepsilon_r^r + j\varepsilon_r^i \tag{2}$$

where ε_r^r is the real part *and* $j\varepsilon_r^i$ is the imaginary part of the relative permittivity and, also referred to as the substrate's dielectric constant.

Equation (3) [92] gives the dielectric material's loss tangent ($tan\delta$).

$$\tan\delta = \frac{\varepsilon_r^r}{\varepsilon_r^i} \tag{3}$$

The dielectric constant of the substrate material determines the flexible antenna's quality (Q) factor. Selecting materials with a low dielectric constant raises the bandwidth by lowering the quality factor. The antenna's dielectric constant and resonant frequency are also connected. The resonant frequency of the antenna can be raised by selecting a substrate with a low dielectric constant, and vice versa. The thickness of the substrate becomes a major factor in the mechanical and electromagnetic performance in the design of flexible antennas using PVDF. This factor is essential for balancing structural flexibility with the electrical efficiency needed for wireless communication systems. Because of their better bending ability, thin PVDF films—typically between 50 and 200 micrometers—are quite popular in flexible antenna applications [93]. Such thin substrates let the antenna fit to curved surfaces, textiles, or biological tissues without notable distortion of the radiating structure or loss in electrical function.

Extreme substrate thickness miniaturization, however, has natural trade-offs. A thinner PVDF layer can lead to more surface wave losses, which then lower the antenna's radiation efficiency and bandwidth. Furthermore, mechanical durability could suffer as ultrathin films are more prone to tearing, weariness, and delamination under repeated mechanical stress. Substrate thickness has to be carefully optimized to provide sufficient mechanical resilience while preserving reasonable electromagnetic performance, therefore meeting these conflicting needs. Often, this optimization calls for repeated design cycles using simulations and mechanical testing to find the best balance between durability in practical operating conditions, dielectric integrity, and flexibility [94].

3.1.2 Loss Tangent and Dielectric Constant

The electrical performance and miniaturization possibilities of flexible antennas are greatly influenced by the dielectric properties of PVDF. Typically ranging from 8 to 12, PVDF helps to design small antennas that use far less surface area than those built on lower-permittivity substrates [95]. For applications in wearable and embedded systems where spatial limits are a key concern, this high dielectric constant is particularly beneficial. Elevated εr is crucial for integrating communication modules into compact and curved platforms since it lets electrically tiny antennas be realised without sacrificing operational frequency. Equally crucial is the low dielectric loss shown by PVDF. Generally, below 0.02 in the RF frequency range, its loss tangent (tan δ) suggests low energy dissipation during signal propagation [96]. Maintaining high radiation efficiency and consistent impedance matching across operating bands depends on a low loss tangent, especially if the antenna experiences mechanical bending or distortion. PVDF-based antennas are guaranteed by this quality to be strong under dynamic circumstances, therefore fitting them for body-mounted, textile-integrated, and deployable communication systems. PVDF is the perfect substrate material for flexible high-performance RF antennas since it combines a moderate to high dielectric constant with low dielectric losses.

3.1.3 Electrode Metallization

A key stage in the production of flexible antennas is the metallization of PVDF substrates since it defines the conductive routes responsible for signal transmission and radiation. The antenna's electrical conductivity, mechanical compliance, and long-term dependability are directly affected by the choice of conductive material and the deposition technique. Excellent electrical conductivity of traditional materials like copper makes them quite popular. Though they usually need the use of specific adhesives or surface treatments to guarantee adequate bonding strength, copper foils can be laminated onto PVDF films to create antenna traces. Although efficient, copper-based constructions might be problematic under continuous mechanical flexing since they are more susceptible to delamination and fracture [19]. On the other hand, printed electronics techniques using silver nanoparticle inks have become more popular for creating fine-featured antenna patterns on flexible PVDF substrates. Moderate conductivity and great patterning resolution are achieved by methods like inkjet printing and screen printing, which allow exact, scalable deposition of conductive traces. These techniques are also compatible with the thermal limits of PVDF since they let more design flexibility and lower processing temperatures. Especially for uses in wearable healthcare devices and bio-integrated antenna systems, these materials present a special mix of electrical conductivity, mechanical flexibility, and biocompatibility. But no matter the metallization material, strong adhesion between the conductive layer and the PVDF substrate is first importance. Under cyclic mechanical loading, poor adhesion can cause delamination and performance decline. Common methods to improve interfacial bonding and guarantee consistent antenna operation over long durations of use are surface treatments including plasma activation, chemical priming, or the application of adhesion promoters.

3.1.4 Mechanical Durability and Bending Radius

One of the defining characteristics of PVDF-based antennas is mechanical flexibility, which allows them to fit into foldable, wearable, or curved surfaces. The minimum bending radius the antenna can sustain without performance loss governs this function most important. A mix of substrate thickness, metallization method, and the ductility of the conductive materials used defines this mechanical threshold. While metallization layers with high ductility—such as silver nanoparticle inks or conductive polymers—are more appropriate for repeated mechanical deformation, thinner PVDF substrates tend to show better bending compliance. Even under bending radii as small as 5–10 millimeters, experimental studies have demonstrated that PVDF-based antennas can preserve stable RF performance, including slight changes in resonant frequency and consistent S-parameter behavior [97]. Applications in which antennas are subjected to regular flexing or movement, such body-worn sensors, rollable communication modules, or deployable aerospace structures, need this durability. Under these circumstances, guaranteeing mechanical durability calls for deliberate material choice and structural design to avoid cracking, delamination, or performance drift over the operational life of the antenna.

3.1.5 Bandwidth Considerations and Impedance Matching

Maintaining constant impedance qualities under mechanical deformation is one difficulty in using flexible antennas. Bending causes slight but important changes in antenna geometry from PVDF's natural flexibility, which could then influence the impedance matching and radiation performance. Designers frequently include broadband impedance matching strategies that improve tolerance to geometrical distortions to offset this. Common methods for stabilizing impedance over a broad frequency range and under different mechanical loads are coplanar waveguide (CPW) feeding structures, tapered transitions, and stub tuning networks. Figure 12 displays the permittivity of a modified PVDF matrix with different weight percentages of PC as a function of frequency. It is evident that adding PC up to 50 weight percent to PVDF enhanced dielectric relaxation. The physical characteristics underwent a significant change at 50PVDF/50PC composition, becoming softer and a little tacky, which suggested an overabundance of PC. This physical state led to instability during dielectric measurements and restricted the material's

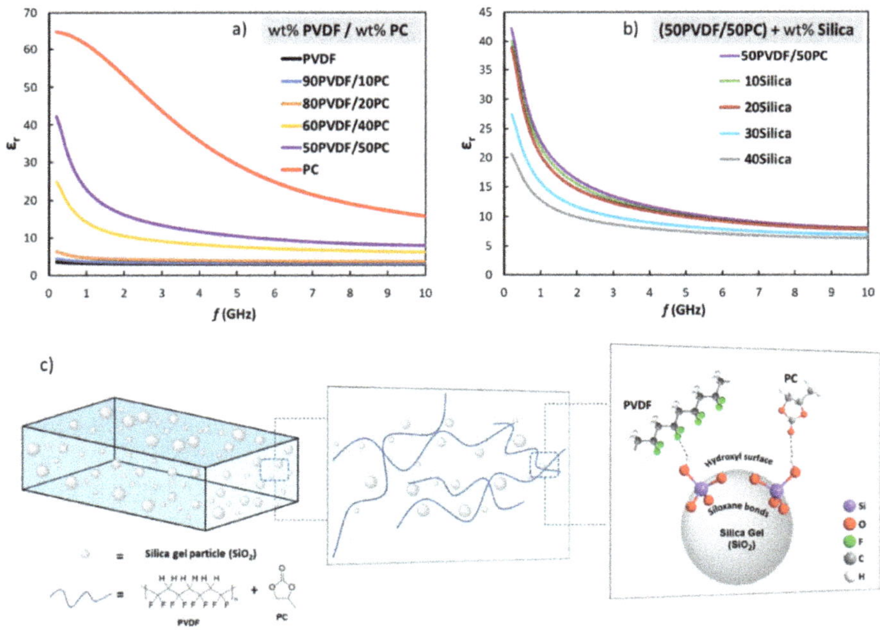

Fig. 12 Dielectric spectra of (**a**) xPVDF(100-x) PC (x in wt%) composite, (**b**) (50PVDF/50PC) + x wt% silica, and c) Silica-filled PVDF/PC composite sketch, showing hydrogen bonding between the acceptor groups of PVDF/PC molecules and the hydroxyl donor groups on the surface of silica particles, indicated by dashed lines [98]

processability. Using nanoparticles of silica as an inorganic filler solves this problem. High-energy ball milling was used to grind the received micron-sized silica gel particles (40–63μm) into nano-sized silica particles, as shown in Fig. S2(ESI). After that, up to 40 percent of the milled silica was added to the 50PVDF/50PC polymeric matrix. Figure 12b displays the permittivity variations.

Nonetheless, the dielectric relaxation, as measured by the power n in 10Silica, is nearly preserved, producing a value comparable to that of the 50PVDF/50PC polymeric matrix, with the 20Silica composite exhibiting the highest relaxation. As seen in figure, this improvement is ascribed to the enhanced intermolecular hydrogen-bonding between the acceptor groups of PVDF (C–F) and PC (C=O) and the hydroxyl donor groups on the surface of silica (–OH). Up to 20 weight percent of silica is added, increasing the degree of hydrogen bonding between them. The permittivity and relaxation strength appeared to decrease with increasing silica addition [98]. This could be because the excess low permittivity silica is now decreasing the overall permittivity, which may be the result of surpassing the threshold of effective interfacial hydrogen bonding formation. The silanol groups on the surface of silica (SiO2) can form hydrogen bonds with the C-F bonds in PVDF, which can function as hydrogen bond acceptors [99–101]. Thus, up to a specific weight percentage, silica can function as a dielectric relaxation enhancer within PVDF and PC. Additionally, it has been reported that when silica is evenly coated on PVDF particles in the form of core-shell, the β- and γ-phases, which are electroactive phases in PVDF,

Apart from strong matching techniques, simulated design optimization is important for guaranteeing consistent antenna performance under deformation. Modern electromagnetic solvers like Ansys HFSS and CST Studio Suite let for parametric modeling of bending situations, where the antenna's response can be examined under several curvature profiles. Before physical prototyping, these tools let designers forecast performance changes including changes in return loss, gain, and bandwidth. Engineers can find and fix possible failure modes by including bending studies into the design phase, therefore guaranteeing that PVDF-based antennas maintain consistent performance even in dynamic settings. Building resilient, application-specific wireless systems depends on this predictive strategy.

3.2 Fabrication Methods for PVDF-Based Flexible Antennas

The fabrication of PVDF-based flexible antennas calls for close attention to maintaining both the electrical performance and the mechanical flexibility of the system. Many manufacturing techniques have been investigated to maximize the integration of conductive patterns onto PVDF substrates, each with benefits depending on the desired application size, cost limits, and performance criteria. Fused filament fabrication (FFF), a kind of additive manufacturing usually known as 3D printing, is among the most promising production techniques. This method creates the wanted antenna structures by extruding PVDF or PVDF-composite filaments through a heated nozzle and depositing them layer-by-layer. FFF's benefit is its capacity to customize complicated three-dimensional shapes, therefore allowing the manufacture of unique and flexible antennas. Combining RF capability with mechanical and biological sensing, researchers have effectively included functional fillers such hydroxyapatite (HAp) and chitosan (CS) into PVDF filaments to manufacture antennas meant for biomedical uses (Fig. 13).

Inkjet printing, which provides a quick, maskless, and affordable way to pattern conductive materials onto PVDF films, is another commonly used technique. With great accuracy, conductive inks—usually based on silver nanoparticles or conductive polymers—are selectively deposited onto the flexible substrate. Low-temperature sintering techniques are then applied to the printed patterns to consolidate the conductive routes without harming the underlying polymer [32]. For prototyping and for uses needing fine feature resolutions, such wearable sensors and flexible communication modules, inkjet printing is especially beneficial. Still, a technical issue that calls for surface treatments or adhesion-promoting interlayers is keeping good adhesion between the printed conductive layer and the PVDF substrate. Mass production of flexible PVDF antennas also uses screen printing. This method creates thicker, more strong conductive tracks than those possible with inkjet printing by pushing a conductive paste through a patterned mesh onto the substrate. Though screen printing is generally low in resolution compared to inkjet-based processes, it is better suited for larger production volumes and provides great mechanical robustness. Lamination and etching methods also offer a high-precision substitute for antenna creation. A laminated PVDF film uses thin metal foil, such copper or silver. The antenna shape is then defined using photolithographic techniques; next, unwanted metal is removed using chemical or plasma etching.

The laser cutting produced complex geometries required for the antenna's conductive components, as seen in Fig. 14. The adhesive-backed conductive layers were attached to the composite film substrate after cutting. First, the monopole antenna was installed on the upper layer of the substrate; the assembly shown in Fig. 10b included several steps. Then, on the underside of the bottom layer of the substrate, the ground plane and circular parasitic patch were placed. A Sub Miniature

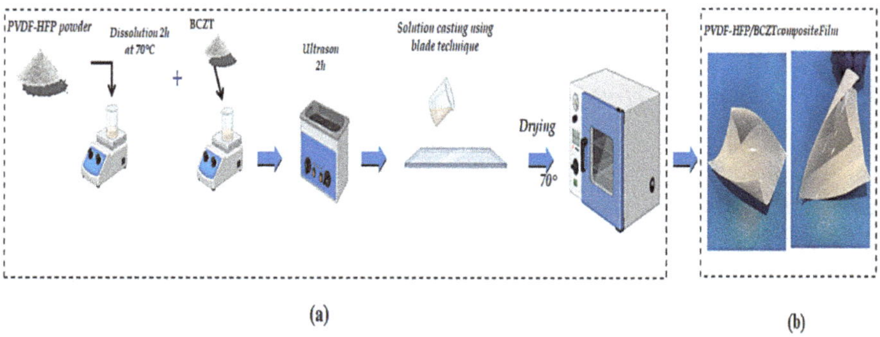

Fig. 13 Schematic representation of the solvent casting method for PVDF-HFP/BCZT films. (**a**) Flowchart of the fabrication process. (**b**) Photograph of the final composite film sample [32]

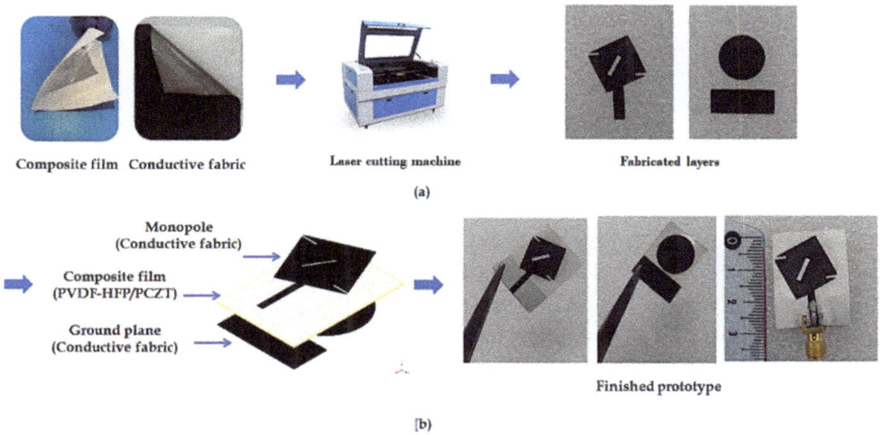

Fig. 14 Antenna prototype assembly process (**a**) precision laser cutting of conductive fabric layers (**b**) integration of antenna components and SMA connector attachment [32]

Version A (SMA) connector was finally fastened to the antenna's rear and transmission line, therefore enabling it to connect with a VNA for performance assessment. Though it requires more complicated and expensive processing, this approach creates very robust antennas with great electrical performance, making it appropriate for high-value aerospace and defense uses where dependability is top priority. To fully exploit PVDF's flexibility without sacrificing the electrical properties of the antenna, every fabrication method calls for parameter optimization including curing temperatures, mechanical stresses, and adhesion qualities. The desired mechanical form factor, operational environment, production scale, and cost factors all have a great influence on the choice of the suitable approach. Incorporating PVDF into antenna design has produced a wealth of creative prototypes and functional systems showing the adaptability and performance of this material across various wireless uses. One especially remarkable design is the dual-band coplanar waveguide (CPW)-fed antenna built on a thin PVDF substrate. Covering standard 3G, 4G, and emerging 5G frequencies, this antenna was designed to run effectively across two vital communication bands: 1.9–2.8 GHz and 3.9–6 GHz. A high-dielectric-constant PVDF substrate allowed notable antenna size miniaturization without loss of bandwidth or efficiency. Especially under moderate bending, the antenna's consistent performance attests to PVDF's resilience. Moreover, its omnidirectional radiation patterns in both frequency bands were perfect for portable and wearable device applications, where user mobility and orientation cannot be controlled.

Another outstanding prototype uses PVDF composite-made patch antennas built by means of FFF 3D printing. PVDF was mixed with hydroxyapatite (HAp) and chitosan (CS) in this instance to produce biocompatible, mechanically stable antenna substrates resonating effectively at 2.45 GHz. Designed especially for implantable or body-worn biomedical devices, these antennas let physiological monitoring and simultaneous communication. The presence of piezoelectric function helped to highlight the multifunctionality of PVDF composite-based antennas by allowing strain and vibration detection.

3.3 Challenges and Design Strategies

Though PVDF-based flexible antennas show great promise, many issues still need to be resolved if they are to be used to their full potential in vital and business uses. Mechanical deterioration with time is one main difficulty. Although PVDF itself has great fatigue resistance, the conductive metallization layers applied on its surface can experience cracking, delamination, or fatigue failure under repeated mechanical deformation including bending, twisting, and stretching. These failures could cause series resistance to rise, radiation efficiency to drop, or perhaps total open-circuit conditions. Researchers have looked at stretchable conductive materials—such as silver nanowire meshes, graphene inks, and conductive elastomers—to offset this problem since they can maintain mechanical deformation without appreciable loss of conductivity. Moreover, creative approaches including serpentine patterning of conductive tracks and pre-strained substrate designs have been used to more properly fit mechanical stresses. The sensitivity of PVDF to moisture absorption is another major difficulty. Though PVDF is chemically resistant and hydrophobic compared to many other polymers, long exposure to high humidity conditions can cause minor water absorption, possibly influencing the dielectric constant and loss tangent. Particularly in precision applications like biomedical telemetry, this might lead to changes in the resonant frequency and deterioration of antenna performance. Protective encapsulation layers like PDMS (polydimethylsiloxane) or parylene coatings are used to shield the PVDF structures without much loss of flexibility, so reducing this effect [68].

Another design restriction is bandwidth limits, especially for ultra-miniaturized PVDF antennas. Though at the cost of lower fractional bandwidth, high dielectric constant materials naturally result in smaller antenna sizes. Advanced design techniques like multi-resonant structures, wideband impedance matching networks, and metamaterial-inspired designs (e.g., split-ring resonators, complementary EBG structures) have been included to increase the operational bandwidth while preserving compact form factors in response to this. A frontier difficulty is finally the combination of flexible PVDF antennas with power management circuits and energy harvesting. Highly efficient, flexible rectifiers, voltage boosters, and energy storage components are required for systems seeking to simultaneously collect mechanical vibration energy and ambient RF energy via PVDF's piezoelectric properties. Current studies are aimed at creating printable, stretchable electronic circuits that can be smoothly co-fabricated with PVDF-based antennas, therefore generating self-sufficient wireless nodes for Internet of Things (IoT) and wearable applications. Essentially, although PVDF-based flexible antennas hold great promise, a multidisciplinary strategy including materials science, mechanical engineering, circuit design, and electromagnetic optimization is needed to fully exploit them.

3.4 *Future Trends in PVDF-Based Flexible Antenna Design*

The investigation of polyvinylidene fluoride (PVDF) in flexible antenna applications has opened interesting new frontiers in wireless communications, wearable electronics, and energy-autonomous systems. Several transforming trends driven by the convergence of material science innovations, new communication standards, and the increasing need for multifunctional devices will shape the future of PVDF-based flexible antenna design as technology progresses.

The Fig. 15 shows several main uses of PVDF-based flexible and conformal antennas, so highlighting their adaptability in current wireless systems. In aircraft applications, these antennas are integrated into curved fuselage surfaces to minimise drag while preserving high-performance communication. PVDF's biocompatibility and capacity to fit to anatomical forms help capsule endoscopy and skin cancer detection in medical diagnostics, therefore enabling accurate imaging and minimally invasive treatments. In the military domain, PVDF-based antennas are incorporated into soldier gear and vehicles for strong, low-profile communication in severe conditions [64]. For smart devices and IoT systems, the material's low dielectric loss and adaptability let smooth integration into small modules for home automation and sensing. Finally, PVDF's skin-friendly and stretchable quality makes it perfect for wearable electronics monitoring physiological parameters in wireless body area networks (WBANs), therefore allowing continuous, real-time health tracking. These various uses show how PVDF can help to create flexible, efficient, and dependable antenna systems in important technology areas. The evolution of multifunctional antennas, which concurrently combine communication, sensing, and energy collecting features inside a single flexible PVDF platform, is one

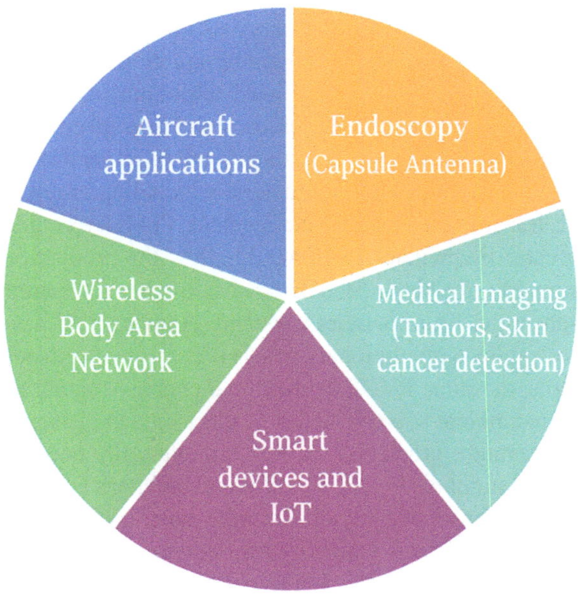

Fig. 15 Applications of a flexible antenna

significant trend. Future antennas could not only be transmitters and receivers but also environmental or physiological sensors and micro-generators of energy by using PVDF's natural ferroelectric and piezoelectric characteristics. Such multi-functionality would be especially useful in Internet of Things (IoT) ecosystems, where devices must run autonomously and sometimes in remote or difficult conditions. Integrating mechanical strain sensors, temperature sensors, or biochemical detectors straight into the antenna structure would allow real-time data collecting without increasing system complexity or weight. Related to this is the expanding area of smart textiles and wearable wireless systems, where PVDF-based antennas are effortlessly integrated into fabrics, clothes, and even skin-conformable patches. Made using printing or embroidery methods, these antennas could allow constant health monitoring, motion tracking, and universal communication without compromising the wearer's comfort or mobility. PVDF's flexibility, mechanical durability, and biocompatibility make it especially appropriate for such uses, surpassing conventional stiff substrates that would otherwise restrict movement or create pain [24]. Future studies are concentrating on improving washability, stretchability, and energy harvesting capacity of PVDF-integrated textiles to produce completely self-sustaining smart clothing systems. The creation of stretchable antennas is yet another interesting trend. While flexible antennas can handle bending and rolling, stretchable antennas can experience notable tensile stresses while preserving electrical continuity and performance. Antennas could stretch and conform to dynamic surfaces, such human joints, soft robotics skins, or inflatable structures, by engineering PVDF films with micro structured conductive patterns such serpentine traces or kirigami-inspired geometries. Real-time wireless communication and sensing in very dynamic and deformable environments would be made possible by these developments, a feature vital for next-generation prostheses, soft robotics, and adaptive wearable technologies. In the field of high-frequency communications, the extension of PVDF-based antennas into the millimeter-wave (mm Wave) and terahertz (THz) domains is gaining traction. Communication systems running at 100 GHz and beyond will need ultra-compact, flexible, low-loss antennas as 5G networks mature and 6G research picks up speed. Though PVDF's dielectric qualities are ideal for sub-6 GHz applications, customizing its microstructure or mixing it with sophisticated nanomaterials including graphene, carbon nanotubes (CNTs), or transition metal dichalcogenides (TMDs) could improve its performance at greater frequencies. This would allow the production of flexible antennas able to support ultra-high-speed data rates, low-latency communication, and high-precision sensing in future wireless systems. Another field where PVDF-based antennas are set to play a revolutionary role is the vision of self-powered IoT nodes. In such systems, the same PVDF element that enables communication can also collect energy from ambient sources including electromagnetic waves, mechanical vibrations, and temperature gradients, thereby removing the need for external batteries or wired power supplies. Future PVDF-based antennas could allow fully autonomous operation for sensors deployed in remote, hostile, or maintenance-free settings by including flexible rectifying circuits, energy storage components, and smart power management systems. Large-scale sensor networks in agriculture, infrastructure monitoring, environmental sensing, and defense applications will find this feature especially appealing.

Future trends will always be led by material innovations. Tunable material characteristics customized to uses will be made possible by composite engineering of PVDF incorporating dielectric, ferroelectric, or conductive nanoparticles. While carbon-based nanomaterials could improve mechanical flexibility and electrical conductivity, for instance, the inclusion of barium titanate (BaTiO3) nanoparticles could greatly increase the dielectric constant, therefore enabling more antenna miniaturization. Moreover, the creation of PVDF copolymers such P(VDF-TrFE) and P(VDF-HFP) provides routes to maximize ferroelectric response, mechanical toughness, and chemical resistance for conditions. New manufacturing techniques will also change what is feasible. Techniques such as laser sintering, aerosol jet printing, electrospinning, and nanoimprint lithography are anticipated to allow more precise control over antenna designs, therefore allowing the creation of complicated, multi-layered, hierarchically organized devices with unmatched performance characteristics. These developments will enable direct printing of flexible antennas with integrated frequency-selective surfaces (FSS), configurable components, and even adaptive beamforming capabilities onto PVDF substrates. From a wider angle, the inclusion of artificial intelligence (AI) and machine learning (ML) methods into the design and optimization of PVDF-based antennas offers yet another interesting prospect. Data-driven models let researchers maximize communication reliability and efficiency by allowing them to dynamically change antenna parameters for varying operating conditions, forecast material behavior under complicated mechanical deformations, and optimize antenna geometries.

Ultimately, PVDF-based flexible antenna design's future is exceptionally bright, marked by the convergence of material innovation, multifunctionality, sophisticated manufacturing, and smart design approaches. PVDF-based antennas are set to be a key enabling technology as wireless communication systems evolve toward more integrated, autonomous, and ubiquitous paradigms, therefore opening new opportunities in healthcare, defense, aerospace, consumer electronics, and beyond.

4 Design of Conformal Antennas Using PVDF

4.1 Design Considerations for PVDF-Based Conformal Antennas

Polyvinylidene fluoride (PVDF), a semi-crystalline piezoelectric fluoropolymer, has attracted much attention as a candidate material for flexible and conformal antenna designs in this changing technological environment. High dielectric strength, extraordinary mechanical flexibility, environmental stability, and electroactive activity define PVDF's unusual combination of desirable qualities. Its capacity to maintain structural and electrical integrity under deformation qualifies it as a perfect substrate and functional layer in antennas meant for wearable, aerospace, biomedical, and reconfigurable platforms. Moreover, PVDF's processability via different scalable methods—from solution casting and electrospinning to additive manufacturing—offers unmatched design freedom and integration possibilities in multi-functional wireless systems. Focusing on their material underpinnings, architectural design concepts, manufacturing techniques, performance assessment, and developing research paths, this chapter offers a thorough and academically grounded investigation of PVDF-based conformal antennas. Conformal body-worn antennas, where PVDF is bonded directly onto textiles or human skin surfaces, have also made good use of this material. Though near high-permittivity biological tissues, which often detune conventional antennas, such antennas have shown their capacity to sustain consistent communication links in body area networks (BANs). Detuning has been reduced and radiation efficiency in these PVDF-based wearable systems has been improved by careful design considerations including the use of artificial magnetic conductors (AMCs) and electromagnetic bandgap (EBG) structures. Collectively, these prototypes show that PVDF-based antennas can be quite competitive substitutes for conventional rigid antennas over a broad spectrum of cutting-edge wireless applications since they can attain high degrees of mechanical flexibility, environmental durability, and electromagnetic performance.

Representing a major development in antenna technology, conformal antennas let RF systems integrate smoothly with curved, flexible, and irregular surfaces. Unlike traditional planar antennas that work best in rigid settings, conformal antennas are meant to work effectively when installed on non-flat substrates including cylindrical, spherical, or anthropomorphic surfaces. Applications where form factor, aerodynamic performance, and spatial adaptability are crucial—such as aircraft fuselages, body-worn communication modules, biomedical implants, helmets, and next-generation wearable electronics—require this capacity. Its unusual mix of electrical and mechanical qualities makes polyvinylidene fluoride (PVDF) a promising substrate material for conformal antennas. A semicrystalline fluoropolymer, PVDF has remarkable dielectric and ferroelectric properties. Its relative dielectric constant (εr) in the range of 8–12 allows a particular design of electrically small antennas, so particularly appropriate for space-limited conformal uses. Even under mechanical deformation or repeated bending, PVDF's low loss tangent (usually <0.02) guarantees high radiation efficiency. Its chemical stability, adaptability, and compatibility with different metallization techniques also help to support long-lasting operation in dynamic settings.

The main benefit of PVDF is its capacity to keep dielectric performance while conforming to the host surface, therefore lowering profile intrusion and improving electromagnetic compatibility.

Designing conformal antennas with PVDF calls for multidisciplinary optimization spanning electromagnetic, mechanical, and materials science fields. Engineers have to consider factors including environmental durability, adhesion quality, metallization methods, permittivity stability, and substrate thickness. Modeling and simulation also have to consider actual deformation situations to forecast antenna performance under bending, curvature, or stretching. Often used in the design stage are full-wave electromagnetic solvers like Ansys HFSS or CST Studio, with multi-physics integration for thermal and mechanical stress analysis. Though it calls for exact engineering, the use of PVDF-based conformal antennas has revolutionary promise for flexible, reconfigurable, smart wireless systems.

4.1.1 Substrate Thickness

Especially with flexible dielectric materials like PVDF, the thickness of the substrate is among the most important factors in conformal antenna design. It directly influences the mechanical bending radius as well as the electromagnetic properties including impedance, bandwidth, and radiation efficacy. Usually, PVDF substrates for conformal antennas are 50 to 150 micrometers thick. At the lower end of this range, the material shows great flexibility, letting it wrap tightly around non-planar shapes without cracking or delaminating. This is perfect for embedded or wearable antennas requiring close surface conformity. On the other hand, very thin films could have disadvantages including lower tear strength, higher sensitivity to mechanical wear, and sensitivity to external environmental damage. Electromagnetically, the substrate thickness affects effective permittivity and surface wave propagation. Very thin PVDF layers have a higher probability of coupling between the radiating component and the ground plane, which can reduce bandwidth and cause losses.

On the other hand, while thicker substrates may sacrifice flexibility and raise the physical height of the antenna an undesirable effect in many conformal applications, they can lower dielectric losses and improve isolation. Thus, the choice of substrate thickness is an optimization issue requiring simultaneous fulfilment of mechanical integrity and RF performance.

Often, advanced simulation methods are used to model the stress-strain response of PVDF under bending and forecast the electromagnetic response in deformed conditions. Long-term dependability of several thickness profiles is also assessed by means of material fatigue testing and environmental stress evaluations. Laminated or multilayer designs using PVDF with intermediate reinforcement layers might also be considered for enhancing structural strength without too compromising flexibility. High-performance, resilient conformal antennas fit for use in flexible and adaptive wireless systems are finally obtained by correctly tuning substrate thickness.

4.1.2 Dielectric Constant and Loss Tangent

The electrical performance of PVDF-based conformal antennas is basically dependent on dielectric constant and loss tangent. The dielectric constant, or relative permittivity, directly affects antenna size, resonance properties, and impedance behavior by determining the wavelength of electromagnetic waves inside the material. PVDF is perfect for small, conformal platforms since it allows for the miniaturization of antenna geometries. Typically, εr ranges from 8 to 12. A key need in wearable, embedded, or morphing antenna structures where real estate is limited, this higher permittivity lowers the effective wavelength, therefore allowing efficient radiators within smaller footprints. Low dielectric loss is equally important, especially when the antenna is deployed in a mechanically dynamic environment. Ranging below 0.02 for PVDF over a broad frequency spectrum, the loss tangent (tan δ) gauges the degree of dielectric energy lost as heat. This guarantees that the antenna radiates efficiently without notable power loss, therefore preserving high gain and efficiency even when conformally wrapped around curved surfaces. Reliable performance depends on the stability of dielectric characteristics under mechanical stress in applications like textile-integrated antennas or conformal antennas on drone fuselages, where surface curvature is pronounced. Unlike many other flexible polymers, PVDF's remarkable quality is its consistent εr and tan δ during deformation. PVDF keeps its polarization stability and dielectric integrity, therefore preserving antenna function, even though bending, stretching, or thermal cycling may distort molecular alignments in low-performance materials. The ferroelectric character of PVDF also enables field-induced modification of dielectric constant, which may be used for dynamic tuning of resonant frequency or beam direction in reconfigurable systems. Antenna matching, impedance control, and bandwidth optimization depend on design engineers' precise awareness of these dielectric characteristics—particularly their change under mechanical stress.

When modeling PVDF-based conformal antennas in simulation settings, it is usual to assign εr and tan δ as functions of bending radius or strain. This guarantees that under reasonable deployment conditions return loss, gain, and efficiency forecasts stay correct. Often used to characterize these qualities in thin PVDF films are measurement techniques including resonant cavity methods, split-post dielectric resonators, and on-wafer probing. These values then guide the structural layout and material choice of the conformal antenna. Therefore, the dielectric characteristics of PVDF not only guarantee reliable operation under flexible, mobile, and conformal configurations but also allow compact and efficient antenna designs.

4.1.3 Electrode Metallization

The production of conformal antennas using PVDF substrates depends much on electrode metallization. The RF performance, mechanical reliability, and environmental robustness of the final antenna structure are directly influenced by the quality, flexibility, and adhesion of the conductive layer. In conformal applications, the metal layer must be able to withstand consistent electrical conductivity and adhesion to the dielectric substrate under repeated mechanical deformation—including bending, folding, or stretching. Widely used in rigid or somewhat flexible antennas, conventional options like copper provide excellent electrical conductivity. True conformal or wearable applications where high mechanical strain is anticipated are less ideal with copper, though, because of its mechanical brittleness and increased stiffness. Although pressure-sensitive adhesives or thermocompression methods allow copper foils to be laminated onto PVDF films, this usually causes problems with long-term adhesion and fatigue resistance. Surface treatments like plasma activation or chemical primers are often used to raise surface energy and enhance bonding strength between copper and PVDF, hence offsetting these constraints.

Silver nanoparticle inks have become a popular option to improve flexibility and compatibility with conformal architectures. Using inkjet or screen-printing methods, these inks can be exactly patterned on PVDF films, therefore offering not only electrical function but also the required mechanical compliance for applications requiring curvature. Though they might oxidize or degrade without appropriate encapsulation, silver inks maintain performance under moderate bending and offer conductivity levels near 10^7 S/m.

Another creative idea is using conductive polymers like PEDOT: PSS, which offer great mechanical flexibility and biocompatibility despite lower conductivity (10^3–10^4 S/m). These materials are especially appropriate for biomedical conformal antennas, where close skin contact or implantation could be necessary. Furthermore, they can be printed or spin-coated onto PVDF surfaces with low cost and little equipment, providing an appealing path to large-area or disposable antenna systems. Adhesion between the metallization layer and the PVDF substrate is still a main worry regardless of the selected material. Especially under cyclic bending, poor adhesion can cause delamination, cracking, or electrical discontinuities. Silane coupling agents, nano-roughening of the PVDF surface, or multi-layer metallization stacks with adhesion-promoting interlayers are among the methods to improve adhesion. These approaches guarantee that the antenna stays mechanically intact and functional all through its lifetime. Thus, efficient electrode metallization is not only about conductivity; it is a foundation of structural resilience and functional dependability in PVDF-based conformal antenna systems.

4.1.4 Mechanical Durability and Bending Radius

For conformal antennas meant for use in real-world settings with ongoing motion, surface curvature, and environmental exposure, mechanical durability is a fundamental need. PVDF-based conformal antennas have to keep their electromagnetic performance when installed on curved or deforming surfaces and withstands repeated mechanical stress without delamination, fracture, or dielectric failure. A key performance measure in this situation is the bending radius—the smallest radius to which the antenna can conform without suffering functional degradation. PVDF shows great mechanical flexibility and fatigue resistance, allowing it to maintain bending radii as low as 5 to 10 mm without notable change of resonant frequency or radiation pattern. Wearable systems, roll-up communication platforms, and deployable aerospace modules all depend on this capacity since curvature is dynamic and sometimes unpredictable. Experiments have demonstrated that antennas printed or laminated on PVDF films can keep S-parameters, return loss, and gain within acceptable limits even after several bending cycles. This durability of PVDF is enhanced by its low modulus of elasticity and high elongation at break, which makes it more suitable than more brittle substrates such FR4 or ceramics in uses requiring regular flexing. Working efficiently with the flexible electronics calls for the antenna to provide a consistent outcome in both conformal and non-conformal configurations. For this reason, the antenna is bent in both the X and Y axes, as indicated in Fig. 16 and Fig. 17, respectively. Both simulations and measurements have a bending radius selected at 10 mm. Figure 17a shows that the return loss of the antenna gets further improved when bent along the X-axis compared to the antenna without any bending. Moreover, the bandwidth stays same as the stiff case. Conversely, when the antenna is bent along the Y-axis, as indicated in Fig. 17b, a small increase in bandwidth and return loss is seen, akin to the results seen in non-conformal conditions. Therefore, one may conclude that the antenna performs nearly the same in both cases, which indicates the performance stability of the suggested work. Moreover, the close correlation between simulated and measured outcomes emphasizes the performance stability of the antenna in relation to return loss.

Mechanical durability, however, is not only dependent on the substrate; it is also greatly affected by the metallization layer's character, the antenna structure's thickness, and the layer-to-layer bonding quality. For instance, while silver nanoparticle inks and conductive polymers can handle closer bending because of their natural

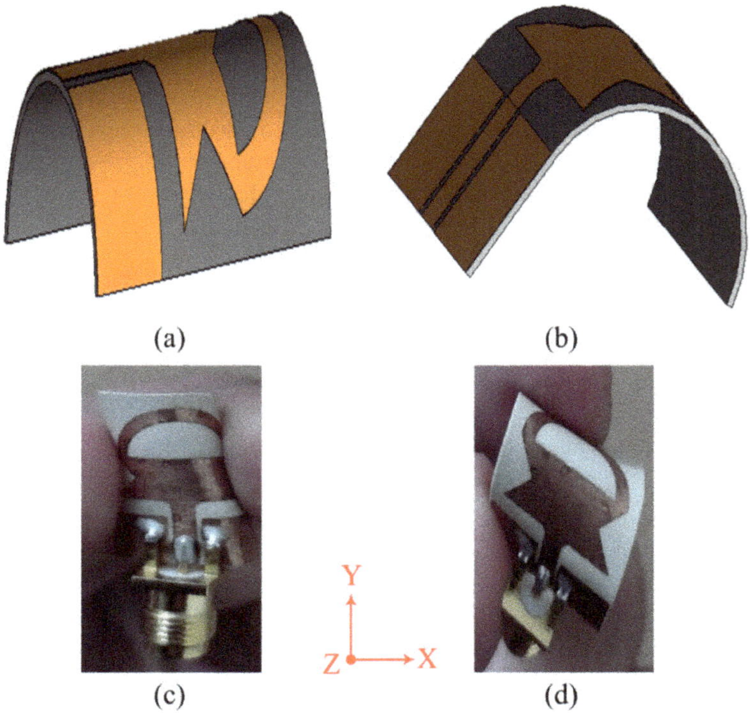

Fig. 16 Simulation setup for conformal structure (**a**) X-axis (**b**) Y-axis; fabricated prototype of the proposed antenna bent along (**c**) X-axis (**d**) Y-axis [102]

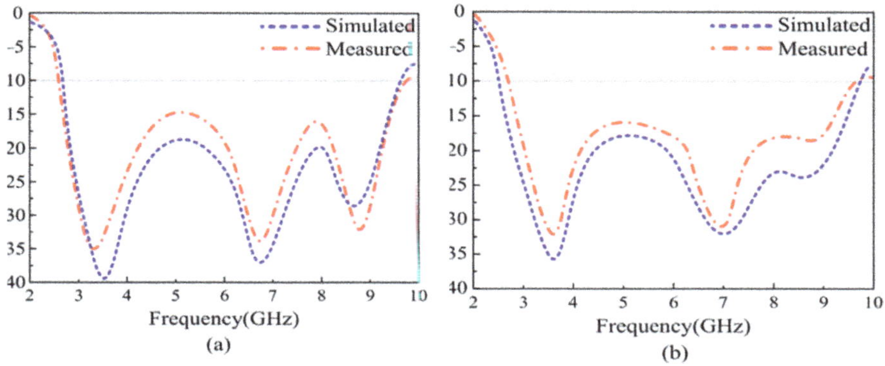

Fig. 17 Comparison between predicted and measured reflection coefficients of the antenna under conformal conditions along the (**a**) *X*-axis and (**b**) *Y*-axis [102]

ductility, copper traces could break or delaminate at large radii if not correctly adhered or strengthened. Long-term performance is significantly influenced by the interfacial adhesion between the PVDF and the metallized layer. While strong bonding, maybe supported by surface roughening or primer layers, can extend operational life, weak adhesion can cause fatigue-induced failures at the material boundary. Mechanical durability is also influenced by environmental factors including temperature, humidity, and UV exposure. PVDF's natural resistance to moisture and chemicals provides a strong protection against environmental deterioration. Thermal cycling, on the other hand, can cause microcracks or start material wear if the coefficient of thermal expansion (CTE) mismatch between layers is not properly handled. Accelerated life tests like cyclic bending, thermal aging, and salt fog testing are run to mimic actual use and measure failure thresholds, therefore evaluating long-term dependability. To qualify PVDF-based conformal antennas for rigorous industries like defense, space, automotive, and medical applications, these mechanical tests are vital.

Ultimately, PVDF is a great candidate for conformal antenna applications given its mechanical compliance, environmental stress resistance, and compatibility with flexible metallization techniques. Successful implementation, therefore, relies on careful consideration of bending radius limits, fatigue tolerance, and interlayer adhesion—qualities that must be thoroughly investigated using simulation, material choice, and rigorous mechanical testing.

4.1.5 Bandwidth Considerations and Impedance Matching

Though their relevance is especially highlighted in conformal antennas, where physical deformation can create parasitic effects and mismatch conditions, bandwidth and impedance matching are fundamental design elements that directly affect the performance of any antenna. Even tiny changes in the geometry of the radiating element or ground plane can cause shifts in input impedance, resonant frequency, and radiation pattern when a PVDF-based antenna is conformed to a curved surface. Unless the antenna is built with enough tolerance to such changes, these consequences could compromise performance. Conformal antenna designers usually depend on broadband impedance matching methods that offer durability under deformation to offset these issues. Common choices are coplanar waveguide (CPW) feeds because of their flat configuration, simple integration, and consistent impedance under curvature. To increase the operating bandwidth or adjust impedance over a broad frequency range, stub tuning, quarter-wave transformers, and parasitic components could also be used. In PVDF-based antennas, inkjet printing or photolithography can be used to exactly implement these structures, therefore enabling small and precisely tuned matching networks. Designing antennas resistant to impedance mismatch brought on by curvature depends much on simulation. Engineers can simulate bending situations by specifying curvature radii, mechanical stress, and surface topology using full-wave electromagnetic modeling tools like CST Studio Suite and Ansys HFSS. Designers may proactively improve the layout to guarantee acceptable performance under expected conformal deployment conditions by watching the effect of these factors on the reflection coefficient (S_{11}), gain, and bandwidth [102]. These simulations also assist in finding mechanical modes that could cause impedance detuning, bandwidth narrowing, or resonance splitting. The comparison between the simulated and measured outcomes of antennas under bending conditions is shown in Fig. 17. Figure 17a shows that the return loss of the antenna gets further improved when bent along the X-axis compared to the antenna without any bending. Moreover, the bandwidth stays the same as the stiff case. Conversely, when the antenna is bent along the Y-axis, as indicated in Figure 17b, a small increase in bandwidth as well as return loss is seen.

The dielectric constant and thickness of the substrate naturally determine bandwidth. Although a higher ε_r, like that of PVDF, allows antenna miniaturization, it can also reduce bandwidth because of more energy storage inside the substrate. Designers might include slotted patches, fractal geometries, or multi-resonant structures to make up for this by improving bandwidth without sacrificing size. The PVDF antenna can also include tunable materials or reconfigurable components to enable dynamic impedance control by means of voltage-controlled devices—an area where the ferroelectric behaviour of PVDF itself could be used for adaptive matching. Repetitive bending or environmental exposure should not compromise impedance matching or bandwidth, which is another crucial factor. Aging, moisture ingress, and thermal cycling can change the dielectric characteristics of PVDF or create delamination at the metal-dielectric interface, therefore influencing impedance. Often, therefore, environmental encapsulation and multilayer barrier coatings are used to maintain the dielectric integrity and stability of the matching network.

In the end, a thorough knowledge of how mechanical deformation affects electromagnetic behaviour will determine the successful design of PVDF-based conformal antennas. Stable impedance, broad operational bandwidth, and consistent performance across a variety of practical use cases may be obtained for conformal antennas by combining broadband matching methods, curvature-aware simulations, and material-specific optimizations.

4.2 *Fabrication Methods for PVDF-Based Conformal Antennas*

Conformal antenna fabrication using polyvinylidene fluoride (PVDF) calls for a synergistic mix of material processing, patterning methods, and structural integration to guarantee mechanical flexibility and RF performance. Unlike rigid antennas, which are usually made by photolithography on flat surfaces, conformal antennas must be made using techniques that preserve electrical characteristics under bending, stretching, and curvature. PVDF is a great dielectric medium for such antennas because of its thermal stability, ferroelectric response, and mechanical compliance. But, to make the most of these qualities, great care must be given to how the PVDF film is prepared, patterned, and integrated with the structural and conductive layers.

The PVDF substrate is prepared to start a typical manufacturing process. Through solution processing, commercial PVDF films can be cast in-house or free-standing sheets. For in-house techniques, PVDF powder is dissolved in a suitable solvent like dimethylformamide (DMF) or N-methyl-2-pyrrolidone (NMP), and the solution is then spin-coated or drop-cast onto a flat surface. The resulting film is dried and annealed to encourage crystallization, especially of the ferroelectric β-phase. Mechanical stretching or electrical poling could be used to raise the proportion of this phase. These post-processing procedures are necessary to guarantee mechanical durability as well as dielectric stability. Once the substrate is set, the following step is patterning the conductive traces. Compatible with PVDF are several metallization methods, each with varying trade-offs in process complexity, conductivity, resolution, and flexibility. Photolithography followed by metal deposition—e.g., sputtering or evaporation of copper, silver, or gold—remains a gold standard for high-fidelity patterns with better conductivity. This technique is costly, though, and not always compatible with roll-to-roll manufacturing or low-temperature settings. Inkjet printing with silver nanoparticle inks is a more easily scalable technique. Inkjet printing lets antenna patterns be directly written onto PVDF without using masks or vacuum techniques. Its low thermal budget makes it compatible with flexible substrates and helps with quick prototyping. Particularly when thicker conductive layers or larger areas are needed, screen printing and aerosol jet printing are also frequent. Thermal or photochemical sintering of the printed traces may follow deposition to improve conductivity. Ultra-flexible or biocompatible designs could use conductive polymers composite inks including carbon nanotubes or graphene. Due to the high dielectric constant and piezoelectric characteristics of the piezoelectric film, the antenna is easy to implement a miniaturized design and can be used in a hybrid energy harvesting system. The conformal antenna in Fig. 18 can cover 2.4G and 5.8G WLAN bands and 4G mobile communication bands, which can be widely used in wireless communication and energy harvesting [103]. In conformal applications, a major issue is the adhesion between the PVDF surface and the printed or deposited metal. Plasma activation, corona discharge, or chemical primers raising surface energy and encouraging improved wetting of the conductive material are commonly used treatments for the PVDF surface to solve this. Without such

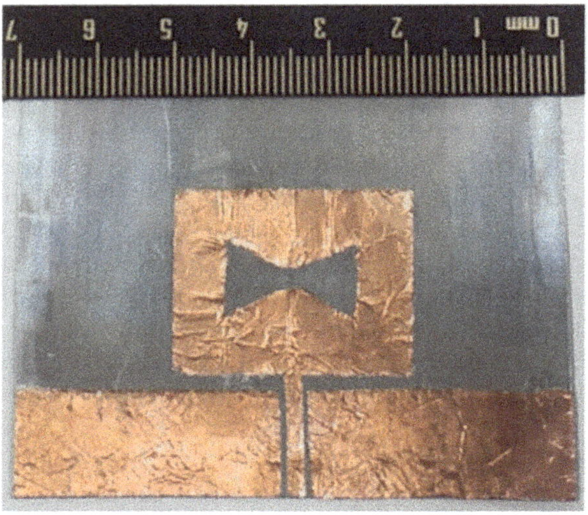

Fig. 18 Fabricated Conformal Antennas work in the 2.4 GHz and 5.8 GHz Wi-Fi frequency bands as well as the 4G mobile communication band, which increases the range and efficiency of RF energy harvesting [103]

treatment, delamination might happen under mechanical stress, therefore impairing antenna performance. Often used to increase durability and environmental resistance are laminated structures, in which the antenna is either embedded within a multilayer stack or sandwiched between flexible barrier films.

Additional manufacturing factors for conformal applications—such as antennas integrated into textiles, attached to helmets, or wrapped around UAV fuselages—include substrate curvature, interconnect routing, and protective encapsulation. While flexible adhesives or thermoplastic bonding layers assist hold the antenna onto non-planar hosts, laser micromachining could be used to exactly shape the PVDF substrate to fit curved surfaces. To pre-shape the substrate for best conformity, thermal forming methods—in which the PVDF is heated and molded over a 3D template—have also been investigated. Sometimes, antennas are included straight into the molding or lamination process of soft shells or wearable textiles, therefore allowing flawless communication without outside devices. Encapsulation to shield the antenna from moisture, dust, and mechanical abrasion is common in the last stage. Without sacrificing flexibility, thin polyurethane coatings, silicone over molding, or parylene deposition can fulfill this function. Radiation-hardened layers could be added in military or aerospace settings. Performance must be validated by post-fabrication testing, including S-parameter measurement, radiation pattern analysis, and mechanical fatigue testing. For long-term deployment, especially under repeated bending cycles, thermal cycling, and humidity exposure, reliability is crucial.

All things considered, the creation of PVDF-based conformal antennas is an interdisciplinary effort combining flexible material science, precision patterning, and mechanical engineering. Antenna development that is not only high-performing but also mechanically robust and conformally integrated into a range of real-world systems can be achieved by using methods such inkjet printing, surface activation, and thermoforming as well as by choosing suitable conductive materials.

PVDF-based conformal antennas are predicted to get even more accessible, scalable, and customizable for next-generation wireless communication systems as manufacturing technology, especially in additive and hybrid manufacturing, develops.

4.3 Future Trends in PVDF-Based Conformal Antenna Design

PVDF-based designs are being more acknowledged as major enablers of next wireless communication systems as conformal antenna technology develops. Antenna design is guided toward multifunctional and adaptive platforms by the need for very integrated, low-profile, and structurally embedded antenna solutions. In this context, PVDF offers a special convergence of environmental resilience, mechanical flexibility, and ferroelectric tunability. Not only are researchers looking at PVDF as a passive dielectric; they are also looking at it as an active medium that enables self-healing, sensing, and reconfigurability in conformal antenna arrays. A significant trend is the use of PVDF in soft electronics and wearable systems where the antenna must not only conform to complicated surfaces but also stretch, compress, or twist without performance loss. In this regard, PVDF composites are being optimized for elongation capacity and fatigue resistance by means of elastomers, low-modulus fillers, or mesh-reinforced layers inclusion [104]. To enable stretchable antennas to maintain dielectric integrity under stress, hybrid laminates combining PVDF with silicone or polyurethane substrates are also being created. These setups are perfect for physiological monitoring systems, e-textiles, and body-area networks. At the same time, notable advances are being made in integrating PVDF antennas with on-board electronics including rectifiers, amplifiers, and impedance matching circuits to enable completely autonomous, ultra-low-power communication modules. The expansion of battery-less IoT networks where energy collecting and efficient signal radiation must coexist depends on such systems. In this concept, PVDF acts as a functional material for sensing or energy harvesting (e.g., via pyroelectric or dielectric modulation effects) as well as a substrate for the antenna. Single-sheet wireless sensor tags that can run independently for years without active maintenance are being made possible by integration with thin-film batteries, organic transistors, and RFID circuitry. The use of PVDF-based conformal antennas in high-frequency bands including Ka, V, and even terahertz domains is another future path. Key to 6G, satellite internet constellations, and ultra-high-speed terrestrial links are these frequency bands. PVDF composites with improved dielectric constants (>20) and low-loss performance at frequencies exceeding 60 GHz are under development to assist such applications. To customize permittivity, lower dielectric loss, and allow smooth film casting on complicated geometries, novel synthesis pathways including in-situ polymerization with high-k ceramic nanoparticles and surface functionalization are being explored. PVDF's natural piezoelectric and pyroelectric qualities motivate academics to include sensory function straight into the antenna substrate. These dual-purpose structures can dynamically change their radiation pattern in reaction to external stimuli including pressure, temperature, or vibration. In security and defense uses, for instance, this could enable smart threat detection or beam reconfiguration. Smart conformal arrays like these could help monitor structural health in space and aerospace while maintaining constant communication under severe environmental conditions. PVDF-based conformal antennas' design and real-time tuning are being increasingly influenced by artificial intelligence and machine learning. Models driven by artificial intelligence are being trained to forecast antenna behavior under several deformation conditions, therefore enabling designers to optimize geometry, feed structure, and matching networks for performance robustness. Future systems could include embedded artificial intelligence cores able to automatically change antenna parameters depending on real-time input, therefore allowing very resilient and context-aware communication nodes.

Ultimately, PVDF-based conformal antennas will evolve from passive, flexible radiators to multifunctional, intelligent modules able of sensing, adapting, and real-time communication. These antennas are set to change what is possible in wearable, embedded, and conformal wireless systems of the next generation with ongoing developments in materials engineering, additive manufacturing, and electronic integration.

5 Applications of PVDF-Based Antennas in 5G and IoT Networks

PVDF-based antennas have advanced from theoretical research to real-world implementation in 5G and IoT environments. The distinct mechanical and electromagnetic characteristics of PVDF materials are opening new design paradigms in communication systems, as illustrated by the thorough real-world application cases presented in this section.

5.1 PVDF Antennas in Smart Gadgets and Mobile Devices

Antenna technology is moving toward novel materials because of consumer electronics' growing need for high-speed data, small form factors, and aesthetic integration. PVDF is a cutting-edge dielectric material that meets the crucial needs of flexibility, compactness, and electromagnetic performance for mobile devices of the future. Smartphones and tablets with 5G capabilities must function across a broad frequency range, frequently supporting both mmWave and sub-6 GHz bands. Because of its modest dielectric constant (P8–10), PVDF makes it easier to reduce the size of antenna components while maintaining adequate gain and bandwidth. In real-world applications, flexible screens, edge frames, and even smartphone back housings incorporate PVDF-based antennas. Better positioning and orientation are made possible by their mechanical compliance without compromising device thickness.

Furthermore, since frequent bending and folding can impair the performance of traditional rigid antennas, the rise of foldable smartphones has presented new difficulties for antenna design. PVDF is appropriate for use in foldable and rollable electronics due to its great mechanical endurance and steady dielectric characteristics under dynamic stress. Multilayer PVDF laminates with conductive traces are now being investigated by manufacturers for application in flexible antenna arrays that are capable of withstanding thousands of bending cycles. PVDF antennas are used to provide Bluetooth, Wi-Fi, and LTE-M communications in smartwatches and earbuds. They are particularly useful in wearables that come into close contact with the human body because of their compact size and low SAR (Specific Absorption Rate) values. Due to PVDF's minimal moisture absorption, antenna performance is guaranteed to be steady even in humid, rainy, or sweaty conditions. Dual- and triple-band PVDF microstrip patch antennas that can combine LTE, Wi-Fi 6, and UWB technologies into a single small unit have been demonstrated in recent prototypes. In these situations, PVDF improves thermal management by dispersing heat produced by RF away from delicate chipsets in addition to acting as a structural element.

Additionally, inkjet or aerosol jet printing techniques can be used to create PVDF antennas. This makes it possible for printed circuits and flexible batteries common in next-generation smart devices—to be seamlessly integrated. PVDF's printability and patternability will facilitate quick prototype and affordable manufacture of RF modules for consumer electronics as additive manufacturing gains traction in the electronics manufacturing industry.

To sum up, PVDF-based antennas are essential to the development of smart devices because they provide dependable, multipurpose, and compact wireless communication. They are essential for attaining smooth, fast communication in the small form factors required by today's mobile consumer electronics due to their mechanical and electromagnetic characteristics.

5.2 Medical and Fitness Wearables Based Antenna

In addition to being small and effective, antennas for wearable medical and fitness equipment must also be long-lasting, biocompatible, and resistant to external influences. PVDF is one of the best substrates for wearable antenna applications because of its exceptional mechanical flexibility and resilience to repeated stress. These antennas send critical health information from the body to distant monitoring systems, including heart rate, oxygen saturation, and glucose levels [98]. PVDF readily conforms to curved body surfaces, including the arms, chest, and wrists, in contrast to conventional rigid materials. It makes it possible to design patches, dipoles, or slot antennas that can be woven into fabric or directly incorporated into armbands, sticky skin patches, or other items. PVDF is safe for extended skin contact because of its non-toxic and inert chemical makeup, which is crucial in medical settings. Real-time data synchronization with mobile devices or cloud systems is made possible by PVDF-based antennas, which support BLE (Bluetooth Low Energy), ZigBee, and Wi-Fi connectivity in fitness applications. For instance, it has been demonstrated that a PVDF-based microstrip patch antenna integrated into a fitness band can sustain steady impedance matching and radiation efficiency while engaging in resistance training, cycling, and running [105]. PVDF performs better than many conventional plastics and composites in the crucial area of antenna robustness to mechanical deformation, which is highlighted by these dynamic use-cases. PVDF antennas are utilized in healthcare settings in respiratory monitoring clothing, smart insoles, and wearable ECG monitors. In these systems, extremely dependable and interference-immune antenna structures are necessary for data transmission via LPWAN (Low Power Wide Area Network) protocols like NB-IoT or LoRa. Devices will continue to function even after extended use or when exposed to external heating or cooling therapies thanks to PVDF's steady dielectric constant across a temperature range of −40 °C to +125 °C. The usage of PVDF-TrFE (copolymerized PVDF) in tunable antenna designs for medical applications has also been highlighted by recent research. By using a tiny electric field to change frequency bands, these antennas allow for dynamic communication with numerous healthcare gateways without requiring hardware development. In settings where bandwidth availability may fluctuate, such as hospitals or senior care facilities, this frequency adaptability is especially helpful [106]. Sterilization compatibility is another crucial feature of PVDF in medical wearables. Ethylene oxide and gamma radiation are two popular sterilizing techniques that PVDF can withstand without losing its material qualities. Because of this, it provides a dependable substrate for medical antennas that are semi-disposable and reusable, which lowers electronic waste and enhances hygiene.

All things considered, the design of wearable health monitoring devices is being revolutionized by PVDF-based antennas, which offer high flexibility, reliable performance, and user comfort. Through dependable, constant wireless connectivity, their incorporation into fitness trackers and medical systems satisfies the rising demand for connected healthcare and preventative medicine.

5.3 IoT for Industry

Wireless connectivity is introduced into the manufacturing, energy, and logistics sectors by the Industrial Internet of Things (IIoT). Antenna materials must withstand physical abuse, electromagnetic interference, and changing ambient conditions in such high-demand settings. The chemical inertness, mechanical robustness, and steady dielectric behavior of PVDF-based antennas under industrial stresses make them ideal for IIoT applications. Conveyor belts, pipelines, mobile robots, robotic arms, and smart sensors mounted on machinery are common IIoT configurations [107]. For these devices to transmit temperature, vibration, flow, or pressure data, embedded antennas are necessary. Laminating PVDF-based antennas onto plastic or metallic housings allows them to continue operating even when subjected to vibrations, heat cycling, or oil splashes. Even in tight areas like junction boxes or rotating shafts, their tiny profile enables non-intrusive insertion.

Additionally, PVDF is appropriate for process sectors like chemical plants, oil refineries, and pharmaceutical manufacture due to its exceptional resistance to oils and chemical solvents. When exposed to these substances, traditional antenna substrates may deteriorate or swell, while PVDF stays stable, extending device life and lowering maintenance frequency. PVDF-based antennas are used in robot control units and machine vision systems in industrial robotics. PVDF substrates aid in maintaining high signal integrity because these systems need real-time data interchange with low latency, particularly in manufacturing settings that are prone to interference or multi-path. Their mechanical dependability guarantees that during robotic activities, flexing or vibrations won't cause signal degradation. Systems for warehouse automation and asset tracking also make use of PVDF antennas. Forklifts, pallets, and containers are equipped with BLE beacons or UHF RFID tags that have PVDF-based antennas. Even when bent, hit, or stored in a dusty or damp environment, these antennas continue to deliver reliable tag reading functionality. Their long-term dependability boosts inventory management systems' productivity and decreases downtime.

PVDF antennas' versatility in wideband and multiband setups is a further benefit. PVDF makes it possible to create small, multi-band antenna designs for IIoT gateways and routers that handle a variety of protocols (such as Zigbee, Wi-Fi, LTE, and LoRa) that may be manufactured with inexpensive printing or lamination methods. Future IIoT infrastructure will depend more on the resilience and adaptability of PVDF-based antenna systems as more industrial settings move toward wireless automation.

5.4 Smart Cities and Smart Infrastructure

The way data is gathered, sent, and used in real time is changing because of the incorporation of PVDF-based antennas into smart infrastructure and urban networks. Antennas need to be low-profile, affordable, and resistant to environmental exposure in smart cities, where thousands of networked devices—from environmental sensors to traffic monitoring cameras and public safety systems—operate simultaneously. These requirements are satisfied by PVDF substrates, which provide flexibility, long-term durability, and compatibility with mass-production techniques like screen printing and inkjet. Installing PVDF-based antennas in interactive kiosks, smart parking meters, intelligent bus stops, and public lighting systems are examples of smart infrastructure. For instance, PVDF antennas are integrated into curved casings of streetlight-mounted air quality monitors to send sensor data using

LPWAN protocols like Sigfox or NB-IoT. These antennas will continue to function even when exposed to rain, UV rays, dust, and seasonal temperature changes thanks to PVDF's environmental durability. PVDF antennas are used in vehicle-to-infrastructure (V2I) communications for traffic management. PVDF-based antennas are used by digital signage systems, traffic light controllers, and emergency communication nodes to transmit data to central management hubs. These antennas, which are frequently mounted on plastic enclosures or metallic poles, function on the LTE, DSRC, or 5G frequency bands. PVDF's adaptability enables these antennas to fit irregular shapes while preserving efficiency and return loss. PVDF antennas are also being utilized in public Wi-Fi networks. They can be subtly placed on walls, ceilings, or inside architectural features due to their tiny form factor and little visible impact. For instance, a solar-powered kiosk can have a flat-panel PVDF antenna running at 2.4 GHz, providing mobile users with smooth access while maintaining visual integrity.

Furthermore, PVDF can be integrated into subterranean infrastructure due to its chemical resistance and lightweight nature. PVDF-based antennas can be integrated into tough enclosures in utility access nodes or smart water meters and still deliver dependable signal strength across soil and concrete layers. This is particularly crucial in nations where automated waste management, gas leak detection, and water usage monitoring depend on subsurface data backhaul. PVDF antennas are used in smart buildings for building automation systems (BAS), which regulate access control, lighting, and HVAC. Signal disruption can occur in these settings due to electromagnetic interference from many devices and dense material constructions like reinforced concrete. With their strong isolation properties, PVDF antennas ensure dependable connection across Wi-Fi, LoRa, or Zigbee protocols.

To sum up, PVDF-based antennas greatly enhance the sustainability and scalability of smart city networks. Their durability, scalability, and manufacturability allow for cost-effective and dependable large-scale deployments—essential qualities for cities looking to improve resource efficiency and connection through smart infrastructure.

5.5 Antenna Based Integration of Automotive and UAV

Some of the most demanding and dynamic situations for antenna placement are found in the automotive and unmanned aerial vehicle (UAV) sectors. Compact, strong, and high-performing antennas that can withstand mechanical, thermal, and electromagnetic stress are needed for both platforms. PVDF-based antennas are particularly well-suited to meet these requirements because of their adaptability, low profile design, and resistance to vibration and temperature changes. Numerous wireless services, such as vehicle-to-vehicle (V2V), vehicle-to-infrastructure (V2I), GPS, Wi-Fi, and in-car connectivity systems, are supported by PVDF antennas in automotive applications. These antennas are frequently included into non-metallic automotive components such interior panels, windshields, mirrors, and bumpers.

PVDF substrates, in contrast to conventional rigid materials, adapt to the aerodynamic curves of contemporary automobiles, allowing for discrete antenna insertion without sacrificing aesthetics. High-frequency communication becomes crucial as cars move toward autonomous systems. Millimeter-wave radar and LiDAR systems that support object detection and situational awareness use PVDF antennas that operate at 28 GHz and higher. For accurate radar resolution, which is essential for real-time navigation and collision avoidance, they exhibit steady dielectric behavior at high frequencies. Another important benefit is thermal durability. Materials that can function dependably at high temperatures are required for under-hood applications and proximity to heating components. PVDF is appropriate for engine-compartment electronics and vehicle gates exposed to sunlight or internal heat since it retains its dielectric qualities up to 150 °C. PVDF-based antennas in UAVs provide high-performance, lightweight options for transmitting payload data, telemetry, and command and control (C2). In drones, where every gram affects flight time and efficiency, weight reduction is crucial. PVDF antennas that are made with printed traces, or thin films add very little mass to the system while offering reliable communication channels. These antennas are commonly mounted on curved surfaces, such as the rotor arms or fuselage. Because of their mechanical compliance, they can withstand aerodynamic stress, abrupt altitude changes, and vibrations while in flight. Longer antenna life during outdoor or high-altitude flights is ensured by PVDF's resistance to UV radiation and environmental deterioration. PVDF antennas are used in beamforming arrays by sophisticated UAV systems to provide real-time video streaming and directional communication. Multiple-element integration is made possible by their small form factor, which enhances interference rejection and spatial resolution. These antennas facilitate mesh networking in drone swarms for cooperative sensing missions and coordinated flight.

Furthermore, autonomous delivery robots and air taxis that are being developed for urban transportation also incorporate PVDF antennas. Multi-band PVDF-based antennas can enable the dependable GPS, LTE/5G, and ultrawideband connectivity needed by these devices. Prototyping and iterative design are made easier by their quick manufacturing, which speeds up development schedules in this quick-paced industry. PVDF antennas are essentially revolutionizing connections in the automobile and unmanned aerial vehicle industries by providing space-saving, high-performance, and robust solutions. PVDF's contributions to dependable and effective wireless communication will only increase in significance as these platforms become more independent and data intensive.

6 Challenges and Prospects for the Future

PVDF-based antennas have a lot to offer 5G and IoT systems, but to reach their full potential, several practical and technological issues need to be resolved. In addition to discussing potential future research avenues and developing trends that could improve the capabilities and uptake of PVDF antennas, this part looks at the present constraints in materials science, manufacturing, and system-level integration. PVDF-based antennas are expected to continue to grow and innovate despite these obstacles. Their special blend of environmental robustness, moderate permittivity, and mechanical flexibility fits very nicely with the changing needs of intelligent, networked ecosystems. PVDF antennas are anticipated to enter new fields such as body-centric networks, vehicular radar, low-earth orbit (LEO) satellite systems, and reconfigurable smart surfaces as research endeavors tackle present constraints in bandwidth, tunability, and sustainability. With joint advancements in manufacturing technology, RF engineering, and materials science, PVDF's place in upcoming wireless architecture is only going to grow.

6.1 Limitations on Bandwidth and Multiband Performance

The trade-off between bandwidth and miniaturization is one of the main issues with PVDF substrates. PVDF can restrict the operating bandwidth, although its moderate dielectric constant permits smaller antennas. For 5G and IoT multiband applications, where a single antenna may be expected to cover sub-6 GHz, Wi-Fi, Bluetooth, LTE-M, and maybe mmWave frequencies, this becomes very difficult. Designers must employ sophisticated strategies like slot loading, meandering radiators, and the incorporation of parasitic parts or stubs to get around this. To achieve impedance matching throughout all desired frequency ranges, these methods may necessitate precise tuning and add complexity to the design. Furthermore, shifting resonance under mechanical deformation makes it especially difficult to achieve consistent multiband performance in small and flexible configurations.

6.2 Surface Adhesion and Printing Technique Compatibility

Although PVDF is compatible with screen and inkjet printing, consistent ink adherence is difficult due to its hydrophobic nature and low surface energy. PVDF film surface preparation is frequently necessary for conductive inks, particularly those based on graphene or silver nanoparticles, to increase wettability and adhesion. Chemical functionalization, corona discharge, and plasma treatment are examples of common techniques. However, the cost and complexity of manufacturing workflows are increased by these extra phases. Furthermore, environmental variables like temperature and humidity might have an impact on the repeatability of conductive layer performance, which could have an impact on mass production yield. Research on embedding adhesion-promoting interlayers or creating PVDF formulations with customized surface properties is still ongoing. Process scalability could be significantly increased by ink chemistry advancements that enhance bonding with PVDF without pretreatment.

6.3 Stability of Heat and Radiofrequency in Adversarial Conditions

Even though PVDF is thermally stable up to about 150 °C, extended exposure to high temperatures or varying humidity levels can still alter its dielectric characteristics. Resonant frequency shifts, decreased gain, or worsened impedance matching can result from these environmental changes. This could affect data throughput and reliability in industrial or outdoor applications. Furthermore, localized heating in PVDF antennas operating at higher RF power levels may impact substrate homogeneity or result in conductive layer delamination. To overcome this, ceramic fillers like $BaTiO_3$ or Al_2O_3, which enhance heat conductivity and dielectric stability, are being included into hybrid PVDF materials. Furthermore, the impacts of UV exposure and temperature cycling can be lessened with the application of thermal spreaders and encapsulating layers.

6.4 Material Lifespan and Environmental Sustainability

Concerns about electronic waste and material sustainability have attracted attention as billions of IoT devices are anticipated to be deployed over the next 10 years. PVDF is durable and resistant to chemicals, however it is not biodegradable, and there are still few recycling solutions. In addition to researching end-of-life disposal and recycling techniques, efforts are being made to create environmentally acceptable PVDF composites. Additionally, in consumer applications where device turnover is significant, life-cycle assessment (LCA) studies are required to estimate the environmental impact of PVDF antenna manufacture.

6.5 Including Emerging Technologies in Integration

PVDF-based antennas must develop in tandem with new communication technologies in order to stay competitive and relevant. For example, new performance needs in phase tuning, beam shaping, and polarization control are brought about by the emergence of reconfigurable intelligent surfaces (RIS), beam-forming arrays, and MIMO (Multiple Input Multiple Output) systems. To allow electrically tunable antennas with varying dielectric constants, research is concentrating on integrating PVDF-TrFE, a ferroelectric copolymer. Without the need for mechanical or MEMS-based switching, these reconfigurable PVDF antennas may dynamically change the direction of radiation or swap frequency ranges. To further enhance performance in adaptive networks, the integration of PVDF antennas with metasurfaces and tunable parts like varactors and tunable RF MEMS is being investigated. These setups hold great promise for upcoming satellite and 6G communication systems, where low-profile design and dynamic reconfiguration are crucial.

6.6 Cost and Scalability of Manufacturing

The capacity to increase PVDF antenna production while preserving consistency and affordability is another important factor. Although printing and laminating methods are scalable, it can be difficult to maintain precise conductor patterning and dielectric thickness tolerances. Antenna performance might vary from batch to batch due to differences in material qualities, particularly in high-frequency or multiband systems. To guarantee reproducible results, PVDF film grades must be standardized, and deposition and curing procedures must have better quality control. Studies on hybrid additive/subtractive manufacturing techniques, like micro-embossing and laser-assisted patterning, could provide improved resolution and reduced variability. These methods can create high-performing antennas while maintaining the low profile and flexibility made possible by PVDF.

7 Conclusion

One of the most important materials in the creation of contemporary antenna systems for wireless applications is polyvinylidene fluoride, or PVDF. In situations where traditional materials are inadequate, its remarkable blend of electrical, mechanical, and environmental characteristics makes it the perfect material for antenna substrates and structures. PVDF's function becomes more crucial as wireless communication technologies advance toward higher frequencies, more compact form factors, and increased integration. Essentially, the dielectric characteristics, environmental stability, and structural constraint compliance of any antenna are critical to its performance in wireless communication. With its modest dielectric constant (εr ~8–10), PVDF allows for size reduction without sacrificing effective radiation properties. More significantly, in wireless systems operating at frequencies ranging from sub-GHz to millimeter-wave, its low loss tangent guarantees little energy dissipation, which is essential for preserving gain and radiation efficiency. Materials that can enable multiband and broadband performance, high data rates, and reliable operation in real-world situations are needed for wireless communication's 5G transition and continuing 6G development. PVDF satisfies these requirements by providing exceptional mechanical durability under dynamic bending or compression, chemical inertness that resists environmental degradation, and customizable dielectric characteristics by copolymerization (e.g., PVDF-TrFE). These characteristics allow PVDF-fabricated antennas to function reliably in embedded systems, wearable technology, vehicle platforms, and unmanned aerial vehicles. PVDF particularly supports the creation of flexible and conformal antennas, two fundamental structures in contemporary wireless systems. PVDF's ability to adapt to different mechanical geometries allows for performance homogeneity across unusual structures, which is advantageous for conformal antennas that are intended to attach to non-flat surfaces. In contrast, PVDF can withstand extended use cycles without sacrificing the electromagnetic response, which is necessary for flexible antennas to function dependably under mechanical stress and deformation. PVDF's compatibility with scalable and economical manufacturing processes is another aspect of its relevance in wireless antenna engineering. High-performance antenna structures on PVDF substrates can be produced on a large scale for commercial wireless devices using roll-to-roll manufacturing, screen printing, and inkjet printing techniques. The application of PVDF is also in line with new developments in fields that are anticipated to rule future wireless networks, such as printable electronics, smart surfaces, and reconfigurable intelligent surfaces (RIS). But there are still difficulties. PVDF's surface energy needs to be carefully treated to ensure strong adhesion with conductive inks, and its comparatively high dielectric constant may restrict bandwidth. Continuous research and innovation are needed in the fields of life-cycle sustainability, recycling, and thermal management. However, by combining surface engineering, hybrid material techniques, and nanofillers, these difficulties are solvable and being actively addressed.

In conclusion, PVDF-based antennas offer a reliable and progressive response to the increasing intricacy and performance requirements of wireless communication. They are essential to the continuous development of wireless technologies due to their mechanical adaptability, stable electromagnetic characteristics, and fabrication flexibility. PVDF is positioned to continue to be a key component in the creation of effective, scalable, and intelligent wireless communication infrastructure as scientists and engineers work to improve its formulations and incorporate them into cutting-edge antenna systems.

References

1. Sharma, A., Das, G., Gupta, S., Gangwar, R.K.: Quad-band quad-sense circularly polarized dielectric resonator antenna for GPS/CNSS/WLAN/WiMAX applications. IEEE Anten Wireless Propag Lett. **19**(3), 403–407 (2020)
2. Concha, V.O.C., Timóteo, L., Duarte, L.A.N., et al.: Properties, characterization and biomedical applications of polyvinylidene fluoride (PVDF): a review. J. Mater. Sci. **59**, 14185–14204 (2024)
3. Sharma, A., Das, G., Gangwar, R.K.: Composite antenna for Ultrawide bandwidth applications: exploring conceptual design strategies and analysis. IEEE Anten Propag Mag. **60**(3), 57–65 (2018)
4. Dwivedi, A.K., Sharma, A., Pandey, A.K., et al.: Two port circularly polarized MIMO antenna design and investigation for 5G communication systems. Wirel. Pers. Commun. **120**, 2085–2099 (2021)
5. Sharma, A., Das, G., Gangwar, R.K.: Dual-band dual-polarized hybrid aperture-cylindrical dielectric resonator antenna for wireless applications. Int. J. RF Microw. Comput. Aided Eng. **27**, e21092 (2017)

6. Das, G., Sahu, N.K., Sharma, A., Gangwar, R.K., Sharawi, M.S.: FSS-based spatially decoupled Back-to-Back four-port MIMO DRA with multidirectional pattern diversity. IEEE Anten Wireless Propag Lett. **18**(8), 1552–1556 (2019)
7. Sharma, A., Das, G., Ranjan, P., Sahu, N.K., Gangwar, R.K.: Novel feeding mechanism to stimulate triple radiating modes in cylindrical dielectric resonator antenna. IEEE Access. **4**, 9987–9992 (2016)
8. Sharma, A., Das, G., Gangwar, R.K.: Design and analysis of tri-band dual-port dielectric resonator-based hybrid antenna for WLAN/WiMAX applications. IET Microwaves, Antennas & Propagation. **12**, 986–992 (2018)
9. Subudhi, P., Punetha, D.: Progress, challenges, and perspectives on polymer substrates for emerging flexible solar cells: a holistic panoramic review. Prog. Photovoltaics Res. Appl. **31**(8), 753–789 (2023)
10. Singh, B., Mishra, R.A., Punetha, D.: Optimizing performance of electrolyte-gated organic field-effect transistors for biosensing: material variations and parameter analysis. Discov. Electron. **2**, 20 (2025)
11. Punetha, D., Kumar, A., Pandey, S.K., et al.: Tertiary nanocomposite-based self-powered E-skin as energy harvester and electronic nose. J. Mater. Sci. Mater. Electron. **35**, 160 (2024)
12. Sharma, D., et al.: A compact wearable textile antenna for NB-IoT and ISM band patient tracking applications. Sensors. **24**(15), 5077 (2024)
13. Mayur, A., Yadav, A., Das, G., Sharma, A.: Performance enhancement of probe fed dielectric resonator antenna by using FSS reflector. In: 2024 IEEE 11th Uttar Pradesh Section International Conference on Electrical, Electronics and Computer Engineering (UPCON), Lucknow, India, pp. 1–3 (2024)
14. Subudhi, P., Sagar, V., Kumar, S., Punetha, D.: Simulation and optimization of substrate layer material for PVDF cantilever based vibration energy harvesting system. IEEE Sensors J. **23**(23), 28689–28695 (2023)
15. Soni, G.K., Yadav, D., Kumar, A.: Design consideration and recent developments in flexible, transparent and wearable antenna technology: a review. Trans Emerging Tel Tech. **35**(1), e4894 (2024)
16. Guggilla, P., Walker, C., Edwards, M., Chilvery, A., Thomas, B.: Dielectric behavior of lithium Tantalate (LiTaO3)/poly(vinylidene fluoride) (PVDF) nanocomposites doped with MWCNTs. Adv. Sci. Eng. Med. **9**(3), 204–208 (2017)
17. Zha, J., Meng, X., Wang, D., Dang, Z., Li, R.K.Y.: Dielectric properties of poly(vinylidene fluoride) nanocomposites filled with surface coated BaTiO3 by SnO2 nanodots. Appl. Phys. Lett. **104**(7), 072906 (2014)
18. Wu, T., Jin, H., Dong, S., Xuan, W., Xu, H., Lu, L., Fang, Z., Huang, S., Tao, X., Shi, L., Liu, S., Luo, J.: A flexible film bulk acoustic resonator based on Β-phase polyvinylidene fluoride polymer. Sensors. **20**(5), 1346 (2020)
19. John, D.M., Vincent, S., Pathan, S., Kumar, P., Ali, T.: Flexible antennas for a Sub-6 GHz 5G band: a comprehensive review. Sensors. **22**(19), 7615 (2022)
20. Yu, Z., et al.: A wearable self-grounding slit antenna for ISM/4G/5G/Bluetooth/WLAN applications. IEEE Access. **11**, 87930–87937 (2023)
21. Martins, P., et al.: Electroactive phases of poly(vinylidene fluoride): determination, processing and applications. Prog. Polym. Sci. **39**(4), 683–706 (2014)
22. Zhu, C., et al.: Enhanced dielectric performance in PVDF-based composites by introducing a transition Interface. Polymers. **17**(2), 137 (2025)
23. Lolla, D., et al.: Polyvinylidene fluoride molecules in nanofibers, imaged at atomic scale by aberration corrected electron microscopy. Nanoscale. **7**(49), 20984–20990 (2015)
24. Zhu, J., Li, W., Huo, X., Li, L., Li, Y., Luo, L., Zhu, Y.: An ultrahigh dielectric constant composite based on polyvinylidene fluoride and polyethylene glycol modified ferroferric oxide. J. Phys. D Appl. Phys. **48**(35), 355301 (2015)
25. Zhang, et al.: “Flexible and high-performance piezoresistive strain sensors based on carbon nanoparticles” @polyurethane sponges. Compos. Sci. Technol. **200**, 108437 (2020)

26. Pradeesh, E.L., Udhayakumar, S., Vasundhara, M.G., et al.: A review on piezoelectric energy harvesting. Microsyst. Technol. **28**, 1797–1830 (2022)
27. Yao, L., Yurui, X., Jiawei, T., Shuang, W., Geyu, L., Hongti, Z., Huibin, Q.: A highly sensitive Piezoresistive electronic skin: coulomb blockade-favored dynamic conduction paths in micellar brush-templated metal Nanoarrays. Chem. Mater. **37**(9), 3274–3283 (2025)
28. Meng, L., Vu, T.V., Criscenti, L.J., Ho, T.A., Qin, Y., Fan, H.: Theoretical and experimental advances in high-pressure behaviors of nanoparticles. Chem. Rev. **123**(16), 10206–10257 (2023)
29. Ashfaq, B., Azeem, I., Sohail, M., Yüce, F.G., Çitoğlu, S., Nayab, S., Abdullah, M., Duran, H., Yameen, B.: Ultraviolet-printing-assisted surface-confined growth of silver nanoparticles on flexible polymer films for Cu2+ and H2S sensing. ACS Appl. Nano Mater. **4**(8), 8200–8216 (2021)
30. Zou, Q., He, K., Ou-Yang, J., Zhang, Y., Shen, Y., Jin, C.: Highly sensitive and durable sea-urchin-shaped silver nanoparticles strain sensors for human-activity monitoring. ACS Appl. Mater. Interfaces. **13**(12), 14479–14488 (2021)
31. Sahoo, R., Mishra, S., Ramadoss, A., Mohanty, S., Mahapatra, S., Nayak, S.K.: Temperature-dependent dielectric properties of metal-doped ZnO nanofiller reinforced PVDF nanocomposites. Mater. Res. Bull. **132**, 111005 (2020)
32. Douhi, S., Labihi, S., Eddiai, A., Lakrit, S., El Achaby, M., Al-Gburi, A.J.A.: Design, characterization, and electromagnetic performance of a flexible wideband RF antenna using composite materials. J. Sci. Adv. Mater. Devices. **10**(1), 100847 (2025) ISSN 2468-2179
33. Han, W.B., Ko, G.-J., Jang, T.-M., Hwang, S.-W.: Materials, devices, and applications for wearable and implantable electronics. ACS Appl Electr Mater. **3**(2), 485–503 (2021)
34. Kumar, R.R., Punetha, D.: Computational prognosis and efficiency augmentation of lead-free and legacy piezoelectric bimorph cantilevers for low-power energy scavenging in IoT and wearable systems. Energy Technol. **13**, 2402193 (2025)
35. Feng, Y., Li, W.L., Hou, Y.F., Yu, Y., Cao, W.P., Zhang, T.D., Fei, W.D.: Enhanced dielectric properties of PVDF-HFP/BaTiO3-nanowire composites induced by interfacial polarization and wire-shape. J. Mater. Chem. C. **3**(6), 1250–1260 (2014)
36. Wang, G., Deng, Y., Xiang, Y., Guo, L.: Fabrication of radial ZnO nanowire clusters and radial ZnO/PVDF composites with enhanced dielectric properties. Adv. Funct. Mater. **18**(17), 2584–2592 (2008)
37. Abolhasani, M.M., Shirvanimoghaddam, K., Naebe, M.: PVDF/graphene composite nanofibers with enhanced piezoelectric performance for development of robust nanogenerators. Compos. Sci. Technol. **138**, 49–56 (2016)
38. Yu, S., Zheng, W., Yu, W., Zhang, Y., Jiang, Q., Zhao, Z.: Formation mechanism of B-phase in PVDF/CNT composite prepared by the sonication method. Macromolecules. **42**(22), 8870–8874 (2009)
39. Cao, X., Ma, J., Shi, X., Ren, Z.: Effect of TiO 2 nanoparticle size on the performance of PVDF membrane. Appl. Surf. Sci. **253**(4), 2003–2010 (2006)
40. Aravindan, V., Vickraman, P.: Characterization of SiO2 and Al2O3 incorporated PVdF-HFP based composite polymer electrolytes with LiPF3(CF3CF2)3. J. Appl. Polym. Sci. **108**(2), 1314–1322 (2008)
41. Chakradhar, R., Prasad, G., Bera, P., Anandan, C.: Stable superhydrophobic coatings using PVDF–MWCNT nanocomposite. Appl. Surf. Sci. **301**, 208–215 (2014)
42. Jaitanong, N., Yimnirun, R., Zeng, H., Li, G., Yin, Q., Chaipanich, A.: Piezoelectric properties of cement based/PVDF/PZT composites. Mater. Lett. **130**, 146–149 (2014)
43. Panda, S., Pasha, S.K.K.: Amplified dielectric properties of PVDF–HFP/SRTIO3 nanocomposites for a flexible film capacitor. Langmuir. **39**(37), 13345–13358 (2023)
44. Bhatt, A.S., Bhat, D.K., Santosh, M.S.: Crystallinity, conductivity, and magnetic properties of PVDF-Fe3O4 composite films. J. Appl. Polym. Sci. **119**(2), 968–972 (2010)

45. Abu-Zurayk, R., Alnairat, N., Bozeya, A., Khalaf, A., Abu-Dalo, D.: Enhanced properties of PVDF membranes using green ag-nanoclay composite nanoarchitectonics. Mater. Res. Express. **11**(4), 045007 (2024)
46. Karan, S.K., Mandal, D., Khatua, B.B.: Self-powered flexible Fe-doped RGO/PVDF nanocomposite: an excellent material for a piezoelectric energy harvester. Nanoscale. **7**(24), 10655–10666 (2015)
47. Islam, M.N., Rupom, R.H., Adhikari, P.R., Demchuk, Z., Popov, I., Sokolov, A.P., Wu, H.F., Advincula, R.C., Dahotre, N., Jiang, Y., Choi, W.: Boosting piezoelectricity by 3D printing PVDF-MOS2 composite as a conformal and high-sensitivity piezoelectric sensor. Adv. Funct. Mater. **33**(42), 2302946 (2023)
48. Dang, Z., Wu, J., Fan, L., Nan, C.: Dielectric behavior of Li and Ti co-doped NiO/PVDF composites. Chem. Phys. Lett. **376**(3–4), 389–394 (2003)
49. Girish, K.M., Lavanya, R., Sowmyshree, N., Jain, K.D., Gurushantha, K., Ramakrihanppa, K.: Preparation and electrochemical properties of Co3O4 as potential electrode materials for supercapacitor and battery applications. Deleted Journal. **6**, 100071 (2025)
50. Arul, N., Suresh, M., Satthiyaraju, M., Babu, N.B.K.: Flexible electrospun PVDF/WO3 nanocomposite fibers for piezoelectric energy harvesting applications. Polym. Compos. **45**(1), 360–372 (2023)
51. Rozana, M.D., Arshad, A.N., Wahid, M.H., Habibah, Z., Ismail, L.N., Sarip, M.N., Rusop, M.: Dielectric constant of PVDF/MgO nanocomposites thin films. In: IEEE Symposium on Business, Engineering and Industrial Applications, pp. 18–22 (2012)
52. Qi, W., Li, L., Han, R., Hou, Y., Zhou, Z., Chen, G., Li, Q.: Enhancing dielectric properties of (CACU3TI4O12 NWS–Graphene)/PVDF ternary oriented composites by hot stretching. *ACS Omega*. **11**, 1 (2024)
53. Zulkefle, H., Hatta, M.H.M., Sin, N.D.M., Kamaruzaman, D., Ahmad, N., Sarip, M.N.: PVDF/SnO2 flexible piezoelectric Nanogenerator films for energy harvesting application. In: 2024 IEEE 4th international conference in power engineering applications (ICPEA), Pulau Pinang, Malaysia, pp. 103–106 (2024)
54. Li, H.P., Liang, J.M., Zheng, R.H., et al.: Influence of β-phase content on the stretching-induced α-β phase transition of highly oriented poly(vinylidene fluoride) ultrathin films. Chin. J. Polym. Sci. **43**, 1406–1414 (2025)
55. Park, M.H., Lee, Y.H., Mikolajick, T., et al.: Review and perspective on ferroelectric HfO2-based thin films for memory applications. MRS Commun. **8**, 795–808 (2018)
56. Zhang, Y., Cui, M.: Porous Y x Fe y Zr 1− x − y O_2 coated TiO_2 solid superacid particles/PVDF hybrid membranes with anti-fouling property. Chem. Eng. J. **301**, 342–352 (2016)
57. Habibpour, S., Zarshenas, K., Zhang, M., Hamidinejad, M., Ma, L., Park, C.B., Yu, A.: Greatly enhanced electromagnetic interference shielding effectiveness and mechanical properties of polyaniline-grafted TI3C2TX MXENE–PVDF composites. ACS Appl. Mater. Interfaces. **14**(18), 21521–21534 (2022)
58. Li, F., Zhan, W., Zhuang, L., Zhou, L., Zhou, M., Bai, G., Zhou, A., Xiao, W., Yang, X., Sui, G.: Acquirement of strong microwave absorption of ZNFE2O4@SIO2@reduced graphene oxide/PVDF composite membranes by regulating crystallization behavior. J. Phys. Chem. C. **124**(27), 14861–14872 (2020)
59. Bhadra, D., Masud, M.G., Sarkar, S., Sannigrahi, J., De, S.K., Chaudhuri, B.K.: Synthesis of PVDF/BiFeO3nanocomposite and observation of enhanced electrical conductivity and low-loss dielectric permittivity at percolation threshold. J. Polym. Sci. B Polym. Phys. **50**(8), 572–579 (2012)
60. Naik, R., S, M., Chavan, S.: Piezoelectric property investigation on PVDF/ZrO2/ZnO nanocomposite for energy harvesting application. Eng. Res. Express. **3**(2), 025003 (2021)
61. Zhao, C., Xu, X., Chen, J., Wang, G., Yang, F.: Highly effective antifouling performance of PVDF/graphene oxide composite membrane in membrane bioreactor (MBR) system. Desalination. **340**, 59–66 (2014)

62. Pishnamazi, M., Koushkbaghi, S., Hosseini, S.S., Darabi, M., Yousefi, A., Irani, M.: Metal organic framework nanoparticles loaded- PVDF/chitosan nanofibrous ultrafiltration membranes for the removal of BSA protein and Cr(VI) ions. J. Mol. Liq. **317**, 113934 (2020)
63. Ramírez, J., Polat, B., Lipomi, D.J.: Metallic Nanoislands on graphene for biomechanical sensing. ACS Omega. **5**(26), 15763–15770 (2020)
64. Fu, R., Chen, S., Lin, Y., Zhang, S., Jiang, J., Li, Q., Gu, Y.: Improved piezoelectric properties of electrospun poly (vinylidene fluoride) fibers blended with cellulose nanocrystals. Mater. Lett. **187**, 86–88 (2016)
65. Li, Y., Huang, X., Hu, Z., Jiang, P., Li, S., Tanaka, T.: Large dielectric constant and high thermal conductivity in poly (vinylidene fluoride)/barium Titanate/silicon carbide three-phase nanocomposites. ACS Appl. Mater. Interfaces. **3**(11), 4396–4403 (2011)
66. Tan, P.D., Ha, D.T., Tung, B.S., Khuyen, B.X., Chi, D.T., Lam, V.D., Lee, Y.: Recoverable broadband absorption based on ultra-flexible meta-surfaces. Crystals. **12**(12), 1817 (2022)
67. Xin, W., Binzhen, Z., Wanjun, W., Junlin, W., Junping, D.: Design, fabrication, and characterization of a flexible dual-band metamaterial absorber. IEEE Photonics J. **9**(4), 1–12 (2017)
68. Nemati, A., Wang, Q., Hong, M., Teng, J.: Tunable and reconfigurable metasurfaces and metadevices. Opto-Electron. Adv. **1**(5), 180009 (2018)
69. Ali, T.-g., Bai, X., Xu, L.-j.: Dual-band energy harvesting antenna based on PVDF piezoelectric material. In: 2020 9th Asia-Pacific Conference on Antennas and Propagation (APCAP), Xiamen, China, pp. 1–2 (2020)
70. Frankhouse, J., Addo-Mensah, E., Wejinya, U.: Self-attenuating graphene-based antennas utilizing polyvinylidene fluoride (PVDF) substrates. In: 2024 International Microwave and Antenna Symposium (IMAS), Marrakech, Morocco, pp. 1–4 (2024)
71. Li, N., Xie, K., Huang, H.: Ferroelectric materials for high energy density batteries: Progress and outlook. ACS Energy Lett. **8**, 4357–4370 (2023)
72. Zhu, G., Zeng, Z., Zhang, L., Yan, X.: Piezoelectricity in β-phase PVDF crystals: a molecular simulation study. Comput. Mater. Sci. **44**(2), 224–229 (2008)
73. Wang, X., Bi, F., Du, H.: Reduction of low frequency vibration of truck driver and seating system through system parameter identification, sensitivity analysis and active control. Mech. Syst. Signal Process. **105**, 16–35 (2017)
74. Kumar, R.R., Punetha, D.: Fermi level modulation for enhanced graphene-based ultrasensitive gas detection. Diam. Relat. Mater. **154**, 112214 (2025)
75. Pi, Z., Zhang, J., Wen, Z., Wu, D.: Flexible piezoelectric nanogenerator made of poly(vinylidenefluoride-co-trifluoroethylene) (PVDF-TrFE) thin film. Nano Energy. **7**, 33–41 (2014)
76. Ji, S.H., Cho, J.H., Jeong, Y.-H., Paik, J.-H., Yun, J.D., Yun, J.S.: Flexible lead-free piezoelectric nanofiber composites based on BNT-ST and PVDF for frequency sensor applications. Sensors and Actuators A: Physical. **247**, 316–322 (2016)
77. Guo, S., Duan, X., Xie, M., Aw, K.C., Xue, Q.: Composites, fabrication and application of Polyvinylidene fluoride for flexible electromechanical devices: a review. Micromachines. **11**, 1076 (2020)
78. Li, Y., Zheng, Y., Liu, Z., Li, J., Zhai, H., Chen, Z., Li, Y.: Design of an Ultrasensitive Flexible Bend Sensor Using a silver-doped oriented poly(vinylidene fluoride) nanofiber web for respiratory monitoring. ACS Appl. Mater. Interfaces. **12**, 1359–1367 (2020)
79. Hu, X., Yan, X., Gong, L., Wang, F., Xu, Y., Feng, L., Zhang, D., Jiang, Y.: Improved piezoelectric sensing performance of P(VDF–TrFE) nanofibers by utilizing BTO nanoparticles and penetrated electrodes. ACS Appl. Mater. Interfaces. **11**, 7379–7386 (2019)
80. Singh, R.K., Lye, S.W., Miao, J.: Measurement of impact characteristics in a string using electrospun PVDF nanofibers strain sensors. Sens. Actuators A Phys. **303**, 111841 (2020)
81. Shi, K., Sun, B., Huang, X., Jiang, P.: Synergistic effect of graphene nanosheet and $BaTiO_3$ nanoparticles on performance enhancement of electrospun PVDF nanofiber mat for flexible piezoelectric nanogenerators. Nano Energy. **52**, 153–162 (2018)

82. Sharma, T., Je, S., Gill, B., Zhang, J.X.: Patterning piezoelectric thin film PVDF–TrFE based pressure sensor for catheter application. Sensors Actuators A Phys. **177**, 87–92 (2011)
83. Dong, K., Peng, X., Wang, Z.L.: Fiber/fabric-based piezoelectric and triboelectric Nanogenerators for flexible/stretchable and wearable electronics and artificial intelligence. Adv. Mater. **32**, 1902549 (2020)
84. Zhang, Y., et al.: A flexible metamaterial absorber with four bands and two resonators. J. Alloys Compd. **705**, 262–268 (2017)
85. Danila, O.: Polyvinylidene fluoride-based Metasurface for high-quality active switching and Spectrum shaping in the terahertz G-band. Polymers. **13**, 1860 (2021)
86. Kumar, G.S., et al.: A wideband flexible antenna utilizing PMMA/PVDF-HFP/PZT composite for wearable applications. Polym. Adv. Technol. **34**(5), 6575 (2023)
87. Martins, P., et al.: Electroactive phases of poly (vinylidene fluoride): determination, processing and applications. Prog. Polym. Sci. **39**(4), 683–706 (2014)
88. Kaur, S., Singh, D.P.: On the structural, dielectric and energy storage behaviour of PVDF-CaCu3Ti4O12 nanocomposite films. Mater. Chem. Phys. **239**, 122301 (2019)
89. Satapathy, S., Gupta, P.K., Varma, K.B.R.: Enhancement of nonvolatile polarization and pyroelectric sensitivity in lithium tantalate (LT)/poly(vinylidene fluoride) (PVDF) nanocomposite. J. Phys. D. Appl. Phys. **42**(5), 055402 (2009)
90. Wang, Y., et al.: Design, characterization, and electromagnetic performance of a flexible PVDF-HFP/BCZT composite antenna. Flexible Printed Electr. **9**(2), 1 (2024)
91. Bai, X., Han, W.-Y., Xu, L.-J.: Multi-band energy harvesting antenna based on PVDF piezoelectric material. In: 2019 International applied computational electromagnetics society symposium—China (ACES), Nanjing, China, pp. 1–2 (2019)
92. Kumar, P., Ali, T., Sharma, A.: Flexible substrate based printed wearable antennas for wireless body area networks medical applications (review). Radioelectron. Commun. Syst. **64**, 337–350 (2021)
93. Kumar, G.S., et al.: Flexible antenna with PMMA/PVDF-HFP/PZT composite. Polym. Adv. Technol. **34**(5), 6575 (2023)
94. Sun, H., et al.: Wearable communications in 5G: challenges and enabling technologies. IEEE Veh. Technol. Mag. **13**(3), 100–109 (2018)
95. Tiwari, R.N., et al.: Design of dual-band 4-port flexible MIMO antenna for mm-wave technologies and wearable electronics. IEEE Access. **12**, 96649–96659 (2024)
96. Paracha, K.N., et al.: Wearable antennas: a review of materials, structures, and innovative features for autonomous communication and sensing. IEEE Access. **7**, 56694–56712 (2019)
97. Panicker, S.S., Rajeev, S.P., Thomas, V.: Impact of PVDF and its copolymer-based nanocomposites for flexible and wearable energy harvesters. Nano-Structures & Nano-Objects. **34**, 100949 (2023)
98. Calisir, I., Yang, X., Bennett, E.L., Xiao, J., Huang, Y.: Enhancing the bandwidth of antennas using polymer composites with high dielectric relaxation. Mater. Today Electron. **3**, 100026 (2023)
99. Jana, S., Garain, S., Sen, S., Mandal, D.: The influence of hydrogen bonding on the dielectric constant and the piezoelectric energy harvesting performance of hydrated metal salt mediated PVDF films. Phys. Chem. Chem. Phys. **17**(26), 17429–17436 (2015)
100. Sarkar, R., Kundu, T.K.: Hydrogen bond interactions of hydrated aluminum nitrate with PVDF, PVDF-TrFE, and PVDF-HFP: a density functional theory-based illustration. Int. J. Quantum Chem. **120**(17), e26328 (2020)
101. Terziyan, T.V., Safronov, A.P.: Solubility and H-bonding of poly (vinylidene fluoride) copolymers in carbonyl liquids: experiment and molecular simulation. J. Mol. Liq. **275**, 378–383 (2019)
102. Ali, E.M., Awan, W.A., Alzaidi, M.S., Alzahrani, A., Elkamchouchi, D.H., Falcone, F., Ghoneim, S.S.M.: A shorted stub loaded UWB flexible antenna for small IoT devices. Sensors. **23**, 748 (2023)

103. Bai, X., Ali, T.-G., Xu, L.-J.: A Dual-frequency Slotted CPW Antenna for 2.45/5.8 GHz RF Energy Harvesting Based on PVDF. In: 2019 International Applied Computational Electromagnetics Society Symposium—China (ACES), Nanjing, China, pp. 1–2 (2019)
104. Punetha, D., Kar, M., Pandey, S.K.: A new type low-cost, flexible and wearable tertiary nano-composite sensor for room temperature hydrogen gas sensing. Sci. Rep. **10**, 2151 (2020)
105. Ali, U., et al.: Design, analysis and applications of wearable antennas: a review. IEEE Access. **11**, 14458–14486 (2023)
106. Douhi, S., Boumegnane, A., Chakhchaoui, N., Eddiai, A., Cherkaoui, O., Mazroui, M.: A wideband flexible antenna utilizing PMMA/PVDF-HFP/PZT polymer composite film and silver-based conductive ink for wearable applications. Polym. Adv. Technol. **35**(9), e6575 (2024)
107. Parlato, S., Centracchio, J., Cinotti, E., Gargiulo, G.D., Esposito, D., Bifulco, P., Andreozzi, E.: A flexible PVDF sensor for Forcecardiography. Sensors. **25**, 1608 (2025)

PVDF in Magnetic and Spintronic Applications

Radha Bayya, Shuvam Pawar, and Vibhor Kumar

Abstract Poly(vinylidene fluoride) (PVDF) and its copolymers have recently gained considerable attention from the research community because they simultaneously exhibit piezoelectric, ferroelectric, and dielectric properties, which are generally not observed together in conventional inorganic materials. Moreover, PVDF is lightweight, flexible, and capable of forming stable polar phases, which makes it a potential alternative to inorganic multiferroics and brittle ceramics such as lead zirconate titanate (PZT) and aluminum nitride (AlN). Due to these advantages, PVDF-based materials are being actively explored for applications in flexible electronics, magnetic random-access memory (MRAM), spintronics, and wearable electronic devices. In this chapter, the fundamental of PVDF and its application in magnetics and spintronics are thoroughly discussed. First, its crystal phases, polarization behaviour and dielectric response are explained. Then, more focus is given on how PVDF forms magnetoelectric composites when combined with ferrites and other magnetic fillers. To improve this multiferroic behaviour, methods like nano-structuring and use of PVDF-TrFE copolymers are also explained. Applications such as sensors, spin valves, tunnelling devices and MRAM based on PVDF have also been included. In each case, PVDF does not just act as a dielectric, but also works as an active layer that helps in spin transport or magnetism. Some new studies even show that mixing PVDF with two-dimensional materials may give better results for next-generation spintronic devices. Finally, a comparison is made between PVDF-based and inorganic materials to highlight their advantages and limitations. Issues like large-scale use, long-term stability and commercialization are also discussed, along with some directions for future research.

R. Bayya · S. Pawar (✉)
Department of Electronics and Communication Engineering, National Institute of Technology, Warangal, Warangal, India
e-mail: pawarshuvam@nitw.ac.in

V. Kumar
College of Science and Engineering, Troy University, Troy, AL, USA

D. Punetha (ed.), *Polyvinylidene Fluoride (PVDF)*, Engineering Materials,
https://doi.org/10.1007/978-981-95-6718-8_6

Keywords Poly(vinylidene fluoride) · Magnetoelectric coupling · Magnetic composites · Spintronics · Flexible sensors · 2D material hybrids

1 Introduction

1.1 Overview of PVDF and Its Multifunctional Properties

Polyvinylidene fluoride (PVDF) is one of the most widely studied electroactive polymers because it combines several useful properties in the same material. It shows piezoelectric, pyroelectric, ferroelectric and dielectric responses, while at the same time having good mechanical strength and high chemical stability [1, 2]. The fact that all these features occur together in a single polymer shown in Fig. 1a is the main reason why PVDF has attracted so much attention, both for fundamental research and for developing practical devices.

From a structural point of view, PVDF is a semi-crystalline fluoropolymer as illustrated in Fig. 1b produced through the polymerisation of vinylidene fluoride (VDF) monomers [3, 4]. Due to its semi-crystalline nature, it can form different crystalline phases, and some of these phases are directly linked to its piezoelectric, ferroelectric and pyroelectric effects. The unusual electrical behaviour in PVDF mostly comes from the strong carbon–fluorine bonds that run along its backbone. Fluorine atoms have very high electronegativity and are responsible for the material's ferroelectric and related properties. The fluorine bonds in PVDF do not break down easily, even with exposure to acids, bases, organic solvents, or radiation. On top of that, the polymer is fairly tough, it resists wear, has decent tensile strength, and lasts for a long time, which is why it finds so many uses in engineering areas [5].

Now, how PVDF is processed makes a big difference. The solvent chosen during fabrication, for instance, has a direct say in how the chains fold and what crystalline phases finally appear. Solvent properties like polarity, boiling point, or even how quickly it evaporates can shift the crystallisation pathway. Because of this, the electroactive behaviour of PVDF is not a fixed thing it can be tuned. By adjusting processing steps and picking the right solvent, researchers can actually shape the electrical response of the polymer to match the needs of a device [6–9].

When PVDF is being processed, the solvent used really changes the way the material turns out. For example, depending on the conditions, the same polymer can come out as a thin film, a nanofibre mat, or even a porous sheet. People generally use a set of common solvent-based methods for making these different forms, which is what Fig. 2 shows. After that, the properties of PVDF can still be improved. Simple steps like stretching the material, applying an electric field to pole it, or even making a copolymer with trifluoroethylene (TrFE) are often done. By adjusting these treatments, the crystalline phases and the overall functional behaviour of PVDF can be tuned. This is why the polymer is considered quite versatile and ends up being used in a wide range of devices [12–14].

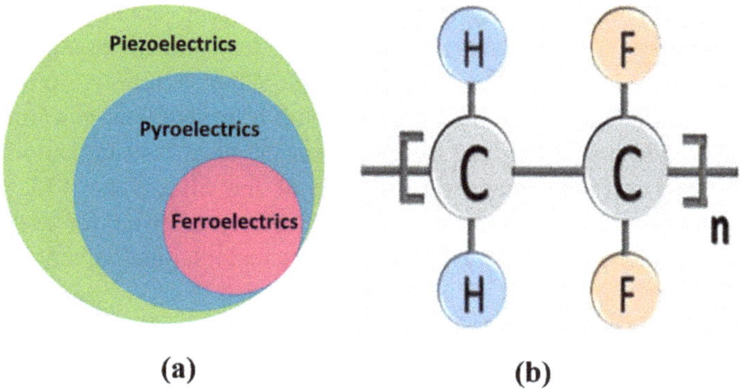

Fig. 1 (**a**) Hierarchy of piezoelectric and (**b**) the repeating unit of polyvinylidene fluoride (PVDF) [10]

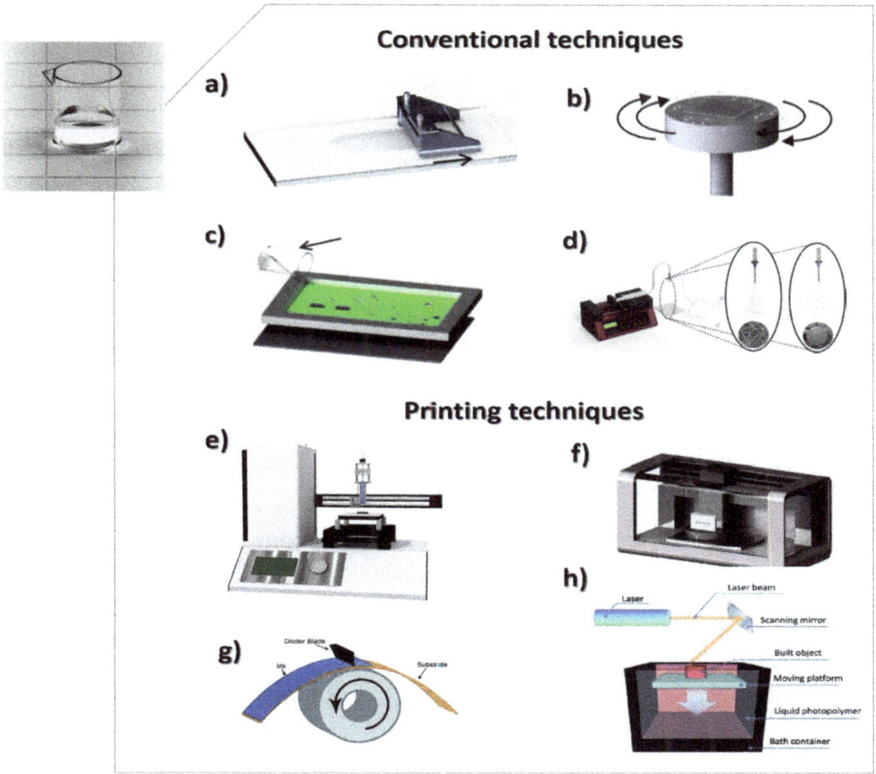

Fig. 2 Solvent-based processing techniques for PVDF: (**a**) doctor blade, (**b**) spin-coating, (**c**) screen-printing, (**d**) electrospinning, (**e**) direct-ink-writing (DIW), (**f**) inkjet printg, (**g**) roll-to-roll printing/coating processes, and (**h**) lithography [7]

Traditionally, PVDF has been employed in sensors, actuators, and energy-harvesting systems. In recent years, however, it has also gained attention in the field of magnetoelectric and spintronic technologies. When PVDF is combined with magnetic nanoparticles or deposited together with ferromagnetic thin films, a magnetoelectric coupling effect is observed, wherein the electric polarization of PVDF interacts directly with the magnetic ordering of the filler phase [15, 16]. Such PVDF-based multiferroic composites are particularly relevant for the design of multifunctional sensors, flexible magnetic detectors, and even biomedical devices [17].

Another key feature of PVDF is its ferroelectric polarization, which can be reversed under an applied electric field. This switchable polarization acts as a tunable barrier for spin-polarized carriers, thereby allowing control of spin transport. Such behaviour provides opportunities for the fabrication of low-power, non-volatile devices like ferroelectric tunnel junctions (FTJs) and hybrid multiferroic structures [18]. These device architectures are being actively explored for advanced applications in non-volatile memory storage, spin-based logic elements, and neuromorphic computing. The mechanical flexibility and chemical stability of PVDF add further advantages for its use in flexible, lightweight, and wearable spintronic systems [19].

Overall, PVDF is recognized as a multifunctional polymer that exhibits piezoelectric, pyroelectric, ferroelectric, dielectric, and even magnetoelectric effects within a single material system. The combination of these properties, together with its ease of processing and good mechanical flexibility, makes PVDF highly suitable not only for conventional electroactive applications but also for emerging areas such as magnetics and spintronics [20]. Because of this versatility, PVDF continues to remain at the forefront of research on multifunctional polymers and their potential in future device technologies [21, 22].

1.2 Why PVDF Is Relevant in Magnetics and Spintronics

Unlike conventional ceramic ferroelectrics, PVDF is lightweight, flexible, and easy to process, making it suitable for thin films, nanofibers, and flexible substrates using simple techniques such as spin-coating, solution casting, or electrospinning, as shown in Fig. 2. This processability makes PVDF particularly suitable for modern device architectures where flexibility, miniaturization, and compatibility with soft substrates are essential. These capabilities, along with PVDF switchable polarization, make it an active material for applications related to magnetics and spintronics (Fig. 3).

In the area of magnetics, PVDF has gained considerable attention when combined with magnetic phases to form flexible composite systems. Traditional magnetic oxides and alloys, though widely used, generally suffer from brittleness and therefore require rigid substrates for support. To overcome this limitation, embedding such magnetic materials in a PVDF matrix provides an effective route to achieve both mechanical flexibility and functional magnetic response. Among the

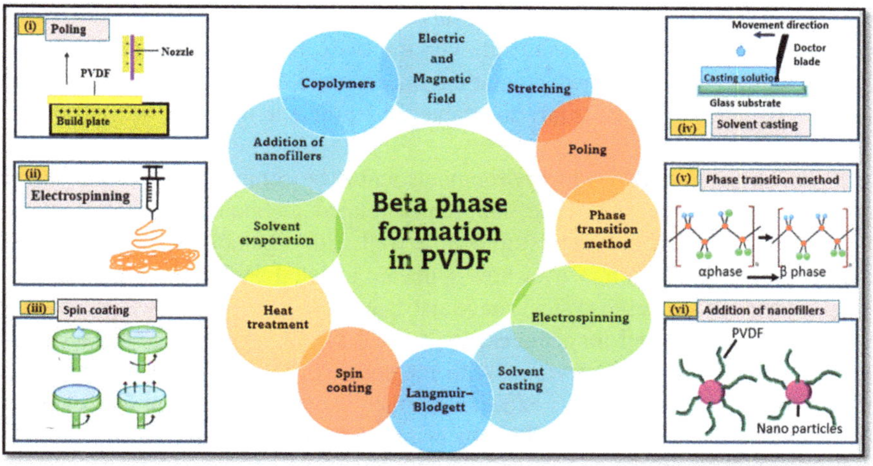

Fig. 3 Different methods to enhance the electroactive phase of PVDF (**i**) Poling. (**ii**) Electrospinning. (**iii**) Spin coating. (**iv**) Solvent casting. (**v**) Phase transition method. (**vi**) Addition of nanofillers [23]

commonly employed fillers, cobalt ferrite ($CoFe_2O_4$), magnetite (Fe_3O_4), and nickel ferrite ($NiFe_2O_4$) are noteworthy, as they are magneto-strictive in nature and respond to external magnetic fields by changing their shape [24, 25]. When such fillers are dispersed within PVDF, the system benefits from a unique combination: the magnetic component ensures field sensitivity, while PVDF, owing to its piezoelectric and ferroelectric properties, enables electromechanical interaction. The fundamental idea of this magnetoelectric effect in these composites, though complex at the microscopic scale, can be explained straightforwardly. An applied electric field across the PVDF matrix can affect the magnetic response of the filler phase, while an external magnetic field can, in turn, alter the polarization of PVDF. This mutual coupling between magnetic and electric parameters makes the system highly attractive for future device applications, particularly in the domain of low-power and multifunctional electronics.

PVDF is quite a useful material because it is soft, lightweight, and can be processed easily into thin films. These features make it well-suited for use with magnetic electrodes and nanostructures, which are important parts of spintronic devices. One of its main advantages is its polarization reversibility can be reversed when an external electric field is applied [26]. This reversible polarization plays a key role in controlling what happens at the interface with magnetic layers, since it can affect how spins align and how they move across the junction. In this way, PVDF links ferroelectric behaviour with spin-dependent conduction, and the advantage is that this can be done on a flexible platform.

Because of this property, PVDF shows promise in devices like spin valves, tunnelling junctions, and non-volatile memories. The spin transport in such systems can be controlled electrically, which reduces the need for strong magnetic fields. PVDF films are flexible, so devices made from them can be bendable and lightweight, which is much harder to achieve with the usual rigid ferroelectric materials. Even though researchers are still working to explore in-depth the finer details of polarisation interaction with spin at the microscopic level [27]. However, PVDF has already shown itself as an adaptable and efficient material for spin-based electronics.

PVDF is stable in different environments and is biocompatible, which makes it attractive for organic spintronic and sensor-related technologies. Strong dipole moment of PVDF further helps it to couple well with organic semiconductors and magnetic oxides, which improves how spins are injected and detected in such systems. Hybrid PVDF-based systems have also been reported to offer dynamic control of spin polarisation and magnetoresistance, which further broadens the scope of multifunctional spintronic devices. With the continuous progress in organic–inorganic hybrid systems, PVDF is expected to play a key role in linking ferroelectricity, magnetism, and spintronics, thereby establishing itself as a significant material in the development of next-generation flexible spintronic devices.

1.3 Scope of the Chapter

This chapter examines the significance of PVDF in the context of magnetic and spintronic applications. It begins with an introduction to the principles of magnetics and the development of spintronics as a technology that utilizes both charge and spin of electrons. The discussion then focuses on the unique properties of PVDF, particularly its ferroelectric polarization, flexibility, and ability to form effective interfaces with magnetic materials. Applications such as ferroelectric tunnel junctions, magnetoelectric heterostructures, and flexible spintronic devices are presented with reference to recent experimental studies. The chapter concludes with an outlook on how PVDF-based systems can contribute to future advances in non-volatile memory, voltage-controlled magnetization, and hybrid multifunctional electronics.

2 Fundamentals of PVDF in Magnetic and Spintronic Context

2.1 Crystalline Phases and Their Role in Coupling Phenomena

The origin of functional properties lies in the molecular arrangement and crystalline phase structure of PVDF, which directly influences dipole orientation and polarization behaviour. Even slight differences in crystallographic phases can alter polarization, dielectric behavior, magnetic ordering, and mechanical response. In ferroelectric and piezoelectric systems, phase transitions enable long-range dipolar alignment, stabilizing electromechanical coupling. Depending on the processing route, such as melt crystallization, stretching, electric poling, or solvent casting, PVDF can crystallize into different polymorphic forms. Among these, the most critical phases are α, β, and γ, each possessing distinct chain conformations and polarization characteristics. Polymeric ferroelectrics such as PVDF illustrate this principle. PVDF exhibits α, β, γ, and δ polymorphs distinguished by chain conformation as shown in the Fig. 4.

PVDF usually forms the α-phase under normal crystallization conditions. In this phase, the polymer chains have a TGTG′ arrangement, where the dipoles of adjacent monomers point in opposite directions. This causes the dipoles to cancel each other out, making the α-phase non-polar and unsuitable for ferroelectric or piezoelectric applications [28].

On the other hand, the β-phase is the most important for practical uses. Here, the chains are in an all-trans (TTTT) configuration, which aligns all the dipoles in the same direction. This alignment produces strong spontaneous polarization, giving the material excellent ferroelectric, piezoelectric, and pyroelectric properties. That is why β-PVDF is widely used in sensors, actuators, energy harvesters, and advanced electronic devices. The α-phase can be converted into the β-phase through mechanical stretching, electrospinning, or applying strong electric fields, which enhances its functionality. Its high remnant polarization makes it particularly suitable for applications like magnetoelectric composites.

The γ-phase is somewhere in between α and β in terms of structure and properties. It has a TTTGTTTG′ chain arrangement, which allows partial dipole

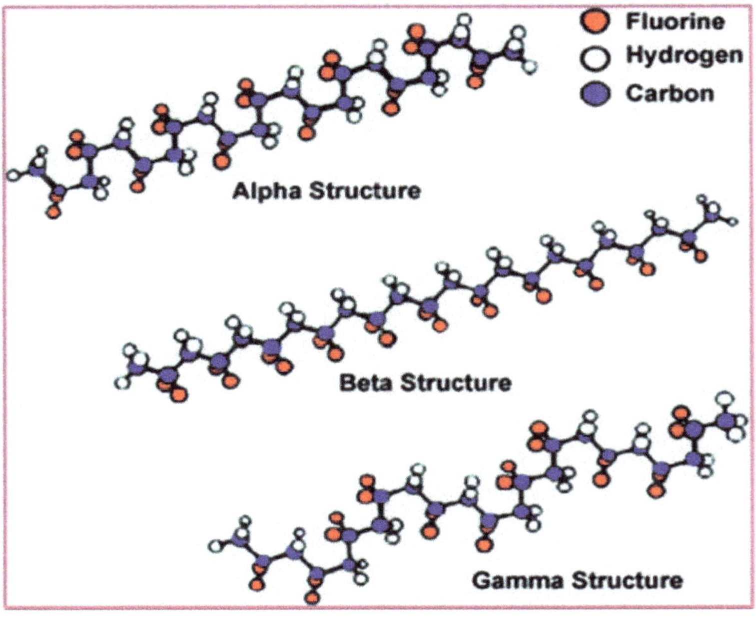

Fig. 4 PVDF phases: nonpolar α-phase, polar β-phase, and polar γ-phase [10]

alignment. This gives it moderate polarization—better than α but not as strong as β. In most applications, having a higher fraction of the β-phase is desirable, which can be achieved through stretching, electrical poling, or careful selection of solvents during processing. Table 1 shows the Controlling which crystalline phase forms is therefore key to tailoring PVDF for different uses, from spintronic devices to multifunctional composite materials [29].

2.2 Ferroelectric and Piezoelectric Properties Relevant to Magnetoelectric Behaviour

The ferroelectric and piezoelectric properties of a material are closely linked to its role in magnetoelectric behaviour, especially when it is used in composite systems with magnetostrictive materials. In the β-phase of PVDF, the molecular dipoles align in the same direction due to the all-trans chain conformation, resulting in spontaneous polarization [29, 30]. This polarization can be reversed by an external electric field, making PVDF ferroelectric in nature. The same alignment also makes PVDF strongly piezoelectric, where mechanical stress generates an electrical signal, and an applied electric field leads to mechanical deformation.

In a magnetoelectric composite, PVDF works as the piezoelectric phase, while the magnetostrictive material acts as the magnetic phase. The coupling between these two takes place through mechanical strain. When a magnetic field is applied, the magnetostrictive material undergoes dimensional change due to magnetostriction and this strain is passed on to the PVDF matrix. Because PVDF is piezoelectric, it converts this mechanical stress into an electrical output. In this way, the magnetic field is linked to the electric response of the composite. The reverse process is also possible. When an electric field is applied to PVDF, the dipoles in the β-phase realign and the polymer undergoes strain. This strain is then transferred to the magnetostrictive layer, which results in a change in its magnetic state. Thus, PVDF establishes a two-way connection between electric and magnetic domains in the composite. This strain-mediated mechanism is very useful for magnetic and

Table 1 Crystallographic data of polymorphs of PVDF [29]

The unit cell (A), the unit (g/cm^3), space group, structure, molecular chain, and electroactive properties						
α-Phase	a = 4.96 b = 6.94 c = 4.62	1.92	$P2_1/c$	Monoclinic	TGTG′	Low polarity
β-phase	a = 8.58 b = 4.91 c = 2.56	1.97	Cm2m	Orthorhombic	TTTT	Largest polarity
γ-phase	a = 4.96 b = 6.94 c = 4.62	1.92	P21/cn	orthorhombic	T3GT3G1	Medium polarity

spintronic applications. In magnetic sensors, small variations in the magnetic field can be converted into electrical signals with good sensitivity when PVDF is used as the matrix. The flexibility of the polymer makes such sensors more adaptable compared to rigid ceramic-based systems.

The overall strength of the magnetoelectric effect depends not only on the intrinsic ferroelectric and piezoelectric coefficients but also on factors such as phase purity, domain orientation, and interface quality. Well-aligned PVDF nanofibers enhance coupling efficiency with magnetostrictive layers, while epitaxial strain in perovskite thin films can boost piezoelectric performance [31]. In this way, ferroelectric polarization supplies a switchable charge background, while piezoelectricity enables strain-mediated pathways. Together, these properties establish a versatile framework for multifunctional magnetoelectric devices, ranging from spintronic memory and logic elements to energy harvesters and flexible sensors.

In spintronic devices, the ferroelectric and piezoelectric properties of PVDF can influence spin transport at the interface with magnetic layers. By applying an electric field to PVDF, the polarization and strain can be tuned, which in turn modifies the spin polarization and magnetization of the adjacent layer. This provides a way of controlling spin-dependent conduction using electric fields rather than large external magnetic fields. For example, in PVDF filled with magnetostrictive nanoparticles, such as $CoFe_2O_4$, the strain generated in the piezoelectric phase is transferred to the magnetic phase [32], leading to a magnetoelectric voltage output, as shown in Fig. 5. Such functionality is essential for devices like magnetic tunnel junctions, spin valves, and non-volatile memory elements. Hence, the combination of PVDF with magnetostrictive phases offers a promising approach for the development of flexible magnetoelectric and spintronic devices.

2.3 Comparison with Other Polymeric Dielectrics

In modern electronic and functional devices, the choice of polymeric dielectrics plays a crucial role in determining device performance. While many polymers such as polyethylene (PE), polypropylene (PP), polystyrene (PS), and polyimides (PI) have been used extensively as insulating materials, they primarily act as passive dielectrics and lack active functional properties are shown in Table 2 [34].

PVDF is a polymer that combines the advantages of processability with intrinsic electroactive properties. Its polar β-phase gives rise to strong ferroelectric and piezoelectric behavior, allowing the material to respond actively to applied electric fields [33]. This switchable polarization distinguishes PVDF from most other polymeric dielectrics, enabling it to be used in applications where electromechanical coupling is required.

Another key advantage of PVDF is its mechanical and thermal stability. Experimental results from literature have shown that PVDF maintains its piezoelectric and ferroelectric performance even under repeated mechanical stress and elevated temperatures, conditions under which conventional polymers such as PE and

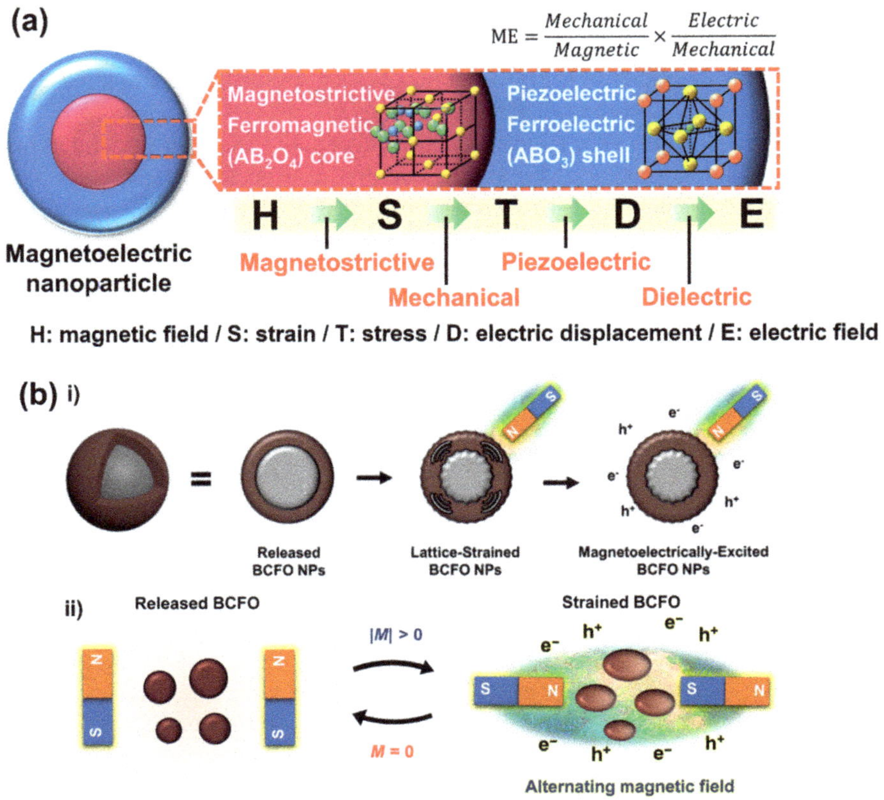

Fig. 5 Magnetoelectric (ME) effect of MENPs. (**a**) Concept of physics; transmission mechanism through ME effect in the MENPs composed of the magnetostrictive core and piezoelectric shell. (**b**) (**i**) The schematic illustration showing that MENPs induce an ME effect when an alternating magnetic field is applied. (**ii**) Repeatedly induced strain and release of the MENPs dependent on the magnetic moments of the applied magnetic field. (Reproduced with permission from [30])

Table 2 Piezoelectric natural and synthetic polymers [34]

Polymer	Dielectric constant (1 kHz; 250 °C)	Piezoelectric coefficient (pC/N)
PLA	3.0–4.0	9.82
Polyhydrox butyrate PHB	2.0–3.5	1.6–2.0
PVDF	6–12	24–34
PVDF-TrFE	18	38
Polamide-11	5	4

PP lose their functional properties [34]. While many polymeric dielectrics are easy to fabricate, they do not provide active or sufficient electro-mechanical functionality as shown in Table 2. Furthermore, a significant amount of research has been focused on achieving the polar β phases of PVDF. Different approaches have been adopted such as the use of nucleating agents and other additives, induction using polar solvents, and the application of external influences like mechanical stress, high pressure, or electric fields. Among these methods, the incorporation of trifluoroethylene (TrFE) units into the PVDF molecular chains, leading to the formation of the copolymer poly(vinylidene fluoride-*co*-trifluoroethylene) [P(VDF-TrFE)], has proven to be a highly effective way to directly obtain the polar β-phase. This copolymer is particularly attractive because of its easy phase accessibility, high remnant polarization, and excellent piezoelectric response. Owing to these properties, PVDF-TrFE thin films have been widely investigated for applications in ferroelectric memory devices [35], energy harvesting systems, pyroelectric sensors, and electrocaloric cooling technologies.

3 Magnetoelectric Coupling and Multiferroic Behavior

3.1 Principles of Magnetoelectric Effect

The magnetoelectric (ME) effect is a fundamental physical phenomenon in which an applied magnetic field induces an electric polarization, or conversely, an applied electric field induces a change in magnetization within a material system (Fig. 6). This coupling between the magnetic and electric order parameters provides an additional degree of freedom in designing multifunctional devices that respond to both electric and magnetic stimuli. The ME effect was first reported in single-phase chromium oxide (Cr_2O_3) by Astro in 1960, where linear coupling between the magnetic and electric fields was experimentally demonstrated [36]. Subsequent theoretical descriptions by Dzyaloshinskii (1960) provided a thermodynamic framework for understanding ME coupling in antiferromagnetic systems shown in Fig. 7.

The magnetoelectric effect represents the coupling between electric and magnetic properties in a material, enabling the conversion of magnetic energy into electrical energy and vice versa. This can be expressed using the linear relations:

$$P_i = \varepsilon_{ij} E_j + \alpha_{ij} H_j$$

$$M_i = \mu_{ij} H_j + \alpha_{ji} E_j$$

In these expressions, Pi represents the electric polarization vector, while Mi represents the magnetization vector. Ej and Hj denote the applied electric and magnetic fields, respectively. The tensors ϵij and μij correspond to the electric permittivity and magnetic permeability of the material, describing its inherent response to electric and magnetic stimuli. The magnetoelectric coupling tensor αij characterizes the interaction between the electric and magnetic subsystems, indicating the ability of a magnetic field to induce electric polarization and an electric field to influence magnetization [37, 38].

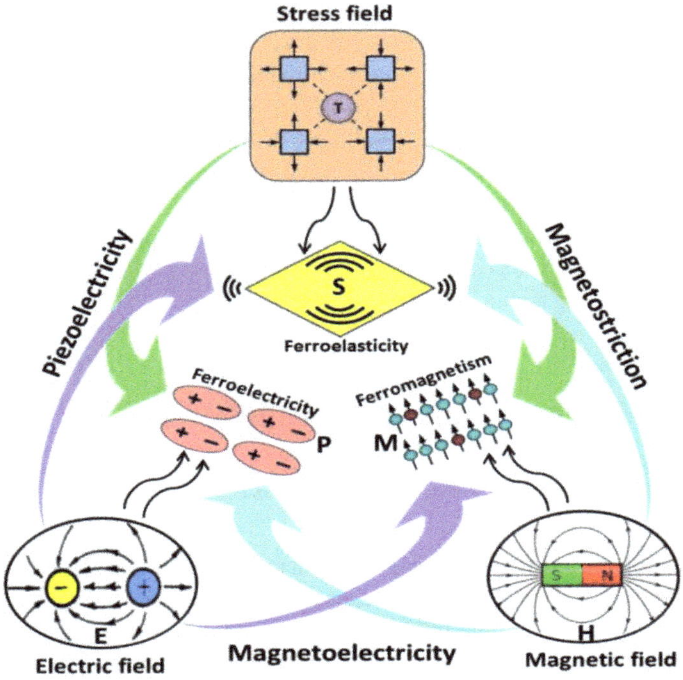

Fig. 6 Coupling mechanisms in multiferroics, illustrating interactions among electric, magnetic, and stress fields through piezoelectricity, magnetostriction, and magnetoelectricity [22]

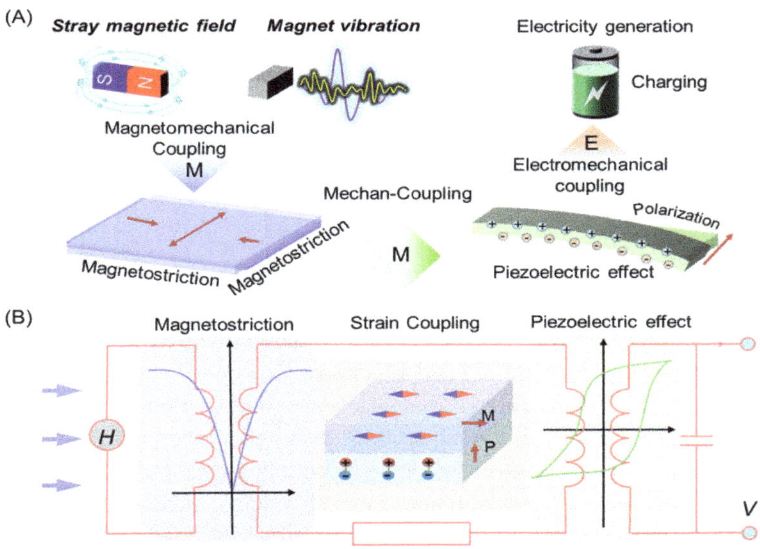

Fig. 7 Schematic of magnetoelectric coupling where magnetic excitation produces magnetostriction, strain is transferred to the piezoelectric layer, and polarization generates an electrical output [38]

This linear coupling is the basis for the functionality of magnetoelectric materials in practical devices. For instance, in sensors, the electric polarization generated by a magnetic field can be measured to detect small magnetic variations. Similarly, in memory and spintronic devices, the electric field can be used to control the magnetic state, enabling energy-efficient switching. The strength and symmetry of the coupling tensor αij determine the efficiency of these effects, making the choice of material and its microstructure critical for achieving high-performance magnetoelectric devices. This interplay between electric and magnetic properties is central to the development of multifunctional materials and devices in modern electronics [39].

The magnetoelectric effect refers to the coupling between electric polarization and magnetization, and it is quite important in the study of PVDF. Because PVDF shows strong ferroelectric and piezoelectric properties, particularly in its polar β-phase, it can work as an active medium in magnetoelectric composites. When PVDF is combined with magnetostrictive materials, the strain created in the polymer by an electric field can alter the magnetization of the magnetic part. In the same way, an applied magnetic field can affect the polarization of PVDF. By adding magnetic fillers into the polymer, a multifunctional material is formed that shows both polarization and magnetic response within a single structure [40]. This two-way interaction makes PVDF-based composites suitable for devices such as sensors, actuators, energy-harvesting systems, and spintronic components. Apart from the coupling behaviour, PVDF has additional benefits like low weight, flexibility, chemical resistance, and easy processability into films or fibres. These advantages make it more practical than brittle ceramic ferroelectrics, especially for flexible electronics and next-generation spintronic devices [41].

3.2 *PVDF-Based Composites with Magnetic Fillers*

PVDF is considered an important polymer because it combines piezoelectric, ferroelectric, and good mechanical properties. When magnetic materials such as nickel ferrite ($NiFe_2O_4$), cobalt ferrite ($CoFe_2O_4$), magnetite (Fe_3O_4), or Terfenol-D are mixed with PVDF, they form magnetoelectric composites. These composites are able to link magnetic and electric responses, which makes them useful for different applications including sensors, actuators, spintronic devices, and systems for energy harvesting.

The efficiency of these composites mainly depends on factors like how much filler is added, how well it is dispersed, the β-phase content of PVDF, and the quality of the interface between the magnetic filler and the polymer matrix. Because of these features, PVDF-based magnetic composites are especially attractive for flexible electronics and multifunctional devices. In general, they offer a simple and practical way of designing materials that show both electric and magnetic behaviour (Fig. 8).

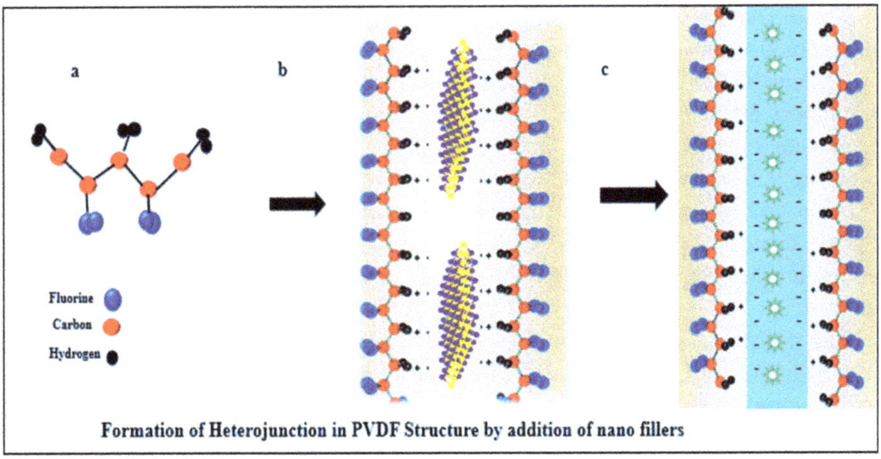

Fig. 8 Mechanism of heterojunction formation in PVDF by addition of nanofillers. (**a**) PVDF matrix. (**b**) Nanofillers. (**c**) Formation of the beta phase in PVDF [42]

Cobalt ferrite ($CoFe_2O_4$, CFO) is widely used as a magnetic filler due to its high magnetostriction coefficient, large magneto-crystalline anisotropy, and chemical stability. CFO nanoparticles embedded in PVDF significantly improve the ME response because the high magneto-strictive strain of CFO can be effectively transferred to the piezoelectric polymer. Experimental studies have reported that the addition of CFO nanoparticles not only enhances ME coupling but also promotes the nucleation of the polar β-phase in PVDF, thereby improving its ferroelectric properties [43]. The particle size and surface modification of CFO play a crucial role in determining the interfacial interaction with PVDF.

Nickel ferrite (NFO) is considered a soft ferrite, while cobalt ferrite (CFO) falls under the hard ferrite category. Because of its lower coercivity and moderate magnetostriction, NFO tends to perform better in applications that require operation under relatively weak magnetic fields. When it is blended with PVDF, the composite not only remains flexible but also maintains stable dielectric properties, making it a practical option for low-frequency magnetoelectric (ME) devices.

Interestingly, researchers have observed that NFO nanoparticles do more than just contribute to the ME effect. They also influence how PVDF crystallizes. Acting as nucleating sites, these particles encourage the formation of the β-phase, which is the phase mainly responsible for the piezoelectric and ferroelectric responses of PVDF. Because of this, NFO/PVDF films usually display a higher dielectric constant along with reduced dielectric loss—both features that are useful for sensing and energy harvesting. Another point worth noting is that, being a soft magnetic material, NFO allows its domains to realign with ease when a magnetic field is applied. This characteristic gives the composite stronger ME responses at relatively lower field strengths compared to CFO/PVDF systems.

Magnetite (Fe_3O_4) nanoparticles are useful because they have high saturation magnetization. In addition, they are biocompatible, and their electrical conductivity can also be adjusted. Composites made of Fe_3O_4 and PVDF have been studied for many applications, such as biomedical sensors, electromagnetic shielding, and flexible electronic devices. When the Fe_3O_4 nanoparticles are properly mixed and spread inside the PVDF matrix, they form good interfacial bonding, which improves the magnetoelectric performance [44, 45]. At very small sizes, Fe_3O_4 shows superparamagnetic behaviour, which reduces hysteresis losses and helps in energy conversion. Compared to other ferrites like CFO and NFO, Fe_3O_4 has better electrical conductivity, which can be useful for adjusting the dielectric properties of PVDF. However, excessive Fe_3O_4 content may lead to leakage currents and reduced breakdown strength of the polymer, emphasizing the importance of optimizing filler concentration for best performance.

The overall ME coupling in PVDF-based composites with magnetic fillers depends on several interrelated factors. Firstly, the filler concentration strongly influences the dielectric constant, loss tangent, and mechanical stiffness of the composites. A low filler loading often results in a weak ME response due to insufficient magnetostrictive contribution, whereas very high loading can deteriorate mechanical flexibility and increase electrical leakage. Secondly, the size and morphology of fillers (nanoparticles, nanorods, or nanosheets) determine the surface area available for stress transfer and the quality of interfacial interaction. Thirdly, processing routes such as hot pressing, melt mixing, or electrospinning significantly affect filler dispersion and PVDF phase content. Optimized processing conditions can maximize the fraction of polar β-phase and improve interfacial stress transfer, both of which are critical for achieving strong ME response.

3.3 *ME Coefficient Improvements Through Nanostructuring*

Nanostructuring has emerged as a promising strategy to significantly enhance the ME coefficient by improving interfacial coupling, optimizing filler distribution, and promoting the ferroelectric β-phase of PVDF. Therefore, a particular emphasis has been placed on exploring the magnetoelectric (ME) effect at the nanoscale, since the preparation and study of nanostructured and nanoscale ME materials provide important insights into their enhanced properties, such as higher ME coefficients [46]. Considerable research has therefore been devoted to the development of different nanoscale structures as shown in Fig. 9.

Among these, spherical core–shell nanoparticles consisting of a magnetostrictive (MS) core surrounded by a piezoelectric (PE) shell have been widely fabricated. In addition, composite nanofibers and core–shell nanofibers, where the MS component forms the inner core and the PE material serves as the outer coating, have also received attention. Another important approach involves the synthesis of polymeric microspheres made of piezoelectric polymers embedded with magnetic nanoparticles. These architectures are being extensively studied with the aim of achieving improved coupling effects at the nanoscale.

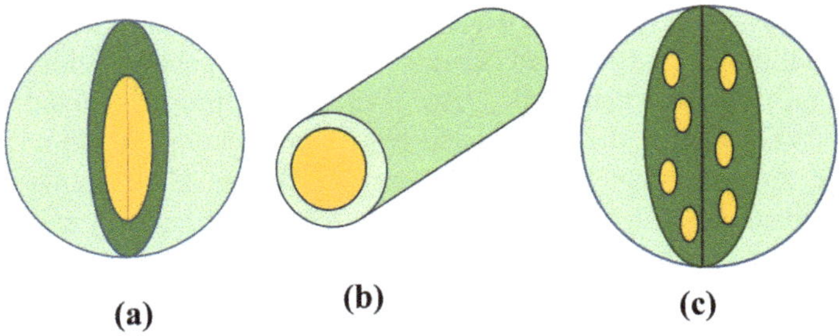

Fig. 9 Different types of nano structure Cross-sectional views of (a) vertical section with a central core, (b) horizontal cross-section of the cylinder, and (c) vertical section showing multiple internal regions

Nanostructured magnetoelectric (ME) materials are expected to display better performance than their bulk forms. This is mainly because of the thin piezoelectric (PE) shell surrounding the magnetostrictive (MS) core, which allows the stress developed in the MS phase to be transferred to the entire PE volume. In bulk ME structures, such stress transfer is often restricted only to the interfacial region. As a result, nanostructures can generate larger dipole moments and achieve stronger ME coupling. In addition, the problem of substrate clamping, which usually limits strain transfer in thin films and laminate structures, can be avoided in nanoscale composites [47]. Better mechanical contact at the MS–PE interface also supports efficient stress transfer between the two phases.

Another important design is the use of superparamagnetic nanoparticles embedded in the PE polymer matrix. Such systems show magnetoelectric activity only when an external magnetic field is applied, remaining inactive otherwise. This switching behaviour helps in conserving energy and also prevents complications like nanoparticle aggregation that are common in magnetic composites.

Geometrical confinement at the nanoscale is also known to affect the physical behaviour of polymers. Studies have shown that crystallization processes and chain dynamics are strongly influenced by confinement effects, which directly modify the piezoelectric response. These aspects make nanoscale ME composites promising candidates for future device technologies, particularly in the areas of sensors, actuators, and biomedical systems.

3.3.1 Types of Nanostructures and Their Impact

Nanoparticles: Spherical nanoparticles of ferrites such as $CoFe_2O_4$, $NiFe_2O_4$, and Fe_3O_4 are the most widely used fillers. When properly surface-functionalized to avoid agglomeration, they provide uniform stress transfer and promote dipole alignment in PVDF. Optimal loading levels (typically 5–20 wt%) are necessary to balance ME coupling with mechanical flexibility and low dielectric losses.

Nanorods and Nanowires: One-dimensional (1D) nanostructures such as $CoFe_2O_4$ nanorods or Fe_3O_4 nanowires exhibit anisotropic magnetostriction, which can be exploited to achieve directional ME coupling. Alignment of such nanorods within the PVDF matrix results in anisotropic strain transfer, enhancing the effective ME coefficient. Electrospinning of PVDF with magnetic nanorods has shown auspicious results for flexible sensor applications (Patro et al. 2015).

Core–Shell Nanostructures: Recently, core–shell nanoparticles have been engineered to optimize stress transfer across the filler–polymer interface. For example, Fe_3O_4 nanoparticles coated with SiO_2 or polymer shells improve dispersion and prevent aggregation, while maintaining magnetostrictive functionality. Such engineered interfaces lead to reduced dielectric losses and higher ME output voltages.

Nanopillars and Thin-Film Architectures: Vertically aligned nanostructures (e.g., nanopillars of CFO embedded in ferroelectric matrices) have been demonstrated in thin-film composites, offering strong coupling due to vertical stress transfer pathways [48]. Though these are more common in inorganic ferroelectric oxides, recent progress suggests potential in polymer-based systems through templated growth and block copolymer-assisted assembly.

Nanostructuring has led to significant improvements in ME response compared to bulk or microstructured composites. For example, PVDF–$CoFe_2O_4$ nanocomposites have exhibited ME voltage coefficients exceeding 200 mV/cm·Oe, which is several times higher than those of their bulk particle counterparts. Electrospun PVDF/Fe_3O_4 nanofiber mats have demonstrated enhanced output voltages under low magnetic fields, demonstrating their suitability for flexible energy harvesting applications. Moreover, the combination of PVDF-TrFE copolymers with nanostructured ferrites has been reported to yield ME coefficients adequate for use in low-power sensors and magnetic field detectors.

3.4 PVDF-TrFE Copolymers in Multiferroics

PVDF-TrFE copolymers represent an essential class of ferroelectric polymers that have gained considerable attention in the field of multiferroics. Unlike neat PVDF, which requires mechanical stretching, poling, or the addition of fillers to stabilize the ferroelectric β-phase, PVDF-TrFE naturally crystallizes into an all-trans configuration that favors the polar β-phase. This intrinsic advantage makes PVDF-TrFE more suitable for device applications where stable and strong ferroelectric behavior is desired. In particular, when combined with magnetostrictive phases such as ferrites or magnetite, PVDF-TrFE-based composites provide a promising pathway for achieving efficient magnetoelectric (ME) coupling.

Incorporation of magnetic fillers such as $CoFe_2O_4$, $NiFe_2O_4$, or Fe_3O_4 into PVDF-TrFE has been widely studied for enhancing ME coupling. Due to its higher inherent ferroelectric polarization, PVDF-TrFE can more effectively convert magneto-strictive strain from fillers into measurable electric signals compared to neat PVDF. For example, composites of PVDF-TrFE with $CoFe_2O_4$ nanoparticles have shown improved ME voltage coefficients, along with stable polarization–electric field (P–E) loops under applied magnetic fields [49, 50]. Similarly, Fe_3O_4/

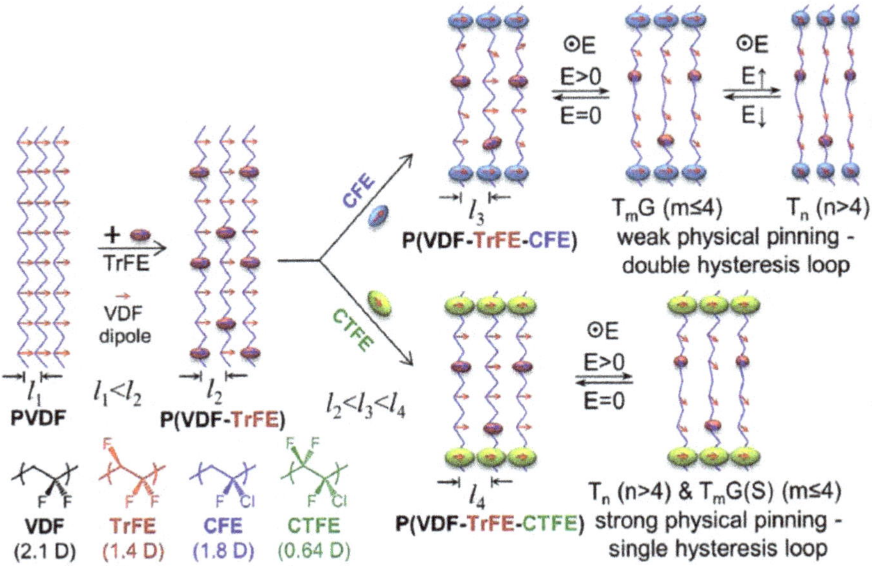

Fig. 10 Molecular structures of PVDF-based copolymers showing dipole alignment and field response, with weak pinning giving double loops and strong pinning giving single hysteresis loops [51]

PVDF-TrFE nanocomposites have demonstrated both ferroelectric and magnetic ordering at room temperature, making them suitable for multifunctional sensor applications (Figs. 10 and 11).

The dispersion of nanoparticles plays a vital role in determining performance. Well-dispersed magnetic fillers increase the interfacial contact area with the copolymer, facilitating strain-mediated coupling. Surface functionalization of nanoparticles and the use of surfactants have been reported to suppress agglomeration and improve filler–matrix interactions. Such modifications lead to higher dielectric constant, reduced loss tangent, and enhanced ME response [47].

3.4.1 Thin-Film Architectures

PVDF-TrFE is especially suitable for thin-film fabrication due to its solubility and compatibility with scalable deposition techniques. Thin films of PVDF-TrFE incorporated with magnetic nanoparticles have been integrated onto flexible substrates for potential use in microelectromechanical systems (MEMS), non-volatile memories, and magnetic field sensors. Studies have also shown that layer-by-layer structures with alternating polymer and magnetic phases result in higher ME output compared to random particulate composites. In such architectures, stress transfer is more efficient, and leakage currents are minimized.

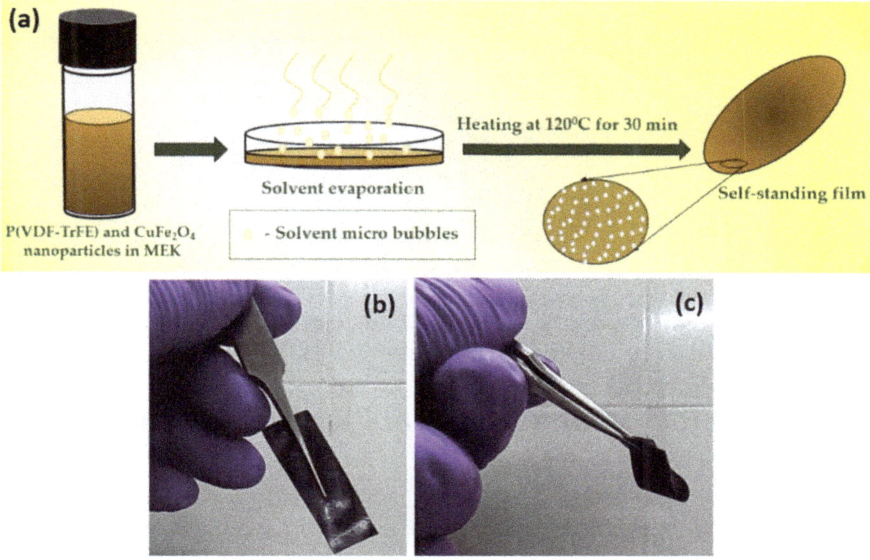

Fig. 11 (**a**) The schematic illustration of the preparation of the P(VDF-TrFE)-$CuFe_2O_4$ composite film, (**b**) and (**c**) digital photographs of the P(VDF-TrFE)-8 composite film before and after bending

3.4.2 Applications and Prospects

PVDF-TrFE-based multiferroic composites find applications in low-power magnetic sensors, flexible transducers, data storage, and energy-harvesting devices. Their lightweight, mechanical flexibility, and ease of processing make them advantageous over ceramic-based ME composites, especially in wearable and biomedical technologies [1]. Moreover, the intrinsic β-phase crystallinity of PVDF-TrFE reduces the need for post-processing treatments, enabling simpler and more reproducible device fabrication [8].

Recent efforts are focused on optimizing copolymer composition (e.g., varying TrFE content between 20–30 mol%) to tune the ferroelectric and dielectric responses. The combination of copolymer matrices with advanced nanostructures, such as nanorods, core–shell fillers, and aligned nanofibers, is expected to enhance the ME coefficients further. With ongoing progress, PVDF-TrFE-based composites are likely to play a significant role in the next generation of soft multiferroic materials.

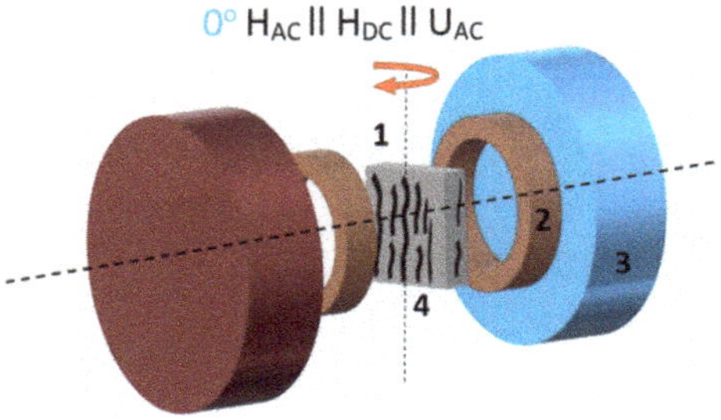

Fig. 12 Scheme of the experiment for direct magnetoelectric (ME) measurements (**1**—sample, **2**—Helmholtz coils, **3**—DC magnetic field source, **4**—aligned chains of particle clusters); the red arrow indicates the direction in which sample was rotated [52]

3.5 Mechanisms of ME Coupling in PVDF-Based Systems

Magnetoelectric (ME) coupling in PVDF and its copolymers emerges from the effective interplay between the piezoelectric nature of the polymer matrix and the magneto-strictive response of the embedded magnetic phase. Unlike single-phase multiferroics, which often suffer from weak intrinsic ME interactions at room temperature, PVDF-based composites employ a **strain-mediated coupling mechanism** to realize strong extrinsic ME effects. This approach provides a pathway to overcome the limitations of conventional ceramics while retaining the mechanical flexibility and processability inherent to polymers illustrated in Fig. 12.

3.5.1 Strain-Mediated Coupling

The dominant mechanism in PVDF-based composites is **mechanical strain transfer** between the magneto-strictive filler and the ferroelectric polymer. Under an applied magnetic field, the magnetic nanoparticles or nanofibers (e.g. $CoFe_2O_4$, $NiFe_2O_4$, Fe_3O_4) undergo magnetostriction, leading to a change in dimension. This stress is transmitted across the filler–polymer interface and induces a deformation in the PVDF matrix. Since PVDF exhibits strong piezoelectricity (particularly in its β-phase or in PVDF-TrFE copolymers), the induced strain directly converts into an electric polarization, thereby generating a measurable ME voltage output. The efficiency of this process depends critically on the filler distribution, interfacial adhesion, and the connectivity pattern (e.g., 0–3 particulate composites, 1–3 fiber-in-matrix systems, or 2–2 laminated structures).

3.5.2 Interfacial Polarization and Maxwell–Wagner Effects

Beyond strain mediation, the **interfacial polarization** mechanism also plays a role, especially when the filler–polymer interfaces exhibit distinct dielectric properties. In heterogeneous systems, charge accumulation at interfaces under an alternating magnetic field can amplify the dielectric response of the polymer, contributing to an enhanced ME coefficient. This **Maxwell–Wagner polarization** is more prominent in nanocomposites [53] where the interfacial area is high, and it acts synergistically with the piezoelectric–magneto-strictive coupling to boost the effective response.

3.5.3 Role of Nanostructuring and β-Phase Enhancement

Nano-structuring of fillers strongly influences the ME coupling pathway. For example, one-dimensional nanofibers and two-dimensional nanosheets facilitate better stress transfer compared to spherical particles, due to larger interfacial contact and anisotropic stress alignment. Simultaneously, many magnetic fillers promote the stabilization of the **polar β-phase of PVDF** during crystallization. Since the β-phase is responsible for the highest piezoelectric activity, the induced phase transformation amplifies the electromechanical response of the polymer. Thus, filler addition not only provides magnetostrictive functionality but also indirectly enhances the ferroelectric and piezoelectric behavior of PVDF.

3.5.4 Direct Magnetoelectric Coupling Pathways

Although strain-mediated mechanisms dominate, certain studies report indications of **direct interfacial magnetoelectric coupling** in PVDF-based systems. These effects are hypothesized to arise from spin–charge interactions at the polymer–filler interface, where surface charges of the polymer dipoles interact with the local magnetic moments of ferrite nanoparticles. While still under debate, such mechanisms could contribute additional pathways to polarization modulation, particularly in composites with conductive or partially oxidized interfaces.

3.5.5 Frequency and Field Dependence

The nature of ME coupling in PVDF composites is also strongly dependent on the applied field conditions. Under low-frequency magnetic fields, strain-mediated processes dominate, leading to a quasi-static ME voltage response. At higher frequencies, however, interfacial polarization effects and mechanical resonance phenomena become more significant, resulting in enhanced ME coefficients. In some laminated PVDF-based heterostructures, resonance conditions have been shown to amplify the output by an order of magnitude, which is promising for energy harvesting and sensing applications.

The ME coupling mechanisms in PVDF-based systems are primarily strain-mediated, with important contributions from interfacial polarization, β-phase stabilization, and possible spin–charge interactions. Optimization of filler morphology, interfacial engineering, and nanostructure alignment remains central to maximizing ME performance. These insights have established PVDF-based composites as versatile candidates for flexible sensors, actuators, memory devices, and energy harvesters, offering significant advantages over rigid ceramic multiferroics.

4 PVDF-Based Magnetoelectric Sensors for Biomedical and Environmental Applications

4.1 *PVDF-Based Magnetoelectric Sensors for Biomedical and Environmental Applications*

Magnetoelectric (ME) sensors have become very popular in recent years because they can directly link magnetic and electric fields through strain. Among the many materials used for ME sensing, polymer-based ferroelectrics like PVDF and its copolymer PVDF-TrFE have a special importance. Unlike ceramic ferroelectrics, which are usually brittle, PVDF is light in weight, flexible, biocompatible, and stable under normal conditions. These features make it very useful for biomedical sensors that need to fit curved and moving body parts, as well as for environmental devices that are exposed to changing stresses [34]. Another advantage is that PVDF has a high piezoelectric voltage coefficient, which helps in effectively converting small mechanical strains into electrical signals, a key requirement in ME devices.

In biomedical applications, these properties are highly useful. For example, monitoring activities like heartbeat, breathing, or muscle movement needs sensors that can sit on soft tissues without causing irritation or mismatch. PVDF films combined with magnetostrictive layers can be made into thin, light patches that attach to the skin and detect very small magnetic field changes linked with body functions. In the case of heart monitoring, such ME patches can measure weak magnetic signals of the heart, giving a safe, non-invasive, and radiation-free alternative to normal imaging methods. Their thin and bendable nature also makes it possible to integrate them into wearables like smart clothing or medical bandages, where ceramic-based sensors cannot be used because of their rigid and fragile nature (Fig. 13).

Another promising biomedical direction is the use of PVDF-based ME devices in implantable sensors. Biocompatibility and mechanical compliance are critical for long-term implantation, and polymers naturally outperform inorganic ferroelectrics in this regard. Within the body, such sensors could be designed to respond to external magnetic stimuli, enabling remote powering or triggering of localized electrical activity. This strategy makes it possible to stimulate specific neurons or release therapeutic agents in a controlled way without relying on large external power sources. In addition, PVDF-based magnetoelectric (ME) devices can be

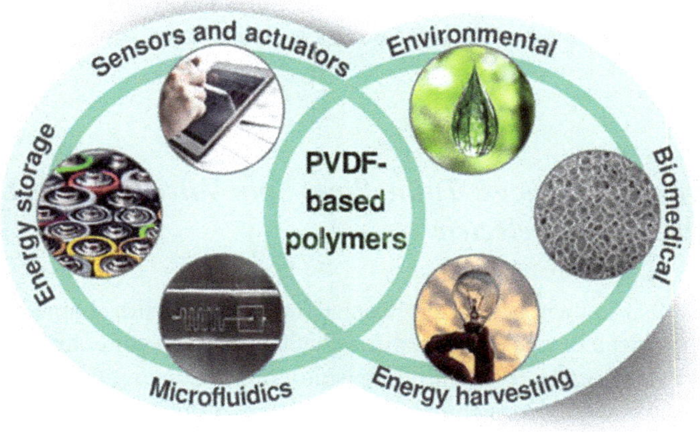

Fig. 13 Application areas of PVDF-based polymers in energy storage, sensors and actuators, microfluidics, environmental, biomedical, and energy harvesting fields [7]

miniaturized using microfabrication or printing techniques, making them easy to integrate into minimally invasive medical tools.

Outside of biomedical applications, PVDF ME sensors are being explored for environmental monitoring. One promising example is in air and water quality assessment, where PVDF membranes containing magnetostrictive nanoparticles can act both as sensors and as active filtration elements. When exposed to a magnetic field, the structure deforms in a controlled manner, changing its porosity or adsorption properties. This allows the membrane not only to detect pollutants but also to assist in their removal, creating a smart, multifunctional system that is compact and energy-efficient.

PVDF-based ME devices are also useful in detecting magnetic anomalies and monitoring structural health. Their thin, flexible design enables embedding into materials such as concrete, bridges, or industrial components. These sensors can detect local magnetic changes or stress-induced variations, providing early warnings of potential damage. Their durability under harsh conditions, combined with simple fabrication methods like spin coating and inkjet printing, makes them suitable for widespread outdoor deployment. Furthermore, their low power requirements make them ideal for remote and wireless sensing applications where battery replacement is challenging.

The fabrication versatility of PVDF and PVDF-TrFE further broadens their utility. Processing routes such as electrospinning, roll-to-roll printing, and flexible lamination allow mass production of sensors at relatively low cost. Control over the crystalline phase, especially the enhancement of the ferroelectric β-phase, is essential for maximizing the piezoelectric response. Techniques such as mechanical stretching, electrical poling, and nanoparticle incorporation have been widely adopted to increase β-phase content. When these optimized polymer films are combined with magnetostrictive components, the resulting ME devices exhibit both high sensitivity and excellent mechanical durability. This scalability is critical for real-world adoption in biomedical wearables and distributed environmental sensor networks.

4.2 Spintronic Devices: Tunnelling, Spin Valves, and PVDF as an Active Dielectric

Spintronics, also known as spin electronics, make use of the charge and spin of electrons to achieve device functionalities that exceed those of conventional semiconductor technologies. By controlling spin-polarised currents across different interfaces, spintronic devices combine magnetic and electronic properties in a single unit. These devices have brought significant advancement in the area of non-volatile memories, magnetic sensors, and logic systems. Among various spintronics devices, magnetic tunnel junctions (MTJs) and spin valves attracted significant attention from the research community due to their direct practical relevance. Recently, functional polymers such as PVDF, whose ferroelectric and dielectric properties provide new ways of controlling spin transport interest have expanded to include the devices in flexible and tunable devices.

4.2.1 Tunneling in Spintronics

Quantum tunneling is a central phenomenon in spintronics, particularly in MTJs. These devices are composed of two ferromagnetic electrodes separated by an ultra-thin insulating barrier. When an external bias is applied, electrons tunnel through the barrier with probabilities that depend strongly on the relative alignment of the magnetization of the electrodes. In the parallel state, the resistance is low due to efficient spin-polarized tunneling, whereas in the antiparallel state, the resistance is higher. This resistance difference is expressed as tunneling magnetoresistance (TMR), which is the key performance metric for MTJs (Fig. 14).

The TMR effect underpins the operation of magnetic random-access memory (MRAM), a non-volatile memory technology in which information is stored in the magnetic alignment of layers, rather than in electric charge. Beyond the traditional MTJ structure, researchers have introduced ferroelectric materials as tunnel barriers, resulting in ferroelectric tunnel junctions (FTJs). These devices exhibit both TMR and tunneling electroresistance (TER), as an applied electric field can switch the ferroelectric polarization of the barrier. The coexistence of TMR and TER in FTJs allows multifunctional operation, merging magnetic and electric control within a single device.

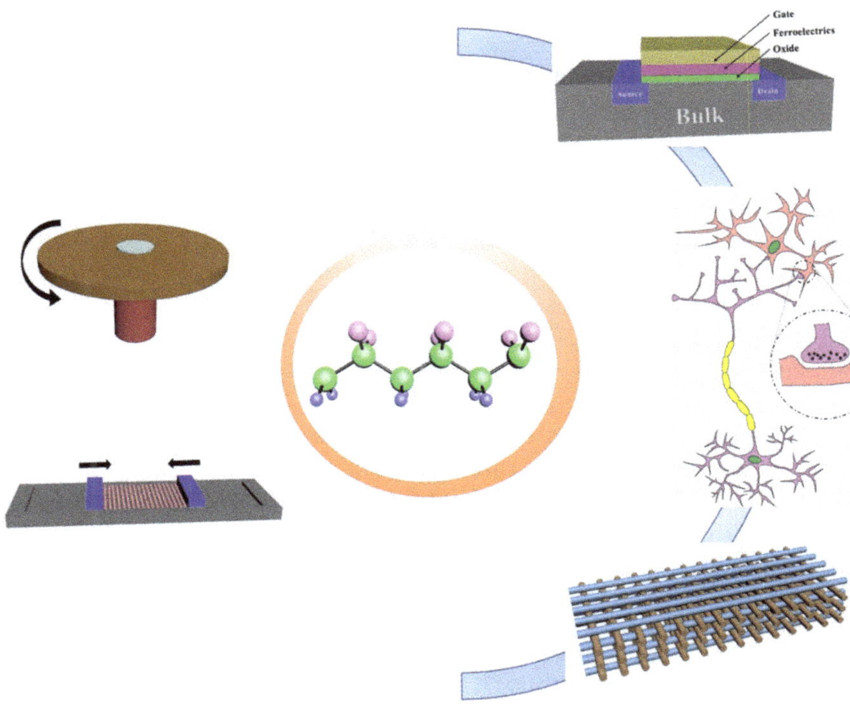

Fig. 14 Illustration of ferroelectric polymer structure, processing methods, and applications in transistors, synaptic devices, and memory arrays [54]

4.2.2 Spin Valves and Magnetoresistance

Spin valves represent another cornerstone of spintronics. A typical spin valve consists of two ferromagnetic layers separated by a thin non-magnetic metallic spacer. One ferromagnetic layer is exchange-pinned by an antiferromagnet, while the other is free to rotate under an applied magnetic field. The device resistance depends on the relative alignment of the two ferromagnetic layers, giving rise to the giant magnetoresistance (GMR) effect.

The discovery of GMR in spin valve structures transformed information storage technology, particularly in hard disk drive read heads, where the effect allowed significant increases in storage density. Spin valves are also widely used in magnetic sensors because of their high sensitivity and scalability. The continued development of spin valve structures has focused on new materials for the non-magnetic spacer and the integration of dielectric layers to improve spin injection efficiency. These refinements expand the functionality of spin valves beyond traditional data storage, toward applications in flexible sensors and spin-based logic devices.

4.2.3 PVDF as an Active Dielectric in Spintronics

PVDF and its copolymer PVDF-TrFE possess a combination of ferroelectric polarization, high dielectric constant, and mechanical flexibility, making them particularly attractive for hybrid spintronic systems. In contrast to conventional oxide dielectrics, PVDF offers electrically switchable dipoles that can directly influence interfacial charge distributions and spin-dependent tunneling processes. When incorporated as a tunnel barrier or dielectric spacer in MTJs and spin valves, PVDF can modulate resistance states through electric-field-induced dipole reorientation.

This property enables spintronic devices with dual tunability: magnetic control through spin alignment and electric control through ferroelectric switching. Such hybrid functionality supports the development of multiferroic systems with potential applications in low-power memory, adaptive logic, and neuromorphic computing. In addition, the organic nature of PVDF allows fabrication of flexible and transparent devices, which extends spintronic applications into wearable electronics, biomedical sensors, and large-area systems. Its compatibility with solution processing further lowers fabrication costs compared to the high-vacuum techniques required for oxide barriers.

The integration of PVDF into spintronic devices illustrates the potential of combining polymer ferroelectrics with magnetic architectures. By leveraging both ferroelectric and spintronic effects, these devices achieve functionalities that are not accessible with conventional materials alone. The use of PVDF not only provides tunability but also broadens the scope of spintronic applications into domains requiring flexibility, transparency, and cost-effective large-scale fabrication.

4.3 Role of PVDF in MRAM and Data Storage Technologies

PVDF and its copolymers, particularly PVDF-TrFE, have emerged as attractive materials for integration in advanced memory and data storage technologies, including magnetoresistive random-access memory (MRAM). The significance of PVDF lies in its unique combination of ferroelectric, piezoelectric, and pyroelectric properties, which allow coupling between electrical and magnetic states when used in hybrid multiferroic systems.

In MRAM devices, data storage relies on controlling the magnetization orientation of magnetic layers, typically through spin-transfer torque or spin–orbit torque mechanisms. By incorporating PVDF as a ferroelectric layer in multiferroic heterostructures, it is possible to achieve voltage-controlled magnetic switching. The strong polarization response of PVDF, combined with its mechanical flexibility, enables efficient strain-mediated magnetoelectric coupling when interfaced with magnetostrictive layers such as $CoFe_2O_4$ or Terfenol-D. This coupling reduces the energy required for magnetic switching, offering a path toward ultra-low-power nonvolatile memories.

Beyond switching, PVDF contributes to enhanced scalability of memory architectures. Unlike conventional oxide ferroelectrics, PVDF films can be processed at relatively low temperatures and deposited over large areas on flexible or unconventional substrates. This makes them suitable for integration into next-generation memory arrays, wearable electronics, and flexible data storage systems. Furthermore, the intrinsic remanent polarization of PVDF can provide additional functionality in hybrid memory devices, combining both ferroelectric and magnetic storage elements in a single platform.

Another promising role of PVDF in data storage lies in its ability to stabilize intermediate polarization states, which can be exploited for multi-level storage. This characteristic allows higher density information storage compared to binary states in conventional MRAM. Moreover, PVDF's chemical stability and mechanical robustness support long retention times and reliability, addressing one of the critical requirements of commercial memory devices.

Overall, PVDF serves as an enabling material that bridges ferroelectricity and magnetism in MRAM technology. Its integration not only reduces switching power and enhances endurance but also broadens the application scope of MRAM beyond rigid platforms, making it a promising candidate for future energy-efficient and flexible data storage technologies.

4.4 Flexible and Wearable Sensor Applications Using PVDF-Based Composites

PVDF and its copolymers are among the most promising materials for flexible and wearable sensors due to their exceptional piezoelectric, pyroelectric, and ferroelectric properties. When combined with functional fillers such as carbon nanotubes, graphene, metallic nanoparticles, or ceramic nanostructures, PVDF-based composites exhibit enhanced electromechanical coupling, improved sensitivity, and tailored flexibility. These characteristics make them highly suitable for next-generation wearable devices that demand lightweight, stretchable, and durable sensing platforms.

In flexible pressure and strain sensors, PVDF composites efficiently convert mechanical deformation into electrical signals, enabling real-time monitoring of body movements, posture, and joint activity. Such sensors have been successfully integrated into wearable bands, smart textiles, and skin-mounted patches for applications in sports, rehabilitation, and healthcare monitoring. The incorporation of nanofillers not only increases the β-phase content of PVDF, which is responsible for strong piezoelectric activity, but also enhances mechanical strength and durability under repeated bending and stretching conditions.

For biomedical applications, PVDF-based wearable sensors offer high sensitivity in detecting subtle physiological signals such as pulse, respiration, and vocal cord vibrations. Their biocompatibility and flexibility allow intimate skin contact without discomfort, while their self-powered nature, arising from the piezoelectric effect, reduces reliance on external batteries. This makes them particularly useful for continuous health monitoring and remote patient care.

In environmental monitoring, PVDF composites have been utilized in wearable devices for detecting acoustic vibrations, impact forces, and thermal changes. Their lightweight and conformal design enables deployment on clothing or flexible substrates without hindering user mobility. Additionally, advances in nanostructured fillers provide opportunities to integrate multifunctional sensing capabilities, such as simultaneous pressure and temperature detection, in a single PVDF-based sensor.

Overall, PVDF-based composites provide a versatile and reliable foundation for flexible and wearable sensor technologies. Their unique combination of electromechanical responsiveness, structural flexibility, and compatibility with large-area fabrication methods positions them as a key material system in the evolution of personalized healthcare, soft robotics, and smart wearable electronics.

5 Recent Experimental Studies

The magnetoelectric coupling in polymer-based composites is strongly influenced by the content, location, and distribution of the magnetic phase within the polymer matrix. Among these factors, the concentration of magnetic fillers plays a vital role, as it governs not only the dispersion and position of particles but also the overall magnetic properties of the composite films. Functionalization of CFO nanoparticles with surfactants or coupling agents has been shown to improve dispersion, reduce agglomeration, and thereby increase ME sensitivity. Additionally, composites fabricated through solution casting and electrospinning techniques have exhibited better alignment of CFO fillers, leading to anisotropic ME behavior (Fig. 15).

Behera et al. reported the incorporation of cobalt ferrite ($CoFe_2O_4$) nanoparticles into a ferroelectric PVDF matrix, which resulted in PVDF–$CoFe_2O_4$ nanocomposite films exhibiting a first-order room-temperature magnetoelectric coupling coefficient of 198 mV/(cm·Oe) under zero DC bias magnetic field. In another study, Gha et al. fabricated flexible CFO/PVDF composite films using a simple solution casting method (Fig. 16), and systematically investigated the effect of CFO particle concentration on the structural, dielectric, and magnetoelectric behavior of the films. With increasing CFO concentration, the magnetic properties increased almost linearly, while the dielectric and ferroelectric responses showed significant improvement [55]. At 10% CFO loading, the dielectric constant reached 64.72 and the remnant polarization was 1.1617 $\mu C/cm^2$. Interestingly, the magnetoelectric coupling coefficient (αME) achieved a maximum value of 52.06 V/(cm·Oe) at 20% CFO content. Moreover, tensile tests verified that the prepared composite films retained good mechanical flexibility Fig. 17, making them suitable for flexible device applications.

The figure illustrates the sol-gel auto-combustion process for synthesizing nickel ferrite ($NiFe_2O_4$) nanoparticles. Initially, Fe^{3+} and Ni^{2+} salt solutions are mixed and

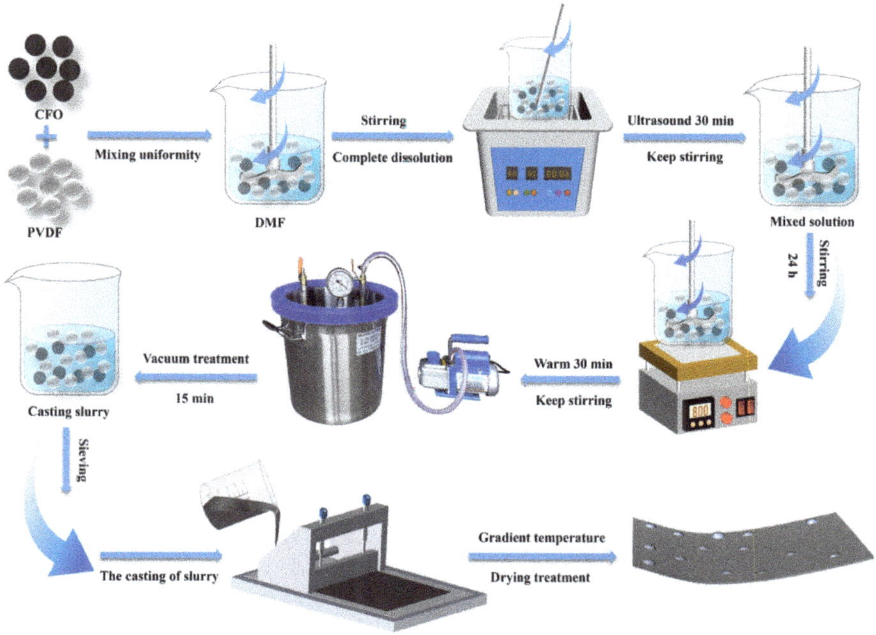

Fig. 15 Preparation route of CFO/PVDF composite films [55]

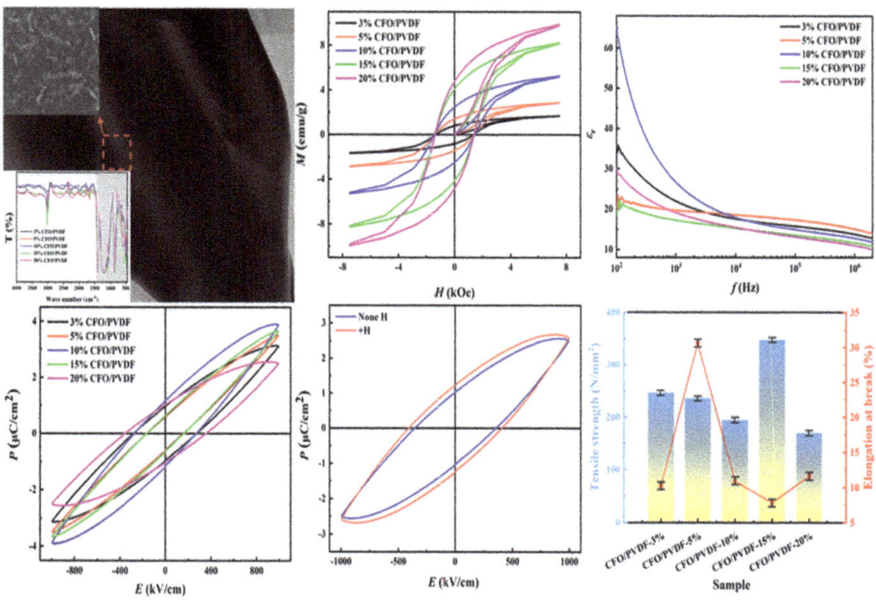

Fig. 16 Photograph image of 20% CFO/PVDF, hysteresis loops and comparison of magnetic parameters, frequency-dependent curves of dielectric constant, P–E loops, P–E loops of CFO/PVDF composite films with CFO particle concentrations under the applied magnetic field and mechanical (tensile) properties of CFO/PVDF composite films with different CFO particle concentrations [55]

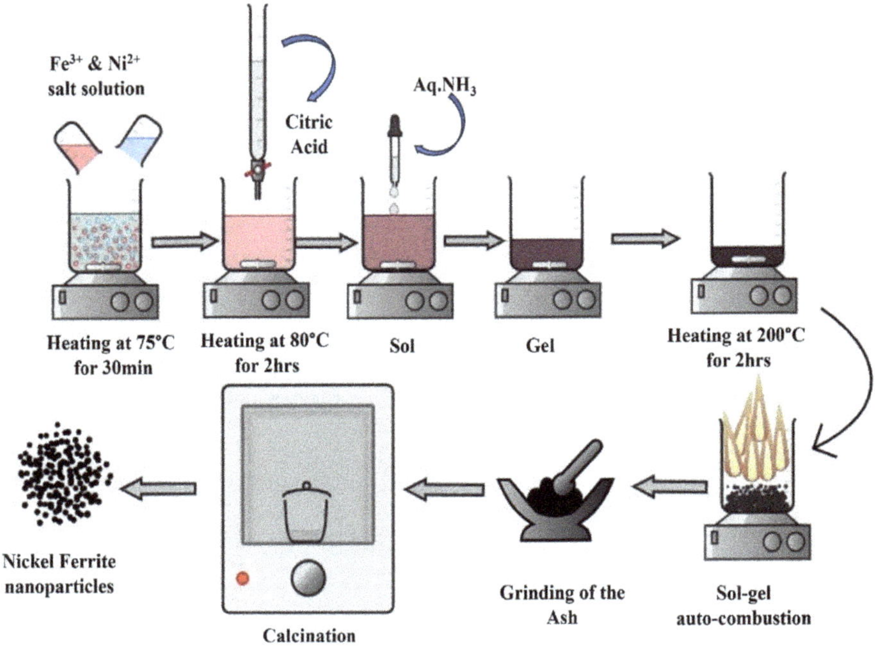

Fig. 17 The pictorial presentation of the synthesis process of $NiFe_2O_4$ nanoparticles by auto-ignited sol-gel method [56, 58]

heated at 75 °C, followed by the addition of citric acid and further heating at 80 °C to form a sol. With the addition of aqueous ammonia, the sol transitions into a gel, which is then heated at 200 °C to undergo drying and decomposition. The gel experiences self-ignition during the sol-gel auto-combustion step, producing an ash-like precursor. This ash is ground into fine powder and subjected to calcination, ultimately yielding nickel ferrite nanoparticles. The process is cost-effective, simple, and allows control over particle size. Moreover, the combustion reaction ensures uniform mixing of metal ions, leading to highly crystalline nanoparticles (Fig. 18).

The photocatalytic activity of the synthesized catalyst was evaluated using the degradation of different organic dyes, as illustrated in Fig. 18. The UV–Vis absorption spectra of methylene blue (MB) (Fig. 18a), congo red (CR) (Fig. 18b), and rhodamine B (Rod B) (Fig. 18c) demonstrate a progressive decrease in absorption intensity with increasing irradiation time, indicating the efficient breakdown of dye molecules. Among the tested dyes, MB exhibited the most rapid and significant reduction in absorbance, followed by CR and Rod B, suggesting a dye-dependent degradation efficiency. The corresponding degradation percentages are summarized in Fig. 18d, where MB achieved nearly complete degradation within 240 min, while

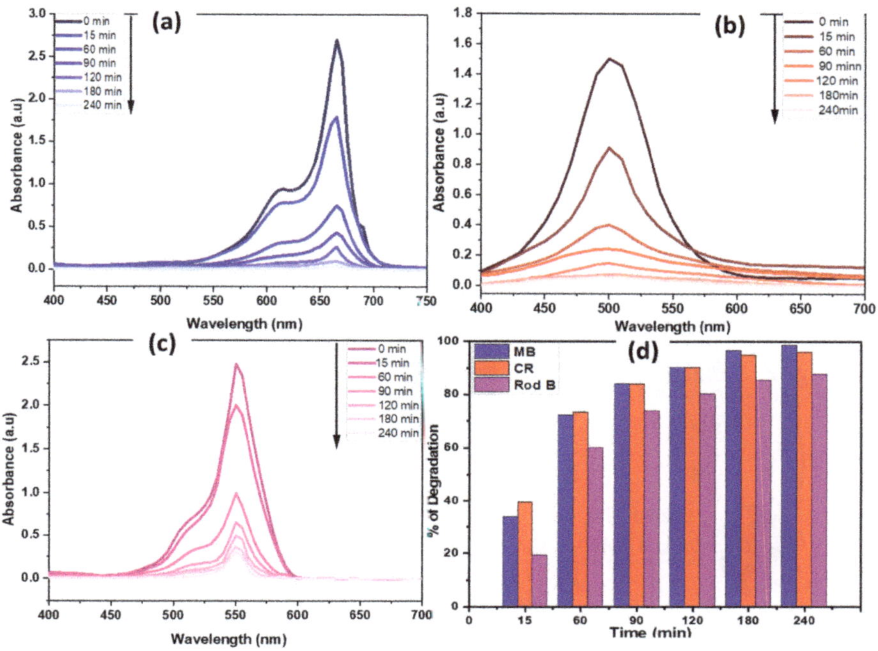

Fig. 18 Chronological UV–Vis spectra of $NiFe_2O_4$ for (**a**) MB, (**b**) CR, (**c**) Rod B, and (**d**) comparison of degradation percentage vs time among MB, CR and Rod B [56, 58]

CR and Rod B also showed high removal efficiencies. These results confirm that the catalyst possesses strong photocatalytic activity toward multiple dyes, highlighting its potential for wastewater treatment applications (Fig. 19).

The figure presents a schematic route for preparing Fe_3O_4/Ag ferrofluids from natural iron sand. Initially, iron sand is treated with hydrochloric acid (HCl) to dissolve the iron content, producing a yellowish iron chloride solution. Ammonium hydroxide (NH_4OH) is then added, which reacts with the iron salt under ultrasonic conditions (400 kHz) to form black Fe_3O_4 precipitates. These precipitates are repeatedly washed with deionized water to remove impurities. Next, silver nanoparticles are introduced by adding sodium borohydride ($NaBH_4$) as a reducing agent, which allows Ag to deposit on the Fe_3O_4 surface, resulting in Fe_3O_4/Ag nanocomposites. The product is cooled to room temperature, separated, and washed to obtain a clean composite. To improve dispersion and long-term stability, the nanocomposites are coated with oleic acid (OA) and dimethyl sulfoxide (DMSO). The final material is a ferrofluid in which Fe_3O_4 provides magnetic behavior, Ag contributes antimicrobial and electrical activity, and the surface coating prevents agglomeration, making the fluid stable and suitable for biomedical, catalytic, and electronic applications.

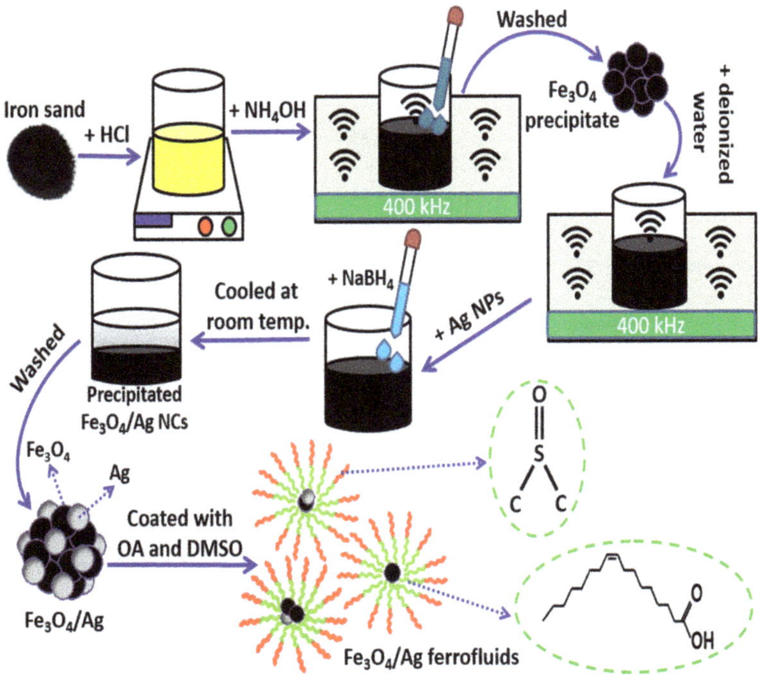

Fig. 19 Synthesis scheme of the Fe_3O_4/Ag nanohybrid ferrofluids [58]

5.1 *Emerging Directions: Hybrid 2D Materials (Graphene, MoS$_2$) with PVDF for Spintronics*

The integration of PVDF with two-dimensional (2D) materials such as graphene and molybdenum disulfide (MoS_2) has opened new opportunities for spintronic technologies. Spintronics, which relies on manipulating the spin of electrons rather than just their charge, requires materials that can efficiently transport spin information and allow low-power control of spin states. PVDF, known for its strong ferroelectric polarization, provides a nonvolatile and energy-efficient means of electrical control, while 2D materials like graphene and MoS_2 offer unique spin transport and spin–orbit characteristics that complement PVDF's properties [57, 58].

When PVDF is placed on graphene, its ferroelectric polarization can be used to control how charges and spins move at the interface. Graphene is ideal for spin transport because electrons can travel long distances without losing their spin, while PVDF acts like a switch that can turn spin currents on or off in a non-volatile way. This makes it possible to build ferroelectric spin field-effect transistors, which require very little energy to operate and perform better than regular charge-based devices.

Things become even more interesting when PVDF is combined with MoS_2 or similar layered materials. Unlike graphene, MoS_2 has strong spin–orbit coupling and valley-dependent electronic states. Using PVDF as a ferroelectric gate on MoS_2 allows simultaneous control of spin and valley states, paving the way for hybrid devices that can store and process information in multiple ways. Such devices could be small, energy-efficient, and work at room temperature.

Hybrid multiferroic structures are another promising direction. In PVDF–graphene or PVDF–MoS_2 stacks with magnetic layers, the polarization of PVDF can directly influence spin injection or even modify magnetization through strain or charge effects. This makes it possible to design low-power memory devices, while the flexibility of PVDF and the thinness of 2D materials allow the construction of bendable and lightweight spintronic systems.

However, there are challenges. The efficiency of these hybrid devices depends on having a clean and well-ordered interface between PVDF and the 2D material. Any defects or trapped charges can reduce performance [59]. Therefore, careful fabrication methods are necessary, along with maintaining the ferroelectric properties of PVDF over long periods. In some cases, combining graphene and MoS_2 under PVDF may help achieve the best balance between long spin coherence and strong spin–orbit effects.

Looking forward, the integration of PVDF with emerging 2D magnetic materials such as CrI_3 and Fe_3GeTe_2 offers an exciting path to directly couple ferroelectric polarization with intrinsic magnetism in atomically thin layers. Such systems could enable highly compact, multifunctional, and energy-efficient spintronic devices [60]. With continued progress in interface engineering and scalable processing, PVDF–2D hybrid structures are expected to play an important role in the future of spin-based memory, logic, and flexible electronic technologies.

5.2 *Future Challenges and Commercialization Aspects*

PVDF is emerging as a strong candidate for next-generation magnetic and spintronic devices, with research moving toward practical memory and sensor technologies. Its ferroelectric nature allows use as a barrier layer in tunnel junctions and as an active element in magnetoelectric RAM, enabling nonvolatile operation at low power. Incorporation of ferrite nanoparticles enhances magnetoelectric coupling, while hybrid designs with graphene, MoS_2, and other two-dimensional materials improve charge transport, spin injection, and mechanical flexibility. These approaches make PVDF attractive for flexible and printable electronics where inorganic multiferroics are less suitable. For real commercialization, emphasis must be placed on scalable fabrication methods such as roll-to-roll coating and wafer-level deposition that can deliver consistent phase quality and device reproducibility. Integration with standard CMOS technology remains essential, requiring careful control of interfaces and processing temperatures. Performance benchmarks must demonstrate stable polarization switching, endurance under repeated cycling, long retention times, and reliability under environmental stresses such as heat, humidity, and mechanical bending. Beyond technical validation, progress depends on establishing robust supply chains for high-purity PVDF and compatible fillers, creating viable cost models that can compete with established MRAM and FeRAM, and securing intellectual property on device architectures and processing routes. Early demonstration units are expected to target applications where PVDF offers unique advantages, such as flexible memory for wearable devices, low-power storage for IoT nodes, and hybrid systems combining sensing and memory. Future adoption will also require close industry collaboration, pilot-scale production, and adherence to reliability standards, while sustainability considerations such as lead-free chemistry and low-energy processing can further strengthen the case for commercial rollout.

5.3 Emerging Directions: Hybrid 2D Materials (Graphene, MoS_2) with PVDF for Spintronics

PVDF in combination with two-dimensional materials such as graphene and MoS_2 offers a versatile platform for developing advanced spintronic devices. The ferroelectric polarization of PVDF or its copolymers provides a nonvolatile control element, while graphene contributes high mobility and efficient spin transport, and MoS_2 adds semiconducting behavior with strong spin–orbit interaction. When integrated together, these hybrid structures allow effective modulation of spin–charge conversion, enhanced interfacial coupling, and improved stability. Experimental studies highlight that ferroelectric gating of graphene with PVDF can achieve controllable spin transport, while PVDF/graphene composites promote higher β-phase crystallinity and stronger piezoelectric or magnetoelectric response. In MoS_2–PVDF systems, dielectric and magnetoelectric properties are significantly improved, expanding the range of applications. Interfaces engineered with magnetic electrodes further strengthen spin injection and reduce resistance losses. In addition, flexible processing methods such as printing, electrospinning, and fiber drawing make these hybrids compatible with emerging wearable and flexible electronics. Together, these developments point toward scalable, low-power, and multifunctional spintronic devices suitable for memory, sensing, and integrated energy-efficient technologies.

6 Conclusion and Future Outlook

6.1 Summary of Key Findings

PVDF is a versatile polymer that combines ferroelectric, piezoelectric, and dielectric properties in a flexible and stable framework. Among its crystalline forms, the β-phase is particularly important, as it provides the intense polarization required for magnetoelectric (ME) coupling. When integrated with magnetic fillers such as

$CoFe_2O_4$, $NiFe_2O_4$, or Fe_3O_4, PVDF-based composites exhibit significant enhancement of ME response compared with conventional polymeric dielectrics. Advances such as nano-structuring, copolymerization with TrFE, and interfacial engineering have been shown to improve ME coefficients and device efficiency. These developments have established PVDF as a promising material for multifunctional devices, including sensors, spin valves, magnetic tunnel junctions, and MRAM. Importantly, its lightweight and flexible nature allows applications in wearable and flexible electronics, where inorganic multiferroics are less practical. Recent studies also highlight the potential of PVDF when combined with two-dimensional materials, such as graphene and MoS_2, offering new pathways for high-performance, energy-efficient spintronic devices. Overall, PVDF provides a robust foundation for next-generation magnetoelectric and spintronic technologies; however, advancements in phase control, long-term stability, and scalable fabrication will be crucial for the practical deployment of these technologies.

6.2 *Potential of PVDF in Next-Generation Spintronic and Memory Devices*

PVDF has been getting a lot of attention in recent years because of its useful properties, especially for spintronic and memory devices. The main reason is that it shows very strong ferroelectric polarization and also has high dielectric strength, while at the same time it is soft and flexible. Because of this mix of properties, PVDF can be added into different device structures without much difficulty. For example, in spintronic systems like magnetic tunnel junctions or spin valves, PVDF can act as a dielectric layer and also help in controlling polarization. This switching of polarization is useful because it can directly affect the flow of spin-polarized current. So, devices based on PVDF can work with lower power and still remain non-volatile, which is one of the big goals in spintronics.

In memory devices such as MRAM (magneto-resistive random-access memory), PVDF gives one more advantage. Here, the ferroelectric switching can be linked with spin transport, which means that the device can store data for a longer time, use less energy during operation, and even combine more than one function in the same system. Compared to rigid oxide materials, PVDF is light and bendable, which makes it more practical for flexible or wearable memories. This is difficult to achieve with the usual inorganic dielectrics.

Researchers are also trying out hybrid PVDF systems where magnetic nanoparticles or 2D materials are added to improve the properties further. Such changes can help at the interfaces and also increase the overall efficiency of the device. If more progress is made in stabilising the right phases, improving interfaces, and scaling up the processing, PVDF can become a very important material for the next generation of spintronic and memory technologies.

6.3 *Research Gaps and Future Opportunities*

Although PVDF and its copolymers have shown promising results in magnetoelectric and spintronic systems, several research gaps remain. One of the significant challenges is the controlled stabilization of the polar β-phase on a large scale, since device performance strongly depends on the fraction of this phase. Achieving consistent phase purity across thin films and composite structures remains challenging. Another gap lies in interfacial engineering, as the weak interaction between PVDF and magnetic fillers often limits the efficiency of magnetoelectric coupling. Long-term reliability also needs attention, as issues such as polarization fatigue, thermal instability, and degradation under ambient conditions affect device stability.

Future opportunities can be explored through hybrid structures that combine PVDF with nanomaterials such as graphene, MoS_2, or other two-dimensional systems, which can improve spin transport and interfacial coupling. Advanced nanostructuring, surface modification, and copolymer design may further enhance coupling strength and device tunability. The flexibility and processability of PVDF also open new directions in stretchable, wearable, and biocompatible electronics, especially for low-power sensors and memory systems. For broader adoption, efforts must focus on scalable fabrication methods, reproducibility, and integration with existing semiconductor processes. Addressing these challenges will allow PVDF to move beyond laboratory demonstrations and establish its role as a key material for future spintronic and memory technologies.

References

1. Dallaev, R., Pisarenko, T., Sobola, D., Orudzhev, F., Ramazanov, S., Trčka, T.: Brief review of PVDF properties and applications potential. Polymers. **14**(22), 4793 (2022)
2. Asandei, A.D.: Photomediated controlled radical polymerization and block copolymerization of vinylidene fluoride. Chem. Rev. **116**(4), 2244–2274 (2016)
3. Marshall, J.E., Zhenova, A., Roberts, S., Petchey, T., Zhu, P., Dancer, C.E., et al.: On the solubility and stability of polyvinylidene fluoride. Polymers. **13**(9), 1354 (2021)
4. Sekhar, M.C., Veena, E., Kumar, N.S., Naidu, K.C.B., Mallikarjuna, A., Basha, D.B.: A review on piezoelectric materials and their applications. Cryst. Res. Technol. **58**(2), 2200130 (2023)
5. Shaukat, H., Ali, A., Bibi, S., Altabey, W.A., Noori, M., Kouritem, S.A.: A review of the recent advances in piezoelectric materials, energy harvester structures, and their applications in analytical chemistry. Appl. Sci. **13**(3), 1300 (2023)
6. Ruan, L., Yao, X., Chang, Y., Zhou, L., Qin, G., Zhang, X.: Properties and applications of the β phase poly(vinylidene fluoride). Polymers. **10**(3), 228 (2018)
7. Costa, C.M., Cardoso, V.F., Martins, P., Correia, D.M., Goncalves, R., Costa, P., et al.: Smart and multifunctional materials based on electroactive poly(vinylidene fluoride): recent advances and opportunities in sensors, actuators, energy, environmental, and biomedical applications. Chem. Rev. **123**(19), 11392–11487 (2023)
8. Yasar, M., Hassett, P., Murphy, N., Ivankovic, A.: β Phase optimization of solvent cast PVDF as a function of the processing method and additive content. ACS Omega. **9**(24), 26020–26029 (2024)

9. Yuan, X., Shi, J., Kang, Y., Dong, J., Pei, Z., Ji, X.: Piezoelectricity, pyroelectricity, and ferroelectricity in biomaterials and biomedical applications. Adv. Mater. **36**(3), 2308726 (2024)
10. Magdy, G., Hassanin, A.H., Kandas, I., Shehata, N.: PVDF nanostructures characterizations and techniques for enhanced piezoelectric response: a review. Mater. Chem. Phys. **325**, 129760 (2024)
11. Kunwar, D., Vazquez, I.R., Jackson, N.: Effects of solvents on synthesis of piezoelectric polyvinylidene fluoride trifluoroethylene thin films. Thin Solid Films. **757**, 139414 (2022)
12. Ramana, E.V., Durairajan, A., Kavitha, D., Tobaldi, D.M., Zavašnik, J., Bdikin, I., Valente, M.A.: Enhanced magnetoelectric and energy storage performance of strain-modified PVDF-$Ba_{0.7}Ca_{0.3}TiO_3$-$Co_{0.6}Zn_{0.4}Fe_2O_4$ nanocomposites. J. Energy Storage. **87**, 111454 (2024)
13. Sarkar, L., Singh, S.G., Vanjari, S.R.K.: Preparation and optimization of PVDF thin films for miniaturized sensor and actuator applications. Smart Mater. Struct. **30**(7), 075013 (2021)
14. Ko, K., Yang, S.C.: Magnetoelectric membrane filters of poly(vinylidene fluoride)/cobalt ferrite oxide for effective capturing of particulate matter. Polymers. **12**(11), 2601 (2020)
15. Kalimuldina, G., Turdakyn, N., Abay, I., Medeubayev, A., Nurpeissova, A., Adair, D., Bakenov, Z.: A review of piezoelectric PVDF film by electrospinning and its applications. Sensors. **20**(18), 5214 (2020)
16. Karageorgou, M.A., Tsakmakidis, K., Stamopoulos, D.: Ferroelectric/piezoelectric materials in energy harvesting: physical properties and current status of applications. Crystals. **14**(9), 806 (2024)
17. Liang, X., Matyushov, A., Hayes, P., Schell, V., Dong, C., Chen, H., et al.: Roadmap on magnetoelectric materials and devices. IEEE Trans. Magn. **57**(8), 1–57 (2021)
18. Majumdar, S.: Ultrafast switching and linear conductance modulation in ferroelectric tunnel junctions via P(VDF-TrFE) morphology control. Nanoscale. **13**(25), 11270–11278 (2021)
19. Hirohata, A., Yamada, K., Nakatani, Y., Prejbeanu, I.L., Diény, B., Pirro, P., Hillebrands, B.: Review on spintronics: principles and device applications. J. Magn. Magn. Mater. **509**, 166711 (2020)
20. Fiebig, M.: Revival of the magnetoelectric effect. J. Phys. D. Appl. Phys. **38**(8), R123 (2005)
21. Ramesh, R., Spaldin, N.A.: Multiferroics: progress and prospects in thin films. Nat. Mater. **6**(1), 21–29 (2007)
22. Palneedi, H., Annapureddy, V., Priya, S., Ryu, J.: Status and perspectives of multiferroic magnetoelectric composite materials and applications. Actuators. **5**(1), 9 (2016)
23. Sun, L.G., Wu, G., Wang, Q., Lu, J.: Nanostructural metallic materials: structures and mechanical properties. Mater. Today. **38**, 114–135 (2020)
24. Lim, B., Alorro, R. D., Aylmore, M., & Grimsey, D. Recycling of Cobalt and Nickel from Nickel Slag by Organic Acid Leaching. (2023)
25. Liu, F., Hashim, N.A., Liu, Y., Abed, M.M., Li, K.: Progress in the production and modification of PVDF membranes. J. Membr. Sci. **375**(1–2), 1–27 (2011)
26. Li, M., Stingelin, N., Michels, J.J., Spijkman, M.J., Asadi, K., Feldman, K., et al.: Ferroelectric phase diagram of PVDF: PMMA. Macromolecules. **45**(18), 7477–7485 (2012)
27. Kang, S.J., Park, Y.J., Bae, I., Kim, K.J., Kim, H.C., Bauer, S., et al.: Printable ferroelectric PVDF/PMMA blend films with ultralow roughness for low voltage non-volatile polymer memory. Adv. Funct. Mater. **19**(17), 2812–2818 (2009)
28. Li, L., Zhang, M., Rong, M., Ruan, W.: Studies on the transformation process of PVDF from α to β phase by stretching. RSC Adv. **4**(8), 3938–3943 (2014)
29. Mohammad pourfazeli, S., Arash, S., Ansari, A., Yang, S., Mallick, K., Bagherzadeh, R.: Future prospects and recent developments of polyvinylidene fluoride (PVDF) piezoelectric polymer; fabrication methods, structure, and electro-mechanical properties. RSC Adv. 13(1), 370–387 (2023)
30. Song, H., Listyawan, M.A., Ryu, J.: Core–shell magnetoelectric nanoparticles: materials, synthesis, magnetoelectricity, and applications. Actuators. **11**(12), 380 (2022)
31. Xin, Y., Zhu, J., Sun, H., Xu, Y., Liu, T., Qian, C.: A brief review on piezoelectric PVDF nanofibers prepared by electrospinning. Ferroelectrics. **526**(1), 140–151 (2018)

32. Koç, M., Dönmez, Ç.E.D., Paralı, L., Sarı, A., Aktürk, S.: Piezoelectric and magnetoelectric evaluations on PVDF/$CoFe_2O_4$ based flexible nanogenerators for energy harvesting applications. J. Mater. Sci. Mater. Electron. **33**(10), 8048–8064 (2022)
33. Xia, W., Zhang, Z.: PVDF-based dielectric polymers and their applications in electronic materials. Iet Nanodielectr. **1**(1), 17–31 (2018)
34. Li, X., Wang, Y., He, T., Hu, Q., & Yang, Y. (2019). Preparation of PVDF flexible piezoelectric film with high β-phase content by matching solvent dipole moment and crystallization temperature. Journal of Materials Science: Materials in Electronics, 30(22), 20174-20180.
35. Park, Y.J., Kang, S.J., Lotz, B., Brinkmann, M., Thierry, A., Kim, K.J., Park, C.: Ordered ferroelectric PVDF-TrFE thin films by high throughput epitaxy for nonvolatile polymer memory. Macromolecules. **41**(22), 8648–8654 (2008)
36. Velev, J.P., Jaswal, S.S., Tsymbal, E.Y.: Multi-ferroic and magnetoelectric materials and interfaces. Philos. Trans. R. Soc. A Math. Phys. Eng. Sci. **369**(1948), 3069–3097 (2011)
37. Wang, Y., Su, Y., Li, J., Weng, G.J.: A theory of magnetoelectric coupling with interface effects and aspect-ratio dependence in piezoelectric-piezomagnetic composites. J. Appl. Phys. **117**(16), 164106 (2015)
38. Liu, S., Liao, S., Wei, K., Deng, L., Zhao, L., Zou, H.: Self-biased magnetoelectric composite for energy harvesting. Batter. Energy. **2**(5), 20230005 (2023)
39. Saleh Medina, L.M., Negri, R.M.: Dielectric behavior and electro-magnetic coupling at room temperature in $BiFeO_3$/PVDF and $CoFe_2O_4$/PVDF composites. J. Phys. Chem. C. **121**(49), 27683–27692 (2017)
40. Liu, Y., Wang, B., Zhan, Q., Tang, Z., Yang, H., Liu, G., et al.: Positive temperature coefficient of magnetic anisotropy in polyvinylidene fluoride (PVDF)-based magnetic composites. Sci. Rep. **4**(1), 6615 (2014)
41. Chen, X., Han, X., Shen, Q.D.: PVDF-based ferroelectric polymers in modern flexible electronics. Adv. Electron. Mater. **3**(5), 1600460 (2017)
42. Siddique, A., Nawaz, H., Razzaque, S., Tabasum, A., Gong, H., Razzaq, H., Umar, M.: PVDF-based piezo-catalytic membranes—a net-zero emission approach towards textile wastewater purification. Polymers. **16**(5), 699 (2024)
43. Gonçalves, R., Martins, P., Moya, X., Ghidini, M., Sencadas, V., Botelho, G., et al.: Magnetoelectric $CoFe_2O_4$/polyvinylidene fluoride electrospun nanofibres. Nanoscale. **7**(17), 8058–8061 (2015)
44. Im, J.S., Park, I.K.: Mechanically robust magnetic Fe_3O_4 nanoparticle/polyvinylidene fluoride composite nanofiber and its application in a triboelectric nanogenerator. ACS Appl. Mater. Interfaces. **10**(30), 25660–25665 (2018)
45. Zhang, Y., Zhang, T., Liu, L., Chi, Q., Zhang, C., Chen, Q., et al.: Sandwich-structured PVDF-based composite incorporated with hybrid Fe_3O_4@BN nanosheets for excellent dielectric properties and energy storage performance. J. Phys. Chem. C. **122**(3), 1500–1512 (2018)
46. Han, B.Q., Lavernia, E.J., Mohamed, F.A.: Mechanical properties of nanostructured materials. Rev. Adv. Mater. Sci. **9**(1), 1–16 (2005)
47. Stolbov, O.V., Raikher, Y.L.: Magnetostrictive and magnetoactive effects in piezoelectric polymer composites. Nanomaterials. **14**(1), 31 (2023)
48. Cao, X., Ma, J., Shi, X., Ren, Z.: Effect of TiO_2 nanoparticle size on the performance of PVDF membrane. Appl. Surf. Sci. **253**(4), 2003–2010 (2006)
49. Rahul, M.T., Chacko, S.K., Vinodan, K., Raneesh, B., Bhadrapriya, B.C., Bose, B.A., et al.: Multiferroic and energy harvesting characteristics of P(VDF-TrFE)-$CuFe_2O_4$ flexible films. Polymer. **252**, 124910 (2022)
50. Rubaiya, F.: High Performance Electromagnetic Interference Shielding from Lightweight-Flexible Polymer Composite Fiber Structures. The University of Texas Rio Grande Valley (2021)
51. Yang, L., Li, X., Allahyarov, E., Taylor, P.L., Zhang, Q.M., Zhu, L.: Novel polymer ferroelectric behavior via crystal isomorphism and the nanoconfinement effect. Polymer. **54**(7), 1709–1728 (2013)

52. Omelyanchik, A., Antipova, V., Gritsenko, C., Kolesnikova, V., Murzin, D., Han, Y., et al.: Boosting magnetoelectric effect in polymer-based nanocomposites. Nanomaterials. **11**(5), 1 (2021)
53. Samet, M., Levchenko, V., Boiteux, G., Seytre, G., Kallel, A., Serghei, A.: Electrode polarization vs. Maxwell–Wagner–Sillars interfacial polarization in dielectric spectra of materials: characteristic frequencies and scaling laws. J. Chem. Phys. **142**(19), 194703 (2015)
54. Yin, Z., Tian, B., Zhu, Q., Duan, C.: Characterization and application of PVDF and its copolymer films prepared by spin-coating and Langmuir–Blodgett method. Polymers. **11**(12), 2033 (2019)
55. Gao, R., Chen, C., Ren, K., Li, H., Sun, G., Deng, X., et al.: Enhanced magnetoelectric properties of flexible $CoFe_2O_4$/PVDF composite films with different $CoFe_2O_4$ particle concentrations. Mater. Today Chem. **42**, 102386 (2024)
56. Biswas, B., Ahmed, M.F., Rahman, M.L., Khanam, J., Bhuiyan, M.H.R., Sharmin, N.: Investigation of structural, optical, and magnetic properties of NiFe2O4 for efficient photocatalytic degradation of organic pollutants through photo fenton reactions. Heliyon. 10(17), e37199 (2024)
57. Choudhuri, I., Bhauriyal, P., Pathak, B.: Recent advances in graphene-like 2D materials for spintronics applications. Chem. Mater. **31**(20), 8260–8285 (2019)
58. Taufiq, A., Saputro, R.E., Susanto, H., Hidayat, N., Sunaryono, S., Amrillah, T., et al.: Synthesis of Fe3O4/Ag nanohybrid ferrofluids and their applications as antimicrobial and antifibrotic agents. Heliyon. 6(12), e05813 (2020)
59. Faraz, M., Singh, H.H., Khare, N.: A progressive strategy for harvesting mechanical energy using flexible PVDF-rGO-MoS_2 nanocomposites film-based piezoelectric nanogenerator. J. Alloys Compd. **890**, 161840 (2022)
60. Wang, H.Q., Wang, J.W., Wang, X.Z., Gao, X.H., Zhuang, G.C., Yang, J.B., Ren, H.: Dielectric properties and energy storage performance of PVDF-based composites with MoS_2@MXene nanofiller. Chem. Eng. J. **437**, 135431 (2022)

PVDF-Based Actuators and Soft Robotics

Prakash Karipoth

Abstract The co-existing remarkable ferroelectric properties and mechanical flexibility of PVDF polymer provides it significant advantages over traditional solid-state ferroelectrics and other elastomers for emerging technological applications. Soft robotics is such a significant avenue where PVDF have shown plenty of opportunities. However, to exploit those potential avenues efficiently, it is important to optimise the material, in particular electroactive properties of PVDF and manufacturing processes. In this chapter, we discuss the key electroactive characteristics of PVDF and optimisation strategies to enable its use in actuator and soft robotic designs. Further, the areas of biomedical engineering where state-of-art PVDF based actuators and soft robots are being explored is presented. This chapter also addresses some of the critical material and fabrication challenges that tend to limit the impact of PVDF in soft robotic development. Finally, the chapter attempts to provide some insight on the possible innovative ways to tackle these challenges and leverage the full potential of PVDF for soft robotics to advance biomedical engineering.

Keywords Soft robotics · Actuators · Biomedical engineering · Elastomers · Electroactive characteristics

1 Introduction

In recent years, several functional materials have emerged as potential stimulus-responsive components for novel applications in actuation and soft robotics. These materials respond to changes such as temperature, humidity, electric field, and magnetic field to yield quantitative and controllable mechanical responses, which are often reversible and repeatable. Based on the type of the stimulus or actuation

P. Karipoth (✉)
School of Design, University of Leeds, Leeds, UK
e-mail: p.karipoth@leeds.ac.uk

D. Punetha (ed.), *Polyvinylidene Fluoride (PVDF)*, Engineering Materials,
https://doi.org/10.1007/978-981-95-6718-8_7

principle, actuators belong to several classes, such as tendon-driven, electric, magnetic, humidity, photothermal, pneumatic, which is summarised in Fig. 1. Based on the actuation principle, the choice of the right material and design is crucial in achieving the desired actuation performance, efficiency and suitability for the desired application. Among these, electrically driven actuators provide significant advantage owing to their ease of design, control, efficiency and reliability.

Compared to piezoelectric ceramics, the piezoelectric response is lower in polymers; however, their low cost, light weight and flexibility make them a better choice for soft robotics applications [1]. There are two main categories of electrically driven actuators, namely electroactive polymers (EAPs) and Ionic Polymer-Metal composites (IPMCs). EAPs provide mechanical deformation in response to an applied electric field, whereas IPMCs undergo ion migration when subjected to an electric field and resulting in actuation response. EAPs in general exhibit larger piezoelectric stress constants [2] than ceramics, which makes them more suitable as electrically driven actuators and soft robots. While there are several EAP candidates [3] such as polypropylene, polystyrene, poly methyl methacrylate, vinyl acetate, which have demonstrated piezoelectric properties, PVDF have been widely explored as a smart electroactive piezoelectric material for application in soft robotics owing to its flexibility, chemical stability, solution-based processing, biocompatibility, and so on. Among these many synthetic polymers, PVDF have demonstrated the highest piezoelectric and pyroelectric properties [4, 5]. Moreover, unlike piezoceramic materials, PVDF possess anisotropic electromechanical properties since their values of d_{20} and d_{21} piezo strain constants are different which is beneficial for mechanical actuation [6].

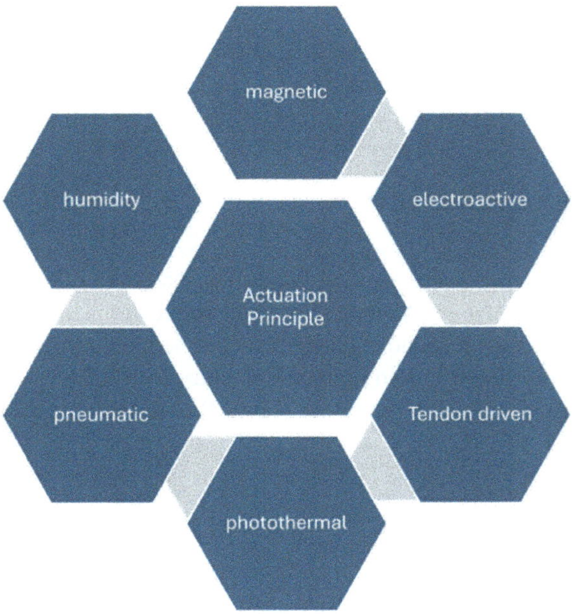

Fig. 1 Various actuation principles employed in soft robotic designs

Recent progress in additive manufacturing, micro/nanoscale fabrication techniques has generated plenty of avenues for PVDF-based actuators and soft robots. While there have been significant developments and practical demonstrations of using PVDF for a wide range of soft robotic applications in biomedical engineering, there are still several materials, manufacturing and optimisation challenges in employing PVDF-based actuators and soft robots for technological innovations to tackle challenging biomedical scenarios.

In this chapter, we explore the key material properties and electroactive behaviour of PVDF, which enables it to provide a mechanical response to applied electric fields. The main aspects of performance optimisation for maximum efficiency and control are also discussed. Based on these optimised properties of PVDF-based actuators, its state-of-art soft robotic applications in biomedical engineering is presented. Finally, various challenges in material processing and device fabrication using PVDF for actuator applications are addressed, and possible manufacturing approaches to overcome these challenges are envisaged.

2 Electroactive Behaviour and Performance Optimization

When exploring the stimulus-responsive behaviour of PVDF, it is important to understand the material characteristics, microscale structure and design architectures which make it an electroactive functional material. The main properties of PVDF are piezoelectricity, ferroelectric polarisation, electrostriction, superior mechanical flexibility and elasticity, chemical stability, and a wide frequency range of operation. We shall explore these properties in detail.

2.1 Piezoelectricity

Conceptually, the piezoelectric effect in materials can be either direct (charge/voltage generated by externally applied force/pressure) or converse (stress/strain induced by applied charge/voltage) [7]. The utilisation of semicrystalline polymer PVDF with the general formula $-(CH_2-CF_2)n$ as actuators is based on the converse piezoelectric effect. The virgin form of PVDF as a raw material is an electric insulator and therefore does not exhibit any piezoelectric property. However, during the material processing or fabrication process, PVDF can be polarised electrically by applying an external electric field. As a result of this polarisation process, the raw PVDF converts from α phase into a β phase, which exhibits piezoelectric properties. Owing to the robustness, flexibility of PVDF polymer, the β phase when constructed as a planar film can provide mechanical deformation perpendicular to their plane under the influence of external applied fields [7]. PVDF in the piezoelectric form constitutes positive and negative dipole regions. While PVDF can exist in 4 different phases, such as α, β, γ, and δ, the α and β phases are the most prominent ones.

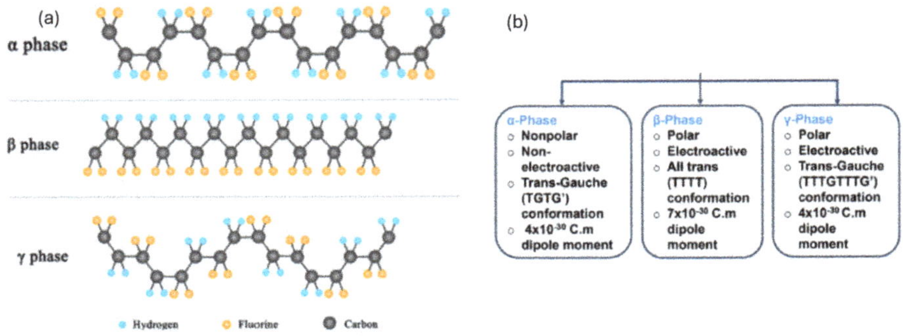

Fig. 2 (**a**) Graphical representation (Reproduced with Permission) [9] and (**b**) key features of α, β, γ phases of PVDF (Reproduced with Permission) [8]

The α phase is characterised by its planar-helical trans-gauche-trans-gauche (TGTG) conformation with the constituent dipoles crystallised in antiparallel orientation. Hence, the net dipole moment will be zero and therefore no electroactive behaviour is exhibited [8]. On the other hand, β phase of PVDF exists as a planar all zigzag trans–trans (TT) configuration. In this configuration, the dipoles are arranged parallel providing a net dipole moment and hence electroactive with remarkable piezoelectric behaviour. The γ phase of PVDF has parallel dipolar arrangement and electroactive, however, the β phase is more desirable in terms of piezoelectric properties as they possess higher dipole moment ($\sim 7 \times 10^{-30}$ C. m) than the γ phase. The three major phases of PVDF and their key features are represented in Fig. 2.

2.2 *Pyroelectric Effect*

Piezoelectric materials also exhibit the pyroelectric effect which refers to the generation of an electrical potential response to thermal mechanisms or temperature changes within the material owing to their polar symmetry [5]. Pyroelectricity is the phenomenon of the conversion of thermal fluctuations into electrical energy. This is different from thermoelectricity which works on the Seebeck effect which requires a sustained temperature gradient [10]. The pyroelectric performance of PVDF also requires enhancement of the β phase.

2.3 *Ferroelectricity*

Many of the pyroelectric materials also exhibit the ferroelectric effect or ferroelectric polarisation in which spontaneous electric polarisation emerges in response to an applied electric field. Hence, it could be generalised that all ferroelectric materials also possess pyroelectric and piezoelectric properties [5].

2.4 Flexibility and Processability

For soft robotic applications in biomedical engineering, it is primarily important that the material is biocompatible. PVDF exhibits excellent biocompatibility and are chemically stable and resistant to oxidation. They also possess hydrolytic stability and does not easily degrade or leach into solutions in normal biomedical pH environments [5]. Apart from these, they have remarkable mechanical properties, such as mechanical flexibility and elasticity, which make them a prominent choice for biomedical engineering applications.

2.5 Performance Optimization Strategies

Like any ferroelectric material, both piezoelectric and pyroelectric responses of PVDF are governed by variation of its electric dipole moment. While PVDF and its composite have demonstrated remarkable properties and performance for applications such as soft robotics, the piezoelectric (d_{33} = ~−30 pC/N) [11] and pyroelectric (p = ~30 μC/m^2/K) [12] coefficients of PVDF are significantly lower than their inorganic or ceramic counterparts, such as $BaTiO_3$ or PZT, limiting their commercial impact and efficiency [5]. Even these specified maximum coefficients are not readily available in PVDF as the most stable phase of PVDF is the α phase, which is non-polar. In order to achieve the performance as close as possible to these maximum reported values, the β phase needs to be enriched by careful manipulation of the material composition, device design and operating parameters. Thus, the need for this performance optimisation at multiple levels makes the efficient use of PVDF for actuators and soft robots for practical deployment in applications such as biomedicine [5] challenging.

Before addressing the performance optimisation strategies, let's look at the equations for piezoelectric and pyroelectric currents in a simple PVDF device architecture. Consider an electroactive PVDF layer sandwiched between two electrodes, thus forming a capacitor arrangement as shown in Fig. 3. Here, the piezoelectric current is given as $I_{piezo} = d_{33}Ad\sigma/dt$ where d_{33} is the piezoelectric coefficient, *A is area of the capacitor, and* σ is the mechanical stress [5, 13]. Similarly, the pyroelectric current can be represented as $I_{pyro} = pAdT/dt$ where I where p is the pyroelectric coefficient, and dT/dt is the rate of change of temperature with time [5, 14]. The piezoelectric coefficient correlates the material's stress and polarisation whereas the

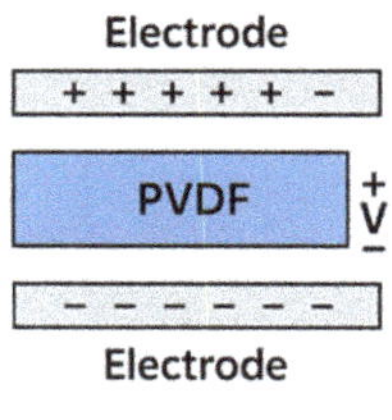

Fig. 3 Schematic of a PVDF active device in a capacitor arrangement

pyroelectric coefficient relates temperature and polarization, and both coefficients are intrinsic parameters. Considering various parameters [5], determining the ferroelectric behaviour of PVDF, maximising the intrinsic coefficients and engineering device design so as to modulate the interplay between these intrinsic parameters, device microstructure and interfaces in a favourable manner is the key to obtaining superior performance of PVDF for actuators and soft robots. By enriching the polar β or γ phase in PVDF, the dielectric characteristics and the intrinsic parameters could be enhanced [5] (Fig. 4).

Various strategies possible to enhance the β phase are primarily aimed at incorporating anisotropy in the microscopic internal configuration of PVDF, which can enable stronger electric polarisation. As summarised in Fig. 4, this can be achieved either by subjecting the material to external fields such as mechanical stress/pressure distorting the molecular conformations [5] or electric field and/or magnetic fields to promote molecular alignment and organisation. Different from using external fields, the other approach involves inducing internal interactions and microstructure refinement by manipulating the material composition via incorporating secondary materials in PVDF at appropriate proportions [5]. These additional materials could be copolymers, inorganic ferroelectrics, nanoparticles, organic/inorganic solvents or plasticizers. The additional materials, when combined with PVDF yield polymer blends or nanocomposites with superior properties as compared to virgin PVDF. It must be noted that achieving PVDF with desired characteristics is often achieved by multi-level optimisation as each of the approaches are significantly different in terms of the process as well as results. For example, the electrospinning process involves a sophisticated instrumental set-up in which the PVDF solution is subjected to controlled spraying through a tiny nozzle and subjected to an applied external electric field. On the other hand, the use of specific solvents or additives provides the crystalline control, particularly in solution casting processes [8]. In many of these enhancement methods, the optimised manufacturing process and composite formation also allow to tune the mechanical and thermal properties of PVDF-based composites to suit the desired performance and applications.

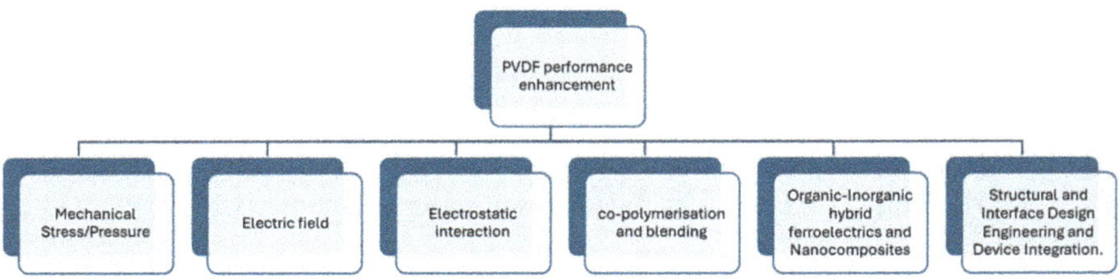

Fig. 4 Various approaches for the enhancement of ferroelectric properties of PVDF

2.5.1 Mechanical Stress/Pressure

Post-extrusion mechanical stretching has been a conventionally followed industrial method to produce piezoelectric PVDF [5]. During extrusion, the PVDF forms films with α phase which, when mechanically strained, aligns the inherent molecular chains into β phase. Upon mechanical stretching, the elongation of the molecular chains takes place. The extent of β phase generated by mechanical stretching is governed by the purity, ordering, stretch ratio and processing temperature [5]. Upto 80% β phase has been achieved by mechanical stretching [15] while further enrichment upto 98% was reported [16] by subjecting to high pressure involving repeated pressing and folding, followed by compaction under higher temperature annealing. While the application of normal pressure could generate higher β phase fraction, this degree of ordering could be depreciated quite easily by thermal fluctuations [5]. While mechanical stretching of PVDF is still useful in generating ferroelectric PVDF films, following the technological progress in microfabrication, miniaturisation and complex designs, the uniaxial stretching became more and more challenging especially when the PVDF films are mechanically coupled with other substrates [17].

2.5.2 Electric Field

The constituent crystal chains of PVDF can respond to an external electric field and orient in the direction of the applied field. The process of subjecting PVDF to a high electric field is often termed as 'poling' [5]. During poling, the random dipoles of the β phase of PVDF molecular chains rotate and align in the direction of the electric field to yield the β phase. When the electric field is removed, the majority of these oriented dipoles do not randomise easily and tend to retain the orientation thereby giving a net dipole moment [18]. The poling could also be done in direct contact mode or separation mode (corona poling) as well as at ambient or elevated temperatures [5].

There are two established fabrication technologies utilizing an electric field, which can effectively enhance the β phase in PVDF namely (1) electrospinning [19] and (2) electrohydrodynamic pulling [20]. Electrospinning technique can produce continuous fibers of nanometer to submicron sizes from a polymer solution or melt. When a high voltage is applied, electrical charges are induced on the surface of the polymer causing the polymer droplets to elongate. In this way, a 'Taylor cone' is formed usually at the tip of the fine nozzle of a syringe in which the polymer is loaded. At optimised voltage and solution parameters, a jet of charged polymer droplets is directed and get deposited on a collector in the form of ultrafine fibers [19]. When PVDF is subjected to the electrospinning process, the polymer droplets are mechanically stretched as well as electrically poled which significantly enhances the β phase [5]. Eventhough high degree of β phase could be achieved by electrospinning, the polarisation takes place in-plane of the fiber film along the fiber axis and polarisation in the normal direction is more or less non-existent. Hence, this method is less suitable for device designs with out-of-plane actuation configurations [5]. This limitation of the electrospinning technique is not present in the electrohydrodynamic pulling. Here, the electric field created across the two electrodes of a parallel plate arrangement, with the PVDF polymer in between. In this way, the polymer from the bottom electrode gets pulled to the top electrode forming an array of fine polymer microfibers with rich β phase. Since the experienced force and hence the generated polarisation on the polymer droplets is normal to the plane of the electrodes, the obtained PVDF film is suitable for out-of-plane actuation applications [5].

2.5.3 Co-polymerisation and Blending

The addition of monomers such as tetrafluorethylene (TrFE), hexafluoropropylene (HFP) and chloride trifluoride ethylene (CTFE) can enhance the crystallinity, thermal and mechanical properties [21]. The enhancement when these additional molecules are incorporated into the PVDF matrix in the β phase could be either due to steric effects or electrostatic effects [5]. In the case of TrFE, with very similar molecular configuration as VDF, the steric effects by TrFE sustains the β phase owing to their energy barriers to rotate their dipoles. The degree of crystallinity and their by polarisation is governed by the VDF:TrFE ratio, and the incorporation of TrFE at 20–50% provides good enhancement in the β phase [5, 22]. The steric effects were found to enhance the piezoelectric and pyroelectric coefficients of PVDF-TrFE. Apart from TrFE, HFP and CTFE also have been greatly explored to improve the β phase in PVDF. Thee HFP and CTFE can provide comparatively larger steric effects than TrFE as they are slightly bulkier molecules. However, they are found to stabilise the non-polar α phase more than the β phase [23, 24] which does not contribute to the desired ferroelectric performance required for actuation. Recently, there have been attempts to acquire more control over the steric effects by co-polymerisation with 3 or 4 constituents rather than just a binary co-polymerisation [5, 25]. For instance, Chen et al. [25] incorporated trace amounts of fluorinated alkyne (FA) monomers along with other 3 monomers to realise P(VDF-TrFE-CFE-FA) tetra-polymer which possessed piezoelectric coefficient (d_{33}) ~> 1000 pm/V at par with traditional inorganic ferroelectric materials (d_{33} > 1000 pm/V) [26]. The presence of comparatively smaller molecules such as FA monomers can facilitate easier chain dynamics of individual polymer units at lower applied voltages and limit dissipation of acquired polarisation. As compared to devices fabricated with PVDF alone, the design of a copolymer of PVDF with other polymers such TrFe also provides cyclic stability and durability in actuators in particularly soft robotic joints.

2.5.4 Electrostatic Interaction

The enhancement of ferroelectric properties utilizing electric charge interactions across the PVDF polymer chains is slightly different from steric interactions where molecular sizes are more relevant. As PVDF constitutes H^+ atoms and highly electronegative F^- atoms, the electrostatic interaction of these charges between these atoms of PVDF itself as well as with different additive molecules could be utilized to enrich the β phase in PVDF. Fillers with strong functional groups or surface charges have been found to facilitate this mechanism. These additives could be either ionic liquids, $-NH_2$ groups or –OH groups [5]. One of the straightforward approaches is to use hydrated salts or solvents to cast PVDF from solution, thereby creating hydrogen bonding between the water molecules and PVDF chains. Upto 80% β phase has been achieved using hydrated or hygroscopic additives with PVDF [27] as polar solvent enhances the crystallisation.

Ionic liquids (IL) are also utilised to manipulate the electrostatic interaction in PVDF to tailor its polar properties. The charged positive or negative ions in these ionic liquids efficiently interact with the electric dipoles in the PVDF. One of the main advantages of using IL is that they are compatible with manufacturing processes [5, 28] such as 3D printing or melt extrusion of PVDF as both PVDF and IL blend together very well and enhanced β-phase of upto 98.3% has been demonstrated. Liquid metals are also studied as an alternative to ILs to enhance the polarisation in PVDF. Liquid metals have the advantage of favouring the electrical conductivity of PVDF while IL tend to compromise the intrinsic electrical properties of PVDF [29]. Nonetheless, one of the main disadvantages of using liquid phase additives for β-phase enhancement is that to retain the liquid within the polymer matrix during operation, the active PVDF layers in the devices often require encapsulation [5].

Ionic polymer composites [30] (IL/PVDF) composites provides an excellent provision of overcoming the challenges related to ionic liquid actuators and PVDF actuators individually and benefit from the solid–liquid architecture. Ionic electroactive polymers are cheap to produce, light in weight and can yield large actuations at low input voltages; however, they are limited by their slow response, repeatability, and difficulty in device integration [31]. These limitations could be overcome by realising IL/polymer composites [30]. IL/PVDF actuators using PVDF-CTFE [32] or PVDF_CTFE/poly(methyl methacrylate) (PMMA) [33] with ILs such as 1-ethyl-3-methylimidazolium bis(trifluoromethylsulfonyl)imide [34], [Emim][TFSI]) or 1-ethyl-3-methylimidazolium tetrafluoroborate [Emim][BF4]) [35], 1-ethyl-3-methylimidazolium trifluoromethanesulfonate [Emim][TfO]) [33] has demonstrated remarkable actuation responses upto 0.53% [36]. In these IL/PVDF devices, the performance depends on anion/cation sizes, device thickness, and IL content. Another type of PVDF-based laminated actuators (Fig. 5a) was demonstrated which, in addition to shape shifting, they can act as flexible charge storage devices [37]. For this, a sandwich design consisting of carbide-derived carbon (CDC), 1-ethyl-3-methylimidazolium tetrafluoroborate, polyvinylidenefluoride-*co*-hexafluoropropylene (PVdF(HFP)) are utilised. Such laminated actuators can

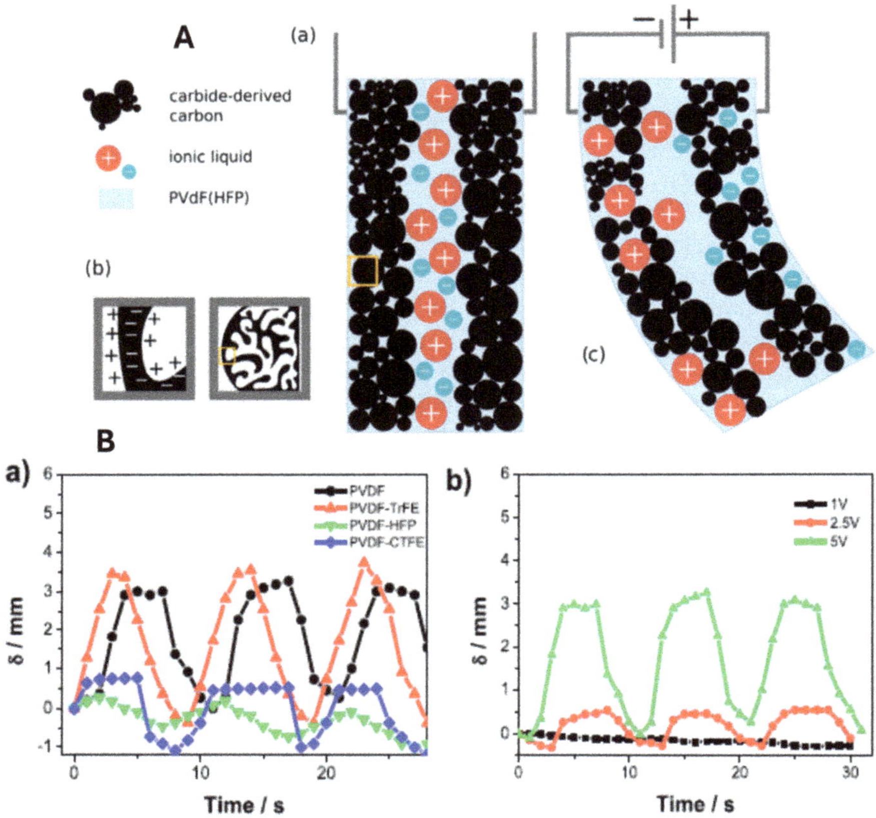

Fig. 5 (**A**) Architecture of CDC based actuator (Reproduced with Permission) [37]. (**B**) Actuation performance of various PVDF copolymers (Reproduced with Permission) [30]

operate at low voltages in the frequency range of 5×10^{-3} Hz and 10 Hz and with a maximum strain change of 0.6% [37].

The performance of various PVDF co-polymers such as poly(vinylidene fluoride) (PVDF), poly(vinylidene fluoride-*co*-trifluoroethylene) (PVDF-TrFE), poly(-vinylidene fluoride-*co*-hexafluropropylene) (PVDF-HFP) and poly(vinylidene fluoride-cochlorotrifluoroethylene) was investigated by Dias et al. [30]. Among these candidates, IL/PVDF-TrFE provided maximum bending actuation response of 3.5 mm at 100 mHz with applied potential difference of 5.0 Vpp. Figure 5b represents the comparative actuation response of PVDFTrFE as compared to other PVDF copolymers. Here, IL/PVDF and IL/PVDF-TrFE does not provide bi-directional actuation while IL/PVDF-HFP and IL/PVDF-CTFE composites is able to actuate in positive and negative directions when the polarity is reversed [30]. These typical behaviours are tunable and determined by the nature and composition of the polymer composites and the driving signals. Hence, while designing an actuator with desired performance, the choice of polymer components as well as applied signal parameters (voltage, frequency) is crucial [30].

2.5.5 Organic–Inorganic Hybrid Ferroelectrics and Nanocomposites

Polymer composites [30] with suitable nanoparticles were found to induce magnetic/magnetoelectric [38] response or enhance electric, dielectric [39] or piezoelectric response [40]. Conductive solid state nano fillers enhance polar phase as well as promote charge mobility within the PVDF matrix. 2D materials with large surface to volume ratio such as graphene [41], graphene oxide [42], and Mxene [43], or high aspect ratio CNTs [44] provide enhanced electrical, mechanical and thermal properties to the PVDF composite as compared to virgin PVDF [5]. Another important elemental additive is Fe which is found to enhance the ferroelectric properties in PVDF. Fe has been incorporated with along with graphene.

Another successful approach to enhance the ferroelectric properties in PVDF is to incorporate inorganic ferroelectric materials into the PVDF matrix. These inorganic components are added in optimised small quantities so that they facilitate the crystallinity and polarisation in the PVDF matrix without degrading the mechanical flexibility which is crucial for soft actuators. Some of the common filler ferroelectric materials are barium titanate [45]and zinc oxide [46]. One of the important criteria to achieve desired properties in PVDF by incorporating inorganic fillers is that these doped entities have to be uniformly dispersed in the PVDF matrix [5]. This could be achieved using nanostructures such as nanowires [47] rather than nanoparticles.

2.5.6 Structural and Interface Design Engineering and Device Integration

While enhancing the polarisation and crystallinity in PVDF by materials engineering is a potential way to obtain high-performing PVDF for actuator applications, there are other design approaches to achieve desired features in these soft robotic devices. For instance, multilayer stacking of fabricated PVDF thin films instead of a single layer can provide desired mechanical responses in PVDF actuators [48]. It is also possible to optimise the PVDF fabrication process to deliberately incorporate porosity in PVDF films and structures [49]. These engineered pores [50, 51] are advantageous to realise deformable structures, surfaces or components for deformable soft robotic devices such as grippers. Similar behaviour and strain/tactile sensitivity could be achieved with surface micropatterning [52, 53] of PVDF films and utilizing these films as building blocks for soft robots or integrating on deformable surfaces. Inorder to integrate PVDF actuators and soft robots with electrical circuits, stretchable electrodes are required to be fabricated on PVDF film surfaces. As traditional copper wires or PCBs are not mechanically compatible with highly flexible PVDF actuators, functional electrodes are made on PVDF films with conductive carbon based (CNT, graphene) nanomaterials [7, 54], silver nanowires [55] or PEDOT:PSS [56]. As far as actuator or soft robotic applications are concerned, silicone elastomers are also utilized as encapsulation [57] on PVDF devices to enhance mechanical properties as well as to protect the active layer from environmental impact.

3 Soft Robotic Applications in Biomedical Engineering

PVDF is biocompatible and can be easily processed into films, membranes, spheres, or fibres [30]. These advantages along with the electromechanical capability, make it a potential choice for actuators and soft robotics devices for biomedical engineering applications. The biocompatibility of piezoelectric PVDF is highly relevant [1] as despite PZT being a high-performance piezoelectric material, the present of toxic Pb makes it unsuitable for biomedical applications inside the body. Moreover, even if they are biocompatible, the rigid inorganic or ceramic ferroelectric materials are not suitable as such for soft robotic applications which require high degree of mechanical flexibility. While PVDF has been found technologically useful for applications and devices (Fig. 6) such as mechanical sensors (strain, pressure), energy harvesting, and soft robotic actuation, this chapter will primarily focus on soft robotic actuation.

The realisation of PVDF actuators and soft robots often require the fabrication of uniform films, scaffolds or 3-D structures of PVDF depending upon the device architecture and desired application. Various fabrication techniques [58] have been

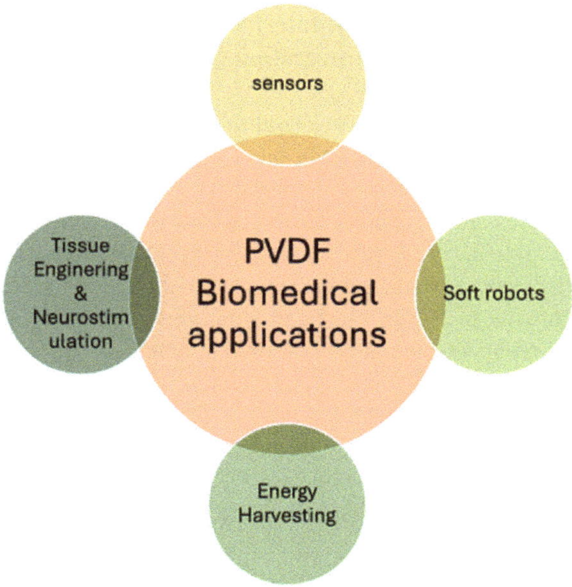

Fig. 6 Overview of biomedical applications of PVDF

explored such as solution casting, electrospinning, electrohydrodynamic pulling, melt electrowriting and 3D printing.

The electroactive characteristics of PVDF are often utilised for intelligent manipulation and tweezing. The piezoelectric and pyroelectric response of PVDF is used to modulate the electric/electrostatic field in tiny regions and the corresponding electric force to potentially manipulate tiny objects [59, 60]. The biocompatible PVDF or PVDF composite films could realise bioactive interactive surfaces and external stimulus fields such as light could be used to generate dielectrophoretic activity on the surface which drives tiny droplets or object in a controllable manner [61]. This is potential for biomedical applications such as biosensing, cell transport, microfluidics and drug delivery [61]. The light induced slippery surfaces generated using pyroelectric [29] or photopyroelectric [62] response extends the capacity to manipulate wide range of materials including biomaterials and adaptable to different physical states and morphologies. The use of liquid metal in the PVDF matrix generates heat on exposure to controlled and localised light which by pyroelectric response facilitates intelligent manipulation.

The electroactive behaviour along with mechanical flexibility, makes PVDF an excellent choice for soft robots for biomedical applications [63]. Apart from being electroactive, the ability of PVDF to respond to environmental stimulus such as temperature, humidity, solvents, enables them to actuate individually or collectively to one or more stimulus variations. PVDF could be also utilized in softrobots with other actuation mechanism for instance magnetic [64]. The selective chemical response of PVDF to acetone was utilized to realise vapomechanical response in a PVDF/PVA bilayer actuator [65]. Insect scale fast moving soft robot 3 cm–by–1.5 cm with a unimorph structure with PVDF layer on PET was fabricated and demonstrated to achieve high speed locomotion of the order of 20 body lengths/s replicating natural organisms [66]. Similarly, graphene-PVDF bimorph actuator designed to realise composite films by solution process were able to generate high actuation stress of the order 312.7 Mpa/g when utilized for fish-like robot. Since graphene is also biocompatible, this combination could be utilized for biomedical applications. The piezoelectric property of PVDF coupled with magnetostrictive property of FeGa in a core shell nanowire architecture was employed for efficient drug delivery [67]. Here, the alternating magnetic field induce actuation of FeGa nanowire generating an electric stimulation in the PVDF shell which then efficiently releases the coated drugs to treat cancer. The PVDF nanotubes were fabricated using a melt-wet technique which is a potential approach to fabricate high aspect ratio PVDF structures [67]. The core-shell design of magnetic-piezoelectric soft robot improvised using a nickel based head-PVDT-TrFE tail to realise a swimming soft robot could be highly potential for targeted drug delivery or in vivo diagnosis and treatment. The advantage of the magnetic effect coupled with piezo electric effect is that magnetic field could be applied remotely and untethered to actuate a magnetically responsive soft structure invivo, whereas this is challenging if using any optical effect coupled with piezoelectric effect.

3.1 Sensors

Soft robotic application for biomedical engineering also includes vital sensing and monitoring of bodily changes [68]. This includes temperature, strain and bodily movements. Owing to the electroactive properties of PVDF, they can provide quantitative and repeatable electrical responses to bodily changes [69]. Apart from these, for intelligent manipulation and control of soft robots and wearables, strain and pressure sensors are also very important. PVDF have been widely explored to design reliable sensors for wearables and soft robots [70]. PVDF-based wearable strain sensor have provided efficient monitoring of heart-beat rate, pulse rate, respiratory rate, and joint or muscle movements [71]. Considering the excellent biocompatibility of PVDF, these sensors could be used either as a wearable or implantable sensor inside the body. This provide the advantage of more reliable and accurate measurement of vital organs and functions [72]. The excellent flexibility of PVDF device allows these implantable devices to be folded and incorporated within ingestible capsules [72]. Among various techniques employed for developing strain sensors using PVDF, electrospinning has been highly efficient in developing highly sensitive sensors as the long-aligned fiber network of the PVDF mesh provides large response to mechanical deformations [73] as compared to solution casted PVDF thin films. Electrospun PVDF-dopamine core-shell nanofibers were demonstrated to be potential implantable sensor for vital biomonitoring [74].

3.2 Energy Harvesting

For continuous body monitoring as wearable or implantable devices, it requires either batteries or self-powering to keep maintenance free operation. These energy harvesters can remain inside the body over a long period of time and assist biomedical interventions such as controlled drug release, wound healing, or cancer therapy without the need for external energy inputs. Either bodily mechanical movements or temperature variations could be exploited with the piezoelectric and pyroelectric properties of PVDF for energy generation. The mechanical stimulus required for the energy generation requires to be synchronous and rhythmic, which is often associated with heart beats and chest expansions during breathing, or walking [26]. Various implantable devices [75–77] such as bioelectronic patches and cardiovascular sensors, have been powered by designed PVDF energy generators which are integrated with the functional devices. Recently Zhang et al. demonstrated a PVDF implantable energy harvester utilizing aorta movement and provide an output power density of 170 nW/cm^3 and charge a 1 μF capacitor to 1.0 V within 40 s which would be sufficient to operate a wide range of implantable devices [78].

3.3 Tissue Engineering and Neural Stimulation

Human cells and tissues often get damages or destroyed owing to various health conditions or accidents. Technological and biomedical advances now allows to repair or grow new tissues to restore the condition to normal using tissue engineering [79]. One of the most promising approach for tissue engineering is using controlled electrical stimulation to facilitate tissue and cell regeneration [80]. While the electrical excitation could be carried out using external energy supply or by external alternating magnetic field [81], it is highly desirable if the energy required could be met by harvesting energy from the body itself [53, 82]. PVDF is found to be suitable for different categories of tissue and biomedical engineering such as cell cycle regulation, migration, proliferation, differentiation, neural communication, cardiomyocyte contraction, and osteogenic differentiation [83].

Piezoelectric neural stimulation [84] has been identified as futurist technology towards treatment of neurological diseases and conditions such as Parkinson's disease and epilepsy. The advantage of PVDF based piezoelectric transducers for neuronal stimulation is that they are biocompatible and can be stimulated remotely or self-powered [85]. Ultrasound-mediated stimulation [86] of PVDF activated calcium channels to facilitate efficient neuronal regeneration. Instead of an ultrasound, magnetic field [87] also has been shown to be potential to remotely activate PVDF/ferromagnetic composite scaffolds for promoting nerve outgrowth and up-regulate gene expression. Similarly photosensitive materials are utilized in combination with PVDF scaffolds to realise artificial photodetectors [88, 89].

Examples of soft robotic, sensing, tissue engineering and neural stimulation application of PVDF is summarised in Fig. 7.

4 Challenges in Material Processing and Device Fabrication

We have seen that PVDF have excellent potential and benefits in actuator and soft robotic applications [63]. However, there are several inherent challenges related to material properties or manufacturing that limit the utilisation of PVDF in realising such applications. The manufacturing challenges are evident either in the material processing stage or in the device fabrication and integration. The major challenges involved in materials engineering and manufacturing of PVDF for actuators and softrobotics are summarised in Fig. 8.

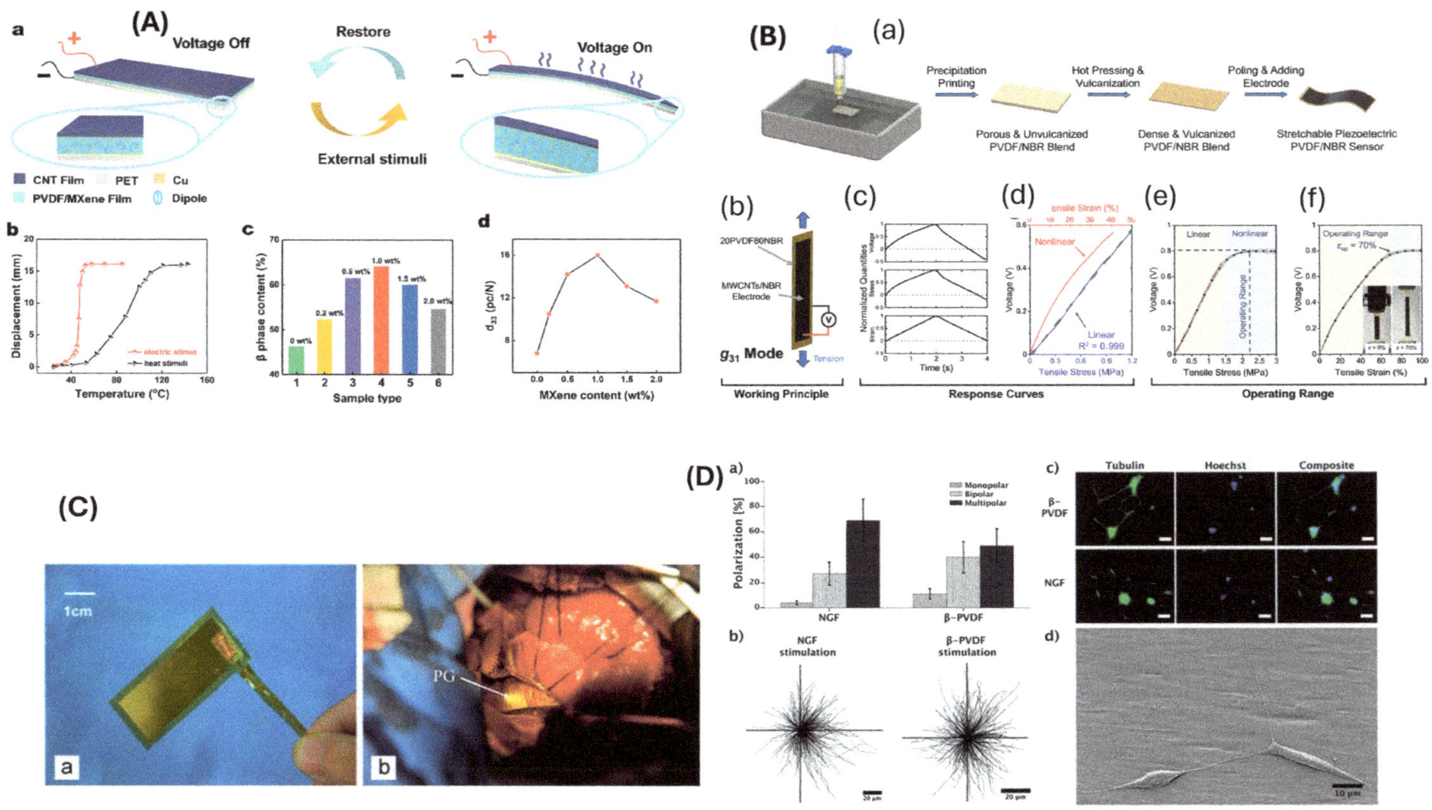

Fig. 7 Examples of (**A**) soft robotic(Reproduced with permission) [130], (**B**) sensing (Reproduced with permission) [131], (**C**) tissue engineering (Reproduced with Permission) [78] and (**D**) neural stimulation (Reproduced with permission [86] application of PVDF

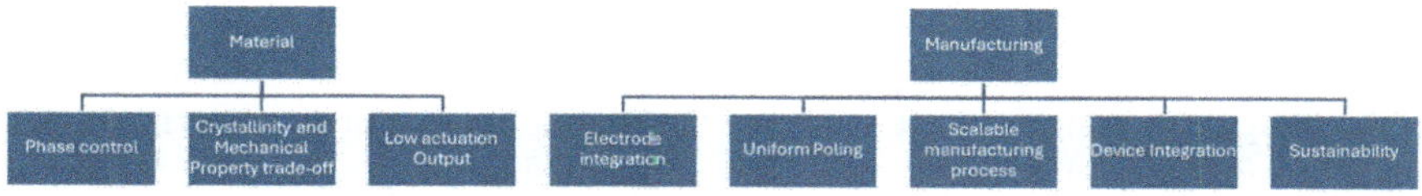

Fig. 8 Overview of challenges in PVDF materials engineering and fabrication for softrobotics

4.1 Material and Material Processing Challenges

It is known that the most stable phase of PVDF is the α-phase which is nonpolar and only β-phase with specific microstructure and crystallinity alone has the desired ferroelectric properties for actuator applications [9]. These β-phase does not originate naturally in PVDF and requires stringent process such as mechanical force, electric poling or optimised solution processes [9]. Even with this enrichment methods, achieving 100% β-phase is nearly impossible and the PVDF obtained after processing is often a mixture of more than one phase and prone to defects in the PVDF films or structures [5]. It is very difficult to maintain proper consistency in the β-phase ratio, degree of crystallinity and microstructure as this phase is not very stable. Even slight variation in the processing conditions can bring drastic variations in the final product. The electroactive characteristics of PVDF is governed by these factors and hence the variations in these adversely affect the performance of the fabricated devices.

We understand that the higher the crystallinity of PVDF, more is the desired electroactive characteristics. However, higher crystallinity makes the PVDF more stiffer and brittle [90]. For actuator and soft robotic applications which requires high degree of softness, mechanical flexibility and elasticity, there is a trade-off between electroactivity and mechanical compliance. Hence, balancing the crystallinity and mechanical properties for specific intended actuator application and desired performance is crucial and often challenging.

In order to mitigate the material challenges and limitations of PVDF, various strategies have been employed [9]. While these strategies have made significant improvements, there are associated comprises or new emerging challenges. For instance, copolymerisation [91] and blending with other monomers, forming nanocomposites with selected nanoparticles [92] or ceramic ferroelectrics [93] have been demonstrated to be effective in enhancing the ferroelectric properties. However, they often compromise the mechanical properties [94], agglomeration [95] and inhomogeneity in the PVDF matrix, or failure during use. There have been new approaches to tackle the material challenges of PVDF using liquid metals [96] or ionic liquids [28] which were effective in enhancing the electroactive performance without much compromise on the mechanical properties. However, as these liquid phase additives form interfaces with solid phase PVDF, encapsulation of the active layers is often necessary to prevent undesirable liquid loss and leaks [97]. It is also possible to enhance the mechanical properties of PVDF by blending with elastomers such as PDMS silicone elastomer [98], however, this could reduce the electrical and electroactive properties drastically. In any case, almost all the mentioned methods to enhance the material properties may add additional steps, complexity and added cost in the material processing as compared to virgin PVDF. While each of the mitigation strategies have the benefit and compromises in the material properties and suitability of actuator applications, optimising the properties according to the desired performance and application is the most important and practical approach.

Another major material characteristics and performance drawback or challenge of PVDF for soft robotics is its much lower actuation output [7] as compared to other actuation technologies such as shape memory alloys [99] which can yield high actuation forces. Supplying extremely high voltage to facilitate enhanced actuation performance would make the control systems unstable. The actuation of PVDF devices is even lower than other electroactive polymers actuators such as ionic polymer-metal composites (IPMC) [100] which makes PVDF less energy efficient. Hence, even though PVDF finds suitability in actuators and softrobots which requires limited deformations [101], they are less desirable in applications which requires large electromechanical responses for shape morphing or complex deformations. Due to the peculiar pyroelectric or thermal properties as well as possible porosity, PVDF material is highly sensitive [102] to environmental temperature variations and humidity with negatively influence the performance of the actuator devices.

4.2 Device Fabrication and Integration Challenges

Apart from material challenges, the realisation of PVDF actuators and soft robots face several challenges during the device fabrication and integration stages.

One of the major challenges in the fabrication of PVDF actuators is the difficulty in achieving robust conductive electrodes. As PVDF is an electroactive material which operates most commonly in a capacitor arrangement, compliant electrodes are necessary. In the case of conventional inorganic or ceramic ferroelectric devices for applications other than actuation, realising conductive electrodes with metal wires, conductive pastes or soldering is quite straight forward. However, as far as highly flexible polymers such as PVDF for actuation applications are concerned, these traditional electrodes will limit flexibility and delaminate under mechanical strain due to interface mismatch. To overcome these, functional materials [7] such as 2D materials (graphene) [103], CNT [44], Ag nanowires [55], PEDOT:PSS [56] have been employed which provides mechanical compliance with PVDF to some extent, however, they are less electrically conductive and have the risk of damage during repeated robotic actuation or environmental variation [104]. There are some alternative approaches to avoid the need for conductive electrodes for PVDF actuation- by combining PVDF with other untethered stimulus response materials such as magnetically [81] or optically [52] active materials to fabricate hybrid actuators and soft robots. The complementary remote actuation of these responsive materials if properly coupled stimulates the PVDF to yield piezoelectric response. However, these approaches are limited to use PVDF as sensors [105], energy harvesters [106] or electrical stimulations [81] and not to be utilized as an independently electric-driven actuator or soft robot.

Inorder to establish and sustain electroactivity in PVDF, electric polarisation [107] of PVDF device with external high voltage supply is necessary. While it is easy to pole the conventional rigid capacitive arrangement of ferroelectric materials, flexible and complex geometry devices cannot be homogenously poled with external voltage [108]. Hence, the possible dimensions, shapes and complexity of PVDF actuators are limited by the electric poling provisions [109] or poling assisted manufacturing [110]. There has been some promising direction in which the PVDF devices could be self-poled [111] without the need of external fields, however, there is no significant progress or performance achieved in such devices.

Manufacturing of actuators and soft robots operating in different mechanisms such as magnetic field [112] are well established and scalable. For instance, 3D [113] and 4D printing [114] have become very versatile and efficient in manufacturing wide variety of soft robots and actuators in varying shapes and sizes. However, as far as PVDF is concerned, such manufacturing opportunities are much limited due to the requirement for phase stabilisation, restricted device architecture and uniformity across large scales [7]. Fabrication methods such as solvent casting [115], electrospinning [19] and to some extend 3D printing [116] has been efficient in prototyping and manufacturing PVDF actuators and soft robots, however, large-area production is not much possible. Due to the defined device configurations, the geometry and size of manufactured PVDF devices are often limited [117]. In-plane or out-of-plane actuations are most commonly achievable while complex designs such as shape-shifting robotics and tasks such as minimal invasive surgeries are quite challenging to realise with PVDF [118].

In many of the actuator or soft robotic applications, functional PVDF layers needs to be integrated with either rigid structures such as silicon wafers [119], or elastomers [120] with drastic elastic mismatch with PVDF. Hence, during deployment, the PVDF layers can easily delaminate damaging the device configuration and functionality. There are also system level integration challenges and assembling various sensors and other components with PVDF actuators for fully embedded devices is quite difficult.

Another important challenge related to PVDF applications in actuators and soft robots is their limited end-of-line recyclability owing to their chemical resistance [121]. The production of raw PVDF is energy intense process [122] and hence longer life and recyclability is highly desirable. While there are some progress [123] in making PVDF material processing and manufacturing more sustainable, more insight and strategies in this aspect are necessary.

While there are many material and manufacturing challenges exist related to PVDF applications in softrobotics, there are several emerging avenues such as roll-to-roll printing [124], electric poling assisted additive manufacturing [110], 4-D printing [125], computational optimisations [126], self-powered [127] energy autonomy, self-healing [128] electrodes, hybrid softrobots [129], improved recyclability [123] and sustainable manufacturing that can leverage the full potential of PVDF based material technologies.

References

1. Lopes, A., Gutiérrez, J., Barandiarán, J.: Direct fabrication of a 3D-shape film of polyvinylidene fluoride (PVDF) in the piezoelectric β-phase for sensor and actuator applications. Eur. Polym. J. **99**, 111–116 (2018)
2. Harvey, J.A.: Smart materials. In: Mechanical Engineers' Handbook: Materials and Engineering Mechanics, vol. 1, p. 439. Wiley (2015)
3. Mani, S.: Surface characterization of polyvinylidene fluoride (PVDF) in its application as an actuator (2009)
4. Fukada, E.: History and recent progress in piezoelectric polymers. IEEE Trans. Ultrason. Ferroelectr. Freq. Control. **47**(6), 1277–1290 (2002)
5. Wang, F., Zhu, X., Du, X.: Intelligent poly(vinylidene fluoride)-based materials for biomedical applications. Adv. Funct. Mater. **35**, 2500685 (2025)
6. Yu, J.Y., Kang, W.Y., Kim, S.J.: Elastic tailoring of laminated composite plate by anisotropic piezoelectric polymers-theory, computation, and experiment. J. Compos. Mater. **29**(9), 1201–1221 (1995)
7. Bae, J.-H., Chang, S.-H.: PVDF-based ferroelectric polymers and dielectric elastomers for sensor and actuator applications: a review. Funct. Compos. Struct. **1**(1), 012003 (2019)
8. Gebrekrstos, A., Muzata, T.S., Ray, S.S.: Nanoparticle-enhanced β-phase formation in electroactive PVDF composites: a review of systems for applications in energy harvesting, EMI shielding, and membrane technology. ACS Appl. Nano Mater. **5**(6), 7632–7651 (2022)
9. Concha, V.O., et al.: Properties, characterization and biomedical applications of polyvinylidene fluoride (PVDF): a review. J. Mater. Sci. **59**(31), 14185–14204 (2024)
10. Sukumaran, S., et al.: Hybrid piezo-, pyro-, and triboelectric nanogenerator based on PVDF and rGO composite fibers for a multifunctional approach to energy harvesting applications. Mater. Des. **254**, 114105 (2025)
11. Mohammadi, B., Yousefi, A.A., Bellah, S.M.: Effect of tensile strain rate and elongation on crystalline structure and piezoelectric properties of PVDF thin films. Polym. Test. **26**(1), 42–50 (2007)
12. Zabek, D., et al.: Micropatterning of flexible and free standing polyvinylidene difluoride (PVDF) films for enhanced pyroelectric energy transformation. Adv. Energy Mater. **5**(8), 1401891 (2015)
13. Wan, C., Bowen, C.R.: Multiscale-structuring of polyvinylidene fluoride for energy harvesting: the impact of molecular-, micro-and macro-structure. J. Mater. Chem. A. **5**(7), 3091–3128 (2017)
14. Whatmore, R.: Pyroelectric devices and materials. Rep. Prog. Phys. **49**(12), 1335 (1986)
15. Sencadas, V., Gregorio Jr., R., Lanceros-Méndez, S.: α to β phase transformation and microestructural changes of PVDF films induced by uniaxial stretch. J. Macromol. Sci. **48**(3), 514–525 (2009)
16. Meng, N., et al.: Ultrahigh β-phase content poly(vinylidene fluoride) with relaxor-like ferroelectricity for high energy density capacitors. Nat. Commun. **10**(1), 4535 (2019)
17. Holz, S., Kiricenko, S., Rissing, L.: Integration of PVDF (polyvinylidene fluoride) in MEMS production. ECS Trans. **64**(1), 189 (2014)
18. Sukumaran, S., et al.: Recent advances in flexible PVDF based piezoelectric polymer devices for energy harvesting applications. J. Intell. Mater. Syst. Struct. **32**(7), 746–780 (2021)

19. Purushothaman, S.M., et al.: A review on electrospun PVDF-based nanocomposites: recent trends and developments in energy harvesting and sensing applications. Polymer. **283**, 126179 (2023)
20. Abdolmaleki, H., et al.: Interfacial engineering of PVDF-TrFE toward higher piezoelectric, ferroelectric, and dielectric performance for sensing and energy harvesting applications. Adv. Sci. **10**(6), 2205942 (2023)
21. Martins, P., Lopes, A., Lanceros-Mendez, S.: Electroactive phases of poly(vinylidene fluoride): determination, processing and applications. Prog. Polym. Sci. **39**(4), 683–706 (2014)
22. Stadlober, B., Zirkl, M., Irimia-Vladu, M.: Route towards sustainable smart sensors: ferroelectric polyvinylidene fluoride-based materials and their integration in flexible electronics. Chem. Soc. Rev. **48**(6), 1787–1825 (2019)
23. Ameduri, B.: From vinylidene fluoride (VDF) to the applications of VDF-containing polymers and copolymers: recent developments and future trends. Chem. Rev. **109**(12), 6632–6686 (2009)
24. Li, Z., Wang, Y., Cheng, Z.-Y.: Electromechanical properties of poly(vinylidene-fluoride-chlorotrifluoroethylene) copolymer. Appl. Phys. Lett. **88**(6), 062904 (2006)
25. Chen, X., et al.: Relaxor ferroelectric polymer exhibits ultrahigh electromechanical coupling at low electric field. Science. **375**(6587), 1418–1422 (2022)
26. Panda, S., et al.: Piezoelectric energy harvesting systems for biomedical applications. Nano Energy. **100**, 107514 (2022)
27. Ghosh, S.K., et al.: Yb^{3+} assisted self-polarized PVDF based ferroelectretic nanogenerator: a facile strategy of highly efficient mechanical energy harvester fabrication. Nano Energy. **30**, 621–629 (2016)
28. Liu, X., et al.: Ionic liquid-assisted 3D printing of self-polarized β-PVDF for flexible piezoelectric energy harvesting. ACS Appl. Mater. Interfaces. **13**(12), 14334–14341 (2021)
29. Wang, F., et al.: Light-induced charged slippery surfaces. Sci. Adv. **8**(27), eabp9369 (2022)
30. Dias, J.C., et al.: Improved response of ionic liquid-based bending actuators by tailored interaction with the polar fluorinated polymer matrix. Electrochim. Acta. **296**, 598–607 (2019)
31. Asaka, K.: Soft actuators. In: Stretchable Electronics, pp. 305–324 (2012)
32. Zhu, Y., et al.: Polar phase formation and competition in the melt crystallization of poly(vinylidene fluoride) containing an ionic liquid. Mater. Chem. Phys. **144**(1–2), 194–198 (2014)
33. Liu, Y., et al.: Enhanced electromechanical response of ionic polymer actuators by improving mechanical coupling between ions and polymer matrix. Macromolecules. **45**(12), 5128–5133 (2012)
34. Asaka, K., et al.: Ionic electroactive polymer actuators based on nano-carbon electrodes. Polym. Int. **62**(9), 1263–1270 (2013)
35. Terasawa, N., et al.: Effect of hexafluoropropylene on the performance of poly(vinylidene fluoride) polymer actuators based on single-walled carbon nanotube–ionic liquid gel. Sens. Actuators B Chem. **160**(1), 161–167 (2011)
36. Mejri, R., et al.: Effect of anion type in the performance of ionic liquid/poly(vinylidene fluoride) electromechanical actuators. J. Non-Cryst. Solids. **453**, 8–15 (2016)
37. Torop, J., et al.: Flexible supercapacitor-like actuator with carbide-derived carbon electrodes. Carbon. **49**(9), 3113–3119 (2011)
38. Martins, P., et al.: Dielectric and magnetic properties of ferrite/poly(vinylidene fluoride) nanocomposites. Mater. Chem. Phys. **131**(3), 698–705 (2012)
39. Lopes, A., et al.: Dielectric relaxation, AC conductivity and electric modulus in poly(vinylidene fluoride)/NaY zeolite composites. Solid State Ionics. **235**, 42–50 (2013)
40. Firmino Mendes, S., et al.: Effect of the ceramic grain size and concentration on the dynamical mechanical and dielectric behavior of poly(vinylidene fluoride)/Pb $(Zr_{0.53}Ti_{0.47})O_3$ composites. Appl. Phys. A. **96**(4), 899–908 (2009)
41. Song, S., et al.: Boosting piezoelectric performance with a new selective laser sintering 3D printable PVDF/graphene nanocomposite. Compos. A Appl. Sci. Manuf. **147**, 106452 (2021)

42. An, N., et al.: Preparation and properties of β-phase graphene oxide/PVDF composite films. J. Appl. Polym. Sci. **132**(10), 41577 (2015)
43. Li, W., et al.: Multilayer-structured transparent Mxene/PVDF film with excellent dielectric and energy storage performance. J. Mater. Chem. C. **7**(33), 10371–10378 (2019)
44. Mishra, S., et al.: Synthesis of PVDF/CNT and their functionalized composites for studying their electrical properties to analyze their applicability in actuation and sensing. Colloids Surf. A Physicochem. Eng. Asp. **509**, 684–696 (2016)
45. Su, Y.P., et al.: Incorporation of barium titanate nanoparticles in piezoelectric PVDF membrane. J. Membr. Sci. **640**, 119861 (2021)
46. Tan, K., et al.: Pyroelectricity enhancement of PVDF nanocomposite thin films doped with ZnO nanoparticles. Smart Mater. Struct. **23**(12), 125006 (2014)
47. Wang, G., et al.: Fabrication of radial ZnO nanowire clusters and radial ZnO/PVDF composites with enhanced dielectric properties. Adv. Funct. Mater. **18**(17), 2584–2592 (2008)
48. Toinet, S., et al.: Design optimization of printed multi-layered electroactive actuators used for steerable guidewire in micro-invasive surgery. Materials. **17**(9), 2135 (2024)
49. Li, X., Zhao, X., Liu, Y.: Research progress of piezoelectric polymer PVDF in sound absorption technology. Text. Res. J. **94**(13–14), 1613–1635 (2024)
50. Djeljadini, S., et al.: Porous PVDF monoliths with templated geometry. Adv. Mater. Technol. **6**(11), 2100325 (2021)
51. Wang, S., et al.: All-fiber volatile organics-perceptive actuators. ACS Nano. **19**(9), 8799–8811 (2025)
52. Won, D., et al.: Digital selective reversible phase control of monolithically integrated heterogeneous piezoelectric polymer for frequency dependent unimorph. Adv. Opt. Mater. **10**(24), 2201206 (2022)
53. Wang, P., et al.: Programmable micropatterned surface for single-layer homogeneous-polymer Janus actuator. Chem. Eng. J. **430**, 133052 (2022)
54. Ünsal, Ö.F., Altın, Y., Çelik Bedeloğlu, A.: Flexible electrospun PVDF piezoelectric nanogenerators with electrospray-deposited graphene electrodes. J. Electron. Mater. **52**(3), 2053–2061 (2023)
55. Zhang, L., et al.: Efficient welding of silver nanowires embedded in a poly(vinylidene fluoride) film for robust wearable electronics. Adv. Mater. Technol. **4**(2), 1800438 (2019)
56. Schmidt, V.H., et al.: Piezoelectric actuators employing PVDF coated with flexible PEDOT-PSS polymer electrodes. IEEE Trans. Dielectr. Electr. Insul. **13**(5), 1140–1148 (2006)
57. Xiao, B., et al.: Design and analysis of a flexible hydrophone based on PVDF with aluminum film substrate and PDMS encapsulation. Sens. Actuators A Phys. **364**, 114822 (2023)
58. Mokhtari, F., et al.: Recent advances of polymer-based piezoelectric composites for biomedical applications. J. Mech. Behav. Biomed. Mater. **122**, 104669 (2021)
59. Jin, Y., et al.: Electrostatic tweezer for droplet manipulation. Proc. Natl. Acad. Sci. **119**(2), e2105459119 (2022)
60. Liu, C., Wang, F., Du, X.: Self-powered electrostatic tweezer for adaptive object manipulation. Device. **2**(10), 100465 (2024)
61. Wang, F., et al.: Light control of droplets on photo-induced charged surfaces. Natl. Sci. Rev. **10**(1), nwac164 (2023)
62. Wang, F., et al.: Photopyroelectric tweezers for versatile manipulation. Innovation. **6**(1), 100742 (2025)
63. Kuznetsov, N., et al.: Electroresponsive materials for soft robotics. Nanobiotechnol. Rep. **18**(2), 189–206 (2023)
64. Lu, H., et al.: Battery-less soft millirobot that can move, sense, and communicate remotely by coupling the magnetic and piezoelectric effects. Adv. Sci. **7**(13), 2000069 (2020)
65. Zhang, L., et al.: Vapomechanically responsive motion of microchannel-programmed actuators. Adv. Mater. **29**(37), 1702231 (2017)
66. Wu, Y., et al.: Insect-scale fast moving and ultrarobust soft robot. Sci. Robot. **4**(32), eaax1594 (2019)

67. Chen, X.-Z., et al.: Hybrid magnetoelectric nanowires for nanorobotic applications: fabrication, magnetoelectric coupling, and magnetically assisted in vitro targeted drug delivery. Adv. Mater. **29**(8), 1605458 (2017)
68. Zhu, M., et al.: Soft, wearable robotics and haptics: technologies, trends, and emerging applications. Proc. IEEE. **110**(2), 246–272 (2022)
69. Xin, Y., et al.: Wearable and unconstrained systems based on PVDF sensors in physiological signals monitoring: a brief review. Ferroelectrics. **500**(1), 291–300 (2016)
70. Liu, W., et al.: PVDF-based biomimetic sensor for application in crawling soft-body mini-robots. In: 2006 IEEE/RSJ International Conference on Intelligent Robots and Systems. IEEE (2006)
71. Panicker, S.S., Rajeev, S.P., Thomas, V.: Impact of PVDF and its copolymer-based nanocomposites for flexible and wearable energy harvesters. Nano-Struct. Nano-Objects. **34**, 100949 (2023)
72. Cheng, X., et al.: Implantable and self-powered blood pressure monitoring based on a piezoelectric thinfilm: simulated, in vitro and in vivo studies. Nano Energy. **22**, 453–460 (2016)
73. Sharma, T., et al.: Aligned PVDF-TrFE nanofibers with high-density PVDF nanofibers and PVDF core–shell structures for endovascular pressure sensing. IEEE Trans. Biomed. Eng. **62**(1), 188–195 (2014)
74. Li, T., et al.: High-performance poly(vinylidene difluoride)/dopamine core/shell piezoelectric nanofiber and its application for biomedical sensors. Adv. Mater. **33**(3), 2006093 (2021)
75. Wang, A., et al.: Piezoelectric nanofibrous scaffolds as in vivo energy harvesters for modifying fibroblast alignment and proliferation in wound healing. Nano Energy. **43**, 63–71 (2018)
76. Han, M., et al.: Three-dimensional piezoelectric polymer microsystems for vibrational energy harvesting, robotic interfaces and biomedical implants. Nat. Electron. **2**(1), 26–35 (2019)
77. Yu, Y., et al.: Biocompatibility and in vivo operation of implantable mesoporous PVDF-based nanogenerators. Nano Energy. **27**, 275–281 (2016)
78. Zhang, H., et al.: A flexible and implantable piezoelectric generator harvesting energy from the pulsation of ascending aorta: in vitro and in vivo studies. Nano Energy. **12**, 296–304 (2015)
79. Chapekar, M.S.: Tissue engineering: challenges and opportunities. J. Biomed. Mater. Res. **53**(6), 617–620 (2000)
80. Chen, C., et al.: Electrical stimulation as a novel tool for regulating cell behavior in tissue engineering. Biomater. Res. **23**(1), 25 (2019)
81. Wang, L., et al.: AC magnetic field-driven wireless charging dual-oriented fibrous magnetoelectric scaffold CFO/PVDF promotes peripheral nerve repair. Colloids Surf. A Physicochem. Eng. Asp. **701**, 134822 (2024)
82. Yue, O., et al.: Biomimetic exogenous "tissue batteries" as artificial power sources for implantable bioelectronic devices manufacturing. Adv. Sci. **11**(11), 2307369 (2024)
83. Sui, B., et al.: Piezoelectric stimulation enhances bone regeneration in alveolar bone defects through metabolic reprogramming of macrophages. Exploration. **4**, 20230149 (2024)
84. Marino, A., et al.: Piezoelectric nanotransducers: the future of neural stimulation. Nano Today. **14**, 9–12 (2017)
85. Cheng, Y., et al.: 3D structured self-powered PVDF/PCL scaffolds for peripheral nerve regeneration. Nano Energy. **69**, 104411 (2020)
86. Hoop, M., et al.: Ultrasound-mediated piezoelectric differentiation of neuron-like PC12 cells on PVDF membranes. Sci. Rep. **7**(1), 4028 (2017)
87. Wu, L., et al.: In situ non-invasive electrical stimulation by magnetically-actuated piezoelectric Fe_3O_4/PVDF alignment scaffolds for synergistically promoting long-distance peripheral nerve regeneration. Chem. Eng. J. **509**, 161308 (2025)
88. Zhu, Y., et al.: Wireless and opto-stimulated flexible implants: artificial retina constructed by ferroelectric $BiFeO_3$-$BaTiO_3$/P(VDF-TrFE) composites. ACS Appl. Mater. Interfaces. **16**(36), 48395–48405 (2024)
89. Peng, B., et al.: A neuron-readable artificial photoreceptor composed of photodeformable liquid crystal polymers and piezoelectric materials. Adv. Funct. Mater. **33**(23), 2214172 (2023)

90. Challier, M., et al.: Damage and fracture of polyvinylidene fluoride (PVDF) at 20 °C: experiments and modelling. Eng. Fract. Mech. **73**(1), 79–90 (2006)
91. Zhang, X., et al.: PVDF-based and its copolymer-based piezoelectric composites: preparation methods and applications. J. Electron. Mater. **51**(10), 5528–5549 (2022)
92. Jaleh, B., Jabbari, A.: Evaluation of reduced graphene oxide/ZnO effect on properties of PVDF nanocomposite films. Appl. Surf. Sci. **320**, 339–347 (2014)
93. Jain, A., et al.: Dielectric and piezoelectric properties of PVDF/PZT composites: a review. Polym. Eng. Sci. **55**(7), 1589–1616 (2015)
94. Vicente, J., et al.: Electromechanical properties of PVDF-based polymers reinforced with nanocarbonaceous fillers for pressure sensing applications. Materials. **12**(21), 3545 (2019)
95. Ismail, A., Ramadan, R., El-Masry, M.M.: The role of nanoparticles inclusion in monitoring the physical properties of PVDF. J. Aust. Ceram. Soc. **59**(2), 333–341 (2023)
96. Cao, C., et al.: Ultrastretchable conductive liquid metal composites enabled by adaptive interfacial polarization. Mater. Horiz. **8**(12), 3399–3408 (2021)
97. Guo, D., et al.: Hydrophilic poly(vinylidene fluoride) film with enhanced inner channels for both water-and ionic liquid-driven ion-exchange polymer metal composite actuators. ACS Appl. Mater. Interfaces. **11**(2), 2386–2397 (2019)
98. He, D., et al.: Significantly enhanced electromechanical performance of PDMS crosslinked PVDF hybrids. Polymers. **10**(7), 714 (2018)
99. Dana, A., Vollach, S., Shilo, D.: Use the force: review of high-rate actuation of shape memory alloys. Actuators. **10**(7), 140 (2021)
100. Bhandari, B., Lee, G.-Y., Ahn, S.-H.: A review on IPMC material as actuators and sensors: fabrications, characteristics and applications. Int. J. Precis. Eng. Manuf. **13**(1), 141–163 (2012)
101. Xiong, J., Chen, J., Lee, P.S.: Functional fibers and fabrics for soft robotics, wearables, and human–robot interface. Adv. Mater. **33**(19), 2002640 (2021)
102. Li, M., et al.: Controlling the microstructure of poly(vinylidene-fluoride) (PVDF) thin films for microelectronics. J. Mater. Chem. C. **1**(46), 7695–7702 (2013)
103. Lee, J.S., et al.: Highly sensitive and multifunctional tactile sensor using free-standing ZnO/PVDF thin film with graphene electrodes for pressure and temperature monitoring. Sci. Rep. **5**(1), 7887 (2015)
104. Takahashi, K., et al.: Mechanical degradation of graphite/PVDF composite electrodes: a model-experimental study. J. Electrochem. Soc. **163**(3), A385 (2015)
105. Costa, C.M., et al.: Smart and multifunctional materials based on electroactive poly(vinylidene fluoride): recent advances and opportunities in sensors, actuators, energy, environmental, and biomedical applications. Chem. Rev. **123**(19), 11392–11487 (2023)
106. Sasmal, A., et al.: PVDF based flexible magnetoelectric composites for capacitive energy storage, hybrid mechanical energy harvesting and self-powered magnetic field detection. Polymer. **281**, 126141 (2023)
107. Mahadeva, S.K., et al.: Effect of poling time and grid voltage on phase transition and piezoelectricity of poly(vinylidene fluoride) thin films using corona poling. J. Phys. D. Appl. Phys. **46**(28), 285305 (2013)
108. Martins, P.M., et al.: Effect of poling state and morphology of piezoelectric poly(vinylidene fluoride) membranes for skeletal muscle tissue engineering. RSC Adv. **3**(39), 17938–17944 (2013)
109. Fan, J., et al.: The effects of additive manufacturing and electric poling techniques on PVDF thin films: towards 3D printed functional materials. In: Smart Materials, Adaptive Structures and Intelligent Systems. American Society of Mechanical Engineers (2020)
110. Lee, C., Tarbutton, J.A.: Electric poling-assisted additive manufacturing process for PVDF polymer-based piezoelectric device applications. Smart Mater. Struct. **23**(9), 095044 (2014)
111. Guo, J., Nie, M., Wang, Q.: Self-poling polyvinylidene fluoride-based piezoelectric energy harvester featuring highly oriented β-phase structured at multiple scales. ACS Sustain. Chem. Eng. **9**(1), 499–509 (2020)

112. Wang, H., et al.: Magnetic soft robots: design, actuation, and function. J. Alloys Compd. **922**, 166219 (2022)
113. Sachyani Keneth, E., et al.: 3D printing materials for soft robotics. Adv. Mater. **33**(19), 2003387 (2021)
114. Hann, S.Y., et al.: 4D printing soft robotics for biomedical applications. Addit. Manuf. **36**, 101567 (2020)
115. Wu, L., et al.: Recent advances in the preparation of PVDF-based piezoelectric materials. Nanotechnol. Rev. **11**(1), 1386–1407 (2022)
116. Koroglu, L., Ayas, E., Ay, N.: 3D printing of polyvinylidene fluoride based piezoelectric nanocomposites: an overview. Macromol. Mater. Eng. **306**(10), 2100277 (2021)
117. Mohammadpourfazeli, S., et al.: Future prospects and recent developments of polyvinylidene fluoride (PVDF) piezoelectric polymer; fabrication methods, structure, and electro-mechanical properties. RSC Adv. **13**(1), 370–387 (2023)
118. Chee, C.Y., Tong, L., Steven, G.P.: A review on the modelling of piezoelectric sensors and actuators incorporated in intelligent structures. J. Intell. Mater. Syst. Struct. **9**(1), 3–19 (1998)
119. Ahbab, N., et al.: A comprehensive review of piezoelectric PVDF polymer fabrications and characteristics. Micromachines. **16**(4), 386 (2025)
120. Shankar, R., Ghosh, T.K., Spontak, R.J.: Dielectric elastomers as next-generation polymeric actuators. Soft Matter. **3**(9), 1116–1129 (2007)
121. Améduri, B., Hori, H.: Recycling and the end of life assessment of fluoropolymers: recent developments, challenges and future trends. Chem. Soc. Rev. **52**(13), 4208–4247 (2023)
122. Memarian, F., et al.: A comprehensive review of nanocomposite PVDF as a piezoelectric material: evaluating manufacturing methods, energy efficiency and performance. Nanofabrication. **8**, 1775 (2023)
123. Husain, M., Singh, R., Pabla, B.: On secondary recycling of PVDF for 3D printing of functional sensors: mechanical, flow, thermal, and dielectric properties. Natl. Acad. Sci. Lett. **47**(4), 385–390 (2024)
124. Zhang, C.: Enhanced piezoelectric and electrodeformation properties of PVDF films fabricated by roll to roll process. J. Appl. Polym. Sci. **141**(45), e56213 (2024)
125. Solnyshkin, A., et al.: Temperature evolution of dielectric characteristics of PVDF and P(VDF-TrFE) films produced by 4D printing. Phys. Chem. Asp. Study Clust. Nanostruct. Nanomater. **16**, 720–728 (2024)
126. Kamath, R.C.: Role of computational material science in improving the properties of piezoelectric smart materials: a review. Eng. Proc. **59**(1), 21 (2023)
127. Yuan, X., et al.: A poling-free PVDF nanocomposite via mechanically directional stress field for self-powered pressure sensor application. Nano Energy. **98**, 107340 (2022)
128. Bonardd, S., et al.: Self-healing polymeric soft actuators. Chem. Rev. **123**(2), 736–810 (2022)
129. Correia, D., et al.: All printed soft actuators based on ionic liquid/polymer hybrid materials. Appl. Mater. Today. **22**, 100928 (2021)
130. Chao, Wang De, Gong Deyuan, Zhang Jun, Cai (2025) A skin-inspired multifunctional soft actuator based on PVDF/MXene with fast response large deformation and high stability Nano Energy 139110994-10.1016/j.nanoen.2025.110994
131. Tu, R., & Sodano, H. A. (2023). Highly stretchable printed poly (vinylidene fluoride) sensors through the formation of a continuous elastomer phase. ACS applied materials & interfaces, 15(18), 22320-22331.

PVDF in Flexible and Wearable Electronics

Harsh Ranjan, Gulshan Kumar, and Saurabh Kumar Pandey

Abstract Polyvinylidene fluoride (PVDF) has become a cornerstone in the advancement of flexible and wearable electronics due to its exceptional mechanical flexibility, chemical stability, and electroactive behavior. This chapter presents a thorough exploration of PVDF's molecular structure, processing methods, and integration strategies that underpin its use in self-powered sensors, energy harvesting devices, and biomedical applications. Special attention is given to the latest progress in fabricating PVDF-based thin films, nanofibers, and composites, as well as their deployment in health monitoring, human–machine interfaces, and smart textiles. The discussion extends to the biocompatibility and long-term durability of PVDF, emphasizing its reliability for continuous operation in real-world environments. The chapter also addresses current challenges such as enhancing stretchability, improving device integration, and advancing sustainable practices while highlighting innovative solutions and future research directions. Altogether, this work underscores PVDF's transformative impact on next-generation wearable technologies and its ongoing role in shaping the future of flexible electronic systems.

Keywords Polyvinylidene fluoride (PVDF) · Flexible electronics · Wearable sensors · Piezoelectric materials · Energy harvesting · Nanofibers · Thin-film composites · Biocompatibility · Self-powered devices · Smart textiles · Membrane technology · Surface modification · Microfabrication · Mechanical durability · Electronic skin (e-skin)

H. Ranjan (✉) · G. Kumar · S. K. Pandey
Department of Electrical Engineering, Indian Institute of Technology Patna, Bihta, Patna, Bihar, India

D. Punetha (ed.), *Polyvinylidene Fluoride (PVDF)*, Engineering Materials,
https://doi.org/10.1007/978-981-95-6718-8_8

1 Introduction

The emergence of flexible and wearable electronics has been significantly accelerated by the integration of advanced polymeric materials, with PVDF standing out due to its unique combination of mechanical flexibility, chemical stability, and outstanding electroactive properties [1]. PVDF's ability to convert mechanical, thermal, and frictional energy into electrical signals via its piezoelectric, pyroelectric, and triboelectric effects has made it a pivotal material for next-generation self-powered sensors, energy harvesters, and health monitoring devices [2, 3]. Its high electroactive β-phase content, achieved through tailored processing techniques, underpins its superior energy conversion efficiency and sensitivity [4, 5]. PVDF is a high-performance, semi-crystalline polymer characterized by strong piezoelectric and dielectric properties, allowing it to transduce mechanical stimuli into an electrical signal and vice versa [6]. This property is particularly valuable in the fabrication of lightweight, flexible, and conformable sensors, actuators, and energy-harvesting devices, key requirements for wearable and skin-mounted electronics [7]. PVDF's robust chemical resistance, low density, and excellent electrical insulation further enhance its reliability in diverse and sometimes harsh environments, including biomedical and industrial applications [5]. Recent advancements in PVDF processing have enabled the fabrication of nanofibers, thin films, and composite structures with optimized β-phase content and enhanced surface area, significantly improving device sensitivity and efficiency [8, 9]. Techniques such as electrospinning, solution casting, and 3D printing have been instrumental in achieving these improvements. The incorporation of functional nanofillers including barium titanate ($BaTiO_3$), MXene, and graphene [10] has further elevated PVDF's piezoelectric performance, supporting applications ranging from self-powered smartwatches to glucose-monitoring patches and smart textiles for real-time health tracking.

PVDF's compatibility with microfabrication and miniaturization aligns with the ongoing trend toward compact, integrated, and multifunctional wearable systems [5]. In healthcare, PVDF's biocompatibility and sterilizability make it ideal for continuous physiological monitoring and implantable devices [11]. The sports and fitness sectors leverage PVDF's sensitivity and flexibility for accurate motion tracking and performance analytics [12]. Additionally, PVDF's role in energy harvesting by converting biomechanical movements into electrical energy enables the realization of self-powered wearable systems, reducing reliance on conventional batteries and enhancing device autonomy. Despite these advantages, challenges remain. Material fatigue under cyclic stress, phase stability in harsh environments, and the need for scalable, cost-effective manufacturing must be addressed for widespread adoption. Strategies such as cross-linked copolymers, surface functionalization, and advanced encapsulation are being explored to enhance durability, safety, and long-term performance. Moreover, the integration of PVDF devices with artificial intelligence and memristive devices [13, 14], soft robotics, and wireless communication modules is paving the way for intelligent, adaptive health monitoring platforms that can deliver personalized, decentralized care [15]. Figure 1 depicts PVDF's phase

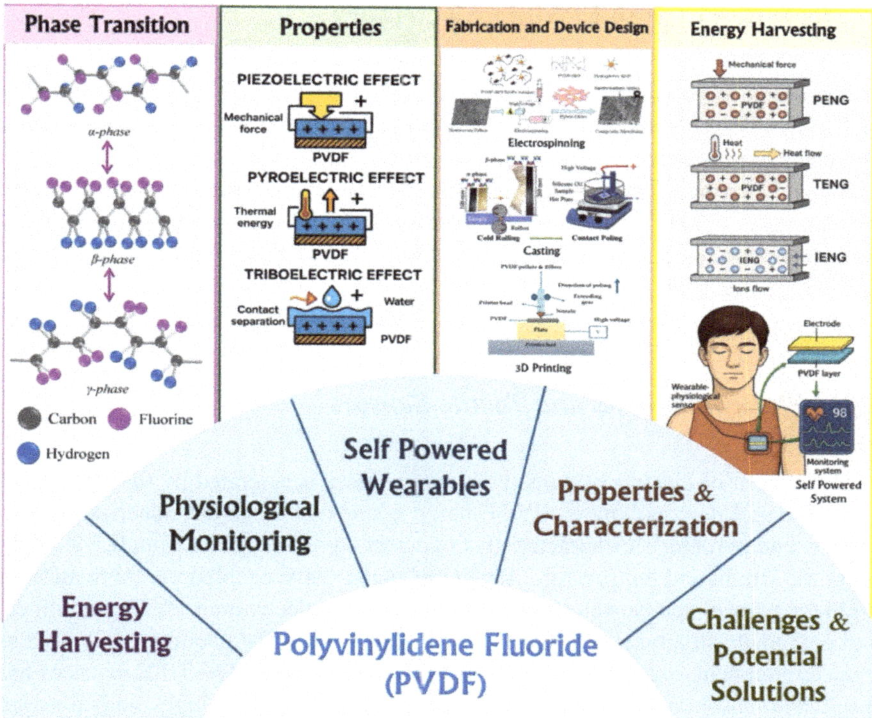

Fig. 1 Overview of PVDF's phase transitions, multifunctional properties, fabrication methods, and its applications in energy harvesting and self-powered wearable sensors [1]

transitions, key functional properties (piezoelectric, pyroelectric, triboelectric effects), advanced fabrication methods (such as electrospinning, casting, and 3D printing), and its integration into energy harvesting devices and self-powered wearable sensors for physiological monitoring. The figure also summarizes major application domains, characterization aspects, and current challenges with potential solutions in the field [1].

In summary, PVDF's adaptability, biocompatibility, and durability make it a key enabler for next-generation flexible and wearable electronics. Advances in material engineering and fabrication techniques continue to unlock new opportunities for PVDF in sensitive health monitoring, autonomous energy harvesting, and multifunctional smart textiles [16]. As research addresses remaining challenges in scalability, integration, and sustainability, PVDF is poised to remain at the forefront of innovation in the rapidly expanding field of wearable technology [17, 18].

2 Role of PVDF in Flexible Electronics

PVDF has become a foundational material in the field of flexible electronics, attributed to its unique blend of mechanical flexibility, pronounced piezoelectric response, and chemical resilience. These attributes are crucial for devices that must conform to dynamic surfaces, endure repeated deformation, and operate reliably in diverse environments. PVDF's suitability for both sensing and energy harvesting applications has positioned it as a material of choice for next-generation wearable and stretchable electronic systems [19].

2.1 Electronic Skins and Tactile Sensors

The evolution of electronic skins (e-skins) has been accelerated by the integration of PVDF and its copolymers. PVDF-based e-skins leverage the material's piezoelectric and pyroelectric characteristics to detect a wide range of stimuli, including pressure, strain, and temperature. Recent advances have enabled the fabrication of highly sensitive, self-powered tactile sensors using electrospun PVDF nanofibers and composite structures. For instance, the incorporation of nanomaterials such as reduced graphene oxide (rGO) or carbon nanotubes into PVDF-TrFE matrices has resulted in sensors with enhanced sensitivity, broader detection ranges, and improved durability under cyclic loading [20]. The fabrication process of the PEDOT@PVDF nanofiber woven fabric is schematically illustrated in Fig. 2. Such tactile sensors are capable of real-time monitoring of physiological signals, including heartbeats, respiratory patterns, and subtle muscle movements. This capability is particularly valuable for non-invasive health monitoring, prosthetic feedback, and human–machine interfaces [22]. Furthermore, the compatibility of PVDF with elastomeric substrates, such as polydimethylsiloxane (PDMS), allows for the creation of stretchable and conformal sensor arrays that can be seamlessly laminated onto skin or integrated into textile.

2.2 Self-Powered Devices and Energy Harvesters

The piezoelectric and triboelectric properties of PVDF are harnessed in the development of self-powered devices, which are increasingly important for autonomous wearable systems. Devices based on PVDF can convert mechanical energy from body motion, ambient vibrations, or even acoustic waves into usable electrical energy. Electrospun PVDF-HFP nanofibers, for example, have demonstrated high mechanical-to-electrical conversion efficiencies, especially when processed to maximize the polar β-phase content [23].

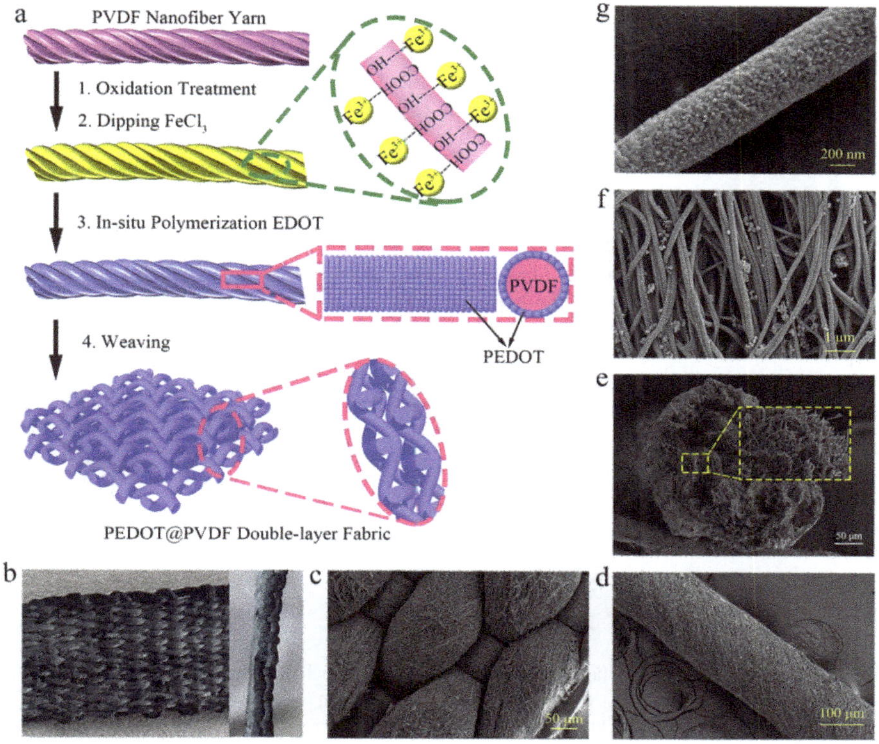

Fig. 2 (**a**) Schematic representation of the fabrication process for PEDOT@PVDF nanofiber woven fabric; (**b**) optical photograph displaying the PEDOT@PVDF nanofiber fabric; scanning electron microscopy (SEM) images showing: (**c**) the surface morphology of the PEDOT@PVDF nanofiber fabric, (**d**) the structure of a nanofiber yarn, (**e**) the alignment of oriented nanofibers, and (**f**) a single PEDOT@PVDF nanofiber, (**g**) enlarged single PEDOT@PVDF nanofiber [21]

Figure 3 illustrates the design and working principle of a multilayer hybrid piezoelectric-triboelectric nanogenerator (BPP-HNG), which integrates BTO-PVDF and PDMS layers for advanced human–machine interaction applications. Hybrid energy harvesters that combine PVDF with conductive polymers or nanomaterials, such as polyaniline or silver nanowires, further enhance charge transfer and device output. These hybrid systems have been shown to power low-energy electronics, such as LEDs and wireless sensors, directly from biomechanical movements [25]. Additionally, the development of porous PVDF nanostructures through template-assisted methods has led to significant improvements in output power density, making these materials suitable for powering next-generation implantable and wearable devices.

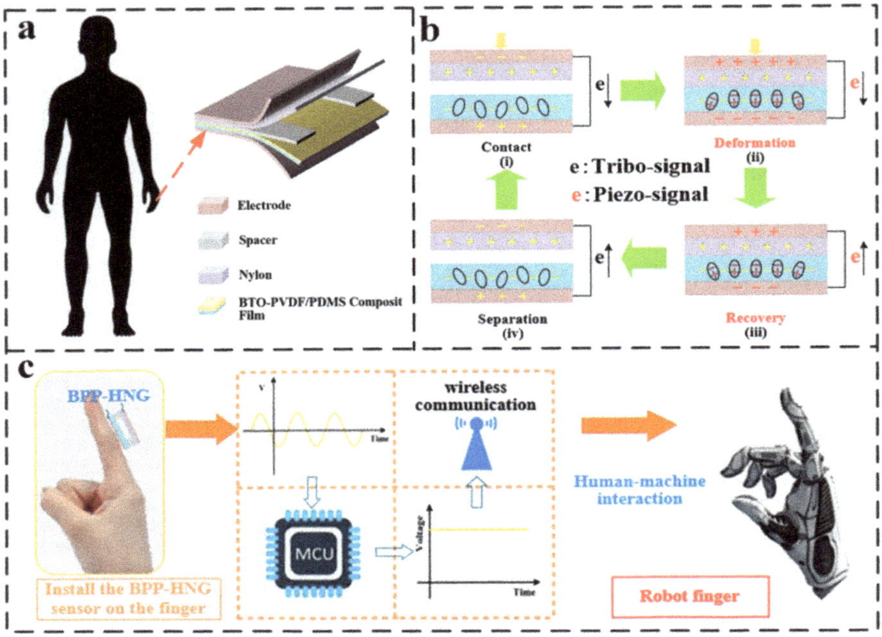

Fig. 3 Illustration of a multilayer hybrid piezoelectric-triboelectric nanogenerator (BPP-HNG) utilizing BTO-PVDF/PDMS, designed for applications in human–machine interaction. (**a**) Depiction of the BPP-HNG's layered architecture, enabling precise gesture detection; (**b**) visualization of the operational mechanism underlying the hybrid nanogenerator; (**c**) diagram demonstrating the use of BPP-HNG to wirelessly control robotic finger movements [24]

2.3 Integration with Stretchable and Textile Substrates

A critical challenge in flexible electronics is ensuring mechanical compatibility between functional materials and stretchable substrates. While PVDF films are inherently flexible, their stretchability can be limited [26]. To address this, researchers have developed composite approaches, blending PVDF with elastomers like thermoplastic polyurethane (TPU) or incorporating PVDF nanofibers into woven textile structures. These strategies yield materials with high elongation at break, robust piezoelectric response, and excellent comfort for wearable applications. Advanced fabrication techniques, such as electrospinning, 3D weaving, and inkjet printing, have enabled scalable production of PVDF-based sensors and energy harvesters [27]. For instance, 3D textile-integrated PVDF fabrics exhibit high tensile strength, breathability, and moisture management, making them suitable for continuous, long-term use in smart garments and medical monitoring systems (Table 1).

Table 1 Key features and applications of PVDF in flexible electronics

Feature	PVDF-based flexible electronics
Flexibility	High; conforms to curved and moving surfaces
Piezoelectricity	Excellent; enables self-powered sensing
Stretchability	Moderate to high (with composites)
Sensitivity	High (detects pressure, strain, temperature)
Self-power capability	Yes (via piezoelectric/triboelectric effects)
Integration methods	Electrospinning, 3D textiles, inkjet printing
Biocompatibility	Good; suitable for skin-contact and implants
Durability	High; maintains performance under cycling

2.4 PVDF-Integrated MEMS Biosensors

Microelectromechanical systems (MEMS) biosensors represent a significant advancement in biomedical and environmental sensing, offering the ability to translate biological interactions into measurable electrical signals with high precision [28]. These devices integrate mechanical components, microstructures, actuators, and electronic circuits on a single silicon substrate, enabling compact and multifunctional platforms. Despite their microscale dimensions often smaller than a human hair MEMS biosensors are highly effective for detecting physical and chemical changes in biological environments. Their development has evolved far beyond mechanical miniaturization, bringing sophisticated sensing capabilities to areas such as point-of-care diagnostics, wearable health monitors, and implantable devices. Figure 4 presents a stepwise illustration of the fabrication process for the ESTENG (Electrospun Triboelectric Nanogenerator) device. One notable development is a MEMS biosensor designed with a network of microresonators for use in digital microfluidic systems.

This sensor comprises three primary components: capacitive sensing elements, electrostatic comb-drive actuators, and a central active region. Such designs allow precise control over biological samples at the microscale level, improving both sensitivity and accuracy. With ongoing improvements in microelectronic manufacturing, modern MEMS devices can now be fabricated with extreme precision, allowing the creation of highly integrated, efficient, and cost-effective systems that can function as part of broader system-on-a-chip (SoC) architectures. The typical dimensions of MEMS biosensors ranging from tens to hundreds of microns are comparable to those used in integrated circuits, enabling their production through similar photolithography and micromachining techniques. This compatibility supports the integration of MEMS devices into existing semiconductor fabrication processes, reducing production costs and improving scalability. In addition, system-level integration enables the seamless inclusion of sensing, data processing, and communication functionalities within a single device. These capabilities are especially critical in developing portable diagnostic tools and lab-on-a-chip systems that require minimal power and provide real-time data acquisition [30].

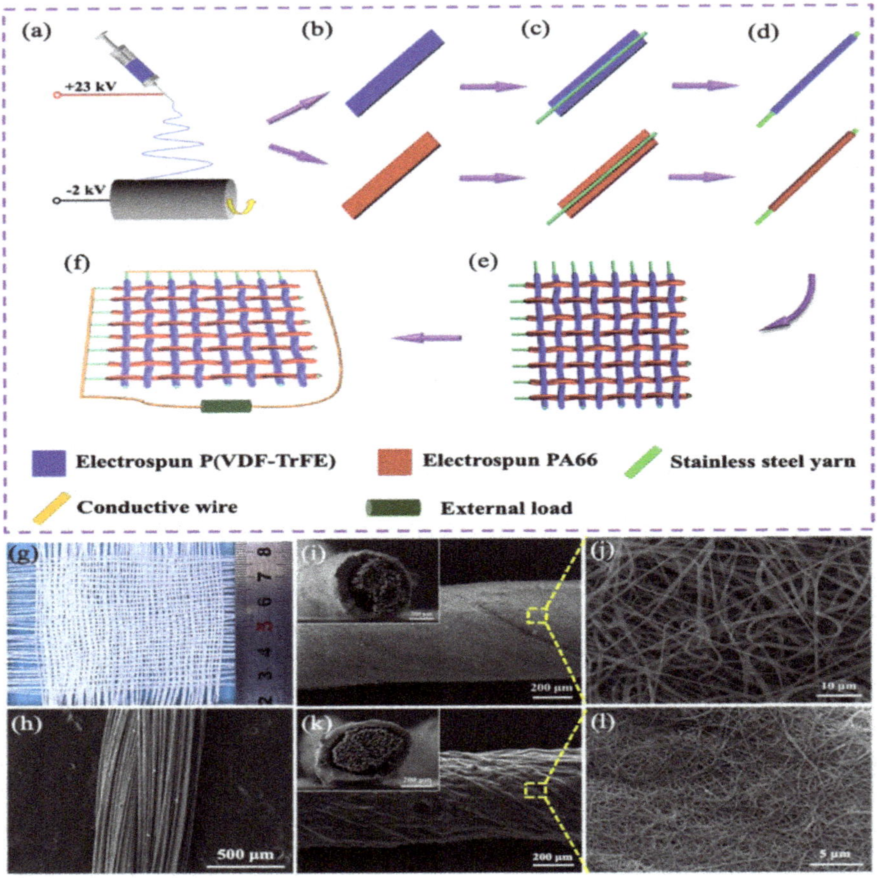

Fig. 4 (**a–f**) Stepwise illustration of the fabrication process for the ESTENG device; (**g**) photograph showing the completed ESTENG; (**h**) SEM image highlighting the structure of the SSL layer; (**i, j**) SEM images of the electrospun P(VDF-TrFE)-SSL surface at low (**i**) and high (**j**) magnification, with the inset in (**i**) displaying the cross-sectional view; (**k, l**) SEM images of the electrospun PA66-SSL surface at low (**k**) and high (**l**) magnification, with the inset in (**k**) showing the cross-section [29]

Figure 5 illustrates various types of MEMS-based biosensors, including microcantilever, magnetic, optical, microfluidic, SAW, and nanomechanical sensors, categorized by their sensing mechanisms and application domains. Various types of MEMS biosensors have been developed to serve specific biomedical functions. Cantilever-based biosensors measure the deflection caused by biomolecular interactions, offering optical, electrical, or piezoelectric detection mechanisms. Microfluidic-based biosensors incorporate miniature channels that guide biological fluids over sensor surfaces, supporting applications like DNA analysis, immunoassays, and cell sorting. Surface Acoustic Wave (SAW) biosensors detect minute changes in surface mass or mechanical properties using acoustic wave propagation

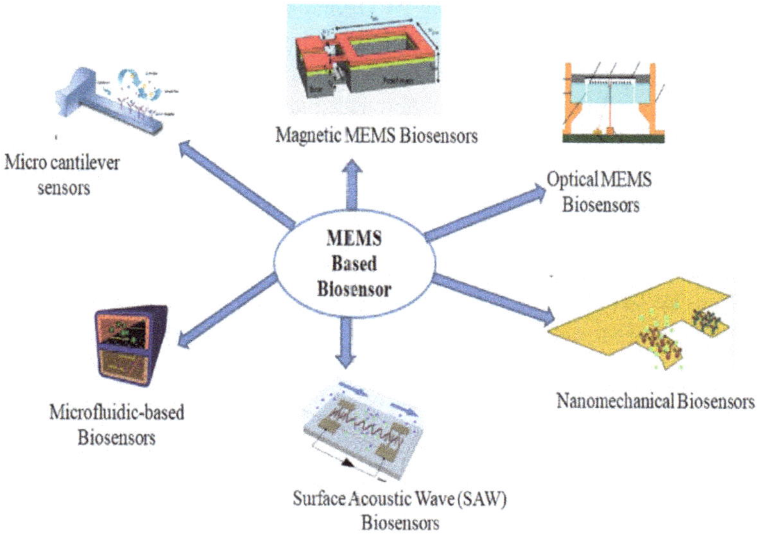

Fig. 5 Different types of MEMS-based biosensors, including microcantilever, magnetic, optical, microfluidic, SAW, and nanomechanical biosensors, categorized by their sensing mechanisms and applications [31]

across piezoelectric substrates [32]. Additionally, microarray biosensors use functionalized spots on a chip surface to enable high-throughput, parallel detection of multiple biomolecules, making them ideal for clinical and research settings.

2.5 *PVDF as a High Dielectric Constant Material in Capacitors*

PVDF has emerged as a highly versatile material in modern electronics, particularly due to its high dielectric constant, which makes it suitable for use in capacitors, transducers, and other energy storage devices. As a semi-crystalline polymer with intrinsic polarity, PVDF offers excellent dielectric behavior that is superior to many other flexible polymers. This property is essential in applications that demand efficient charge storage, such as flexible and wearable electronic systems where traditional rigid components are unsuitable [33]. Capacitors built with PVDF-based dielectric films can achieve high energy density while maintaining mechanical flexibility. PVDF's molecular structure, containing alternating $-CH_2$ and $-CF_2$ units, provides strong dipolar polarization when subjected to an electric field [34]. This results in enhanced dielectric permittivity, enabling the material to store more energy per unit volume compared to conventional plastics. Furthermore, its lightweight and flexible form factor allows integration into bendable substrates, which is critical for next-generation stretchable electronics and compact energy systems.

One of the key advantages of using PVDF in capacitors is its low dielectric loss, which ensures minimal energy dissipation during operation. This makes PVDF ideal for high-frequency and high-efficiency applications, including power conditioning circuits, embedded sensors, and flexible printed circuit boards (PCBs). Moreover, PVDF-based dielectrics maintain stable performance under dynamic mechanical stress, making them well-suited for flexible and rollable electronics where materials must endure bending, twisting, and stretching without performance degradation [35].

In addition to pure PVDF, composites incorporating ceramic fillers such as barium titanate ($BaTiO_3$) or lead zirconate titanate (PZT) are frequently used to further enhance the dielectric constant [36]. These hybrid materials combine the flexibility of the polymer matrix with the high permittivity of inorganic components, achieving superior energy storage capabilities. Such PVDF-based nanocomposites are now being actively explored in the design of flexible energy storage modules and microelectronic systems where space and weight are limited. Figure 6 provides a comprehensive comparison of dielectric capacitors and their advantages.

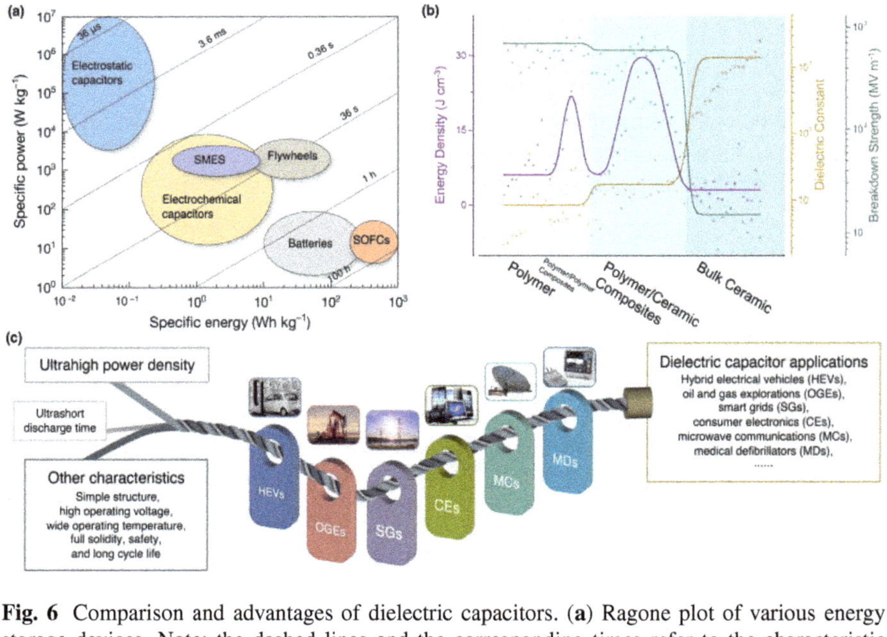

Fig. 6 Comparison and advantages of dielectric capacitors. (**a**) Ragone plot of various energy storage devices, Note: the dashed lines and the corresponding times refer to the characteristic times. (**b**) Dielectric constant, breakdown strength, and energy density of various dielectrics, (**c**) unique properties and various application fields of dielectric capacitors [37]

2.6 *PVDF in Energy Storage and Battery Technologies*

PVDF plays a crucial role in modern energy storage systems, particularly in lithium-ion batteries, supercapacitors, and dielectric capacitors. Its chemical stability, thermal resistance, and electrochemical inertness make it a preferred binder and separator material in battery technologies. PVDF not only enhances the structural integrity of electrodes but also improves cycling performance and energy density in rechargeable systems [38]. In lithium-ion batteries, PVDF is widely used as a polymeric binder in both cathode and anode formulations. Figure 7 depicts the charge/discharge mechanism of a lithium-ion battery, showing the reversible movement of lithium ions between the cathode and anode during energy storage and release. It helps maintain electrode cohesion by binding active material particles (such as $LiCoO_2$ or graphite) to conductive additives and the current collector [40]. The strong bonding capability of PVDF, combined with its resistance to common organic electrolytes, ensures long-term mechanical and electrochemical stability. This results in better cycle life and minimizes capacity fade during prolonged battery use.

Apart from its role as a binder, PVDF is also being explored in separator membranes and solid polymer electrolytes. Its excellent dielectric properties and flexibility enable the design of thin, durable membranes that separate electrodes while allowing controlled ion flow. Modified PVDF films doped with ceramic fillers or ionic liquids have shown improved ionic conductivity and thermal safety, which are critical for preventing battery short-circuits and thermal runaway in high-performance applications. PVDF also shows promise in supercapacitors and flexible energy storage devices, where its piezoelectric and ferroelectric properties support the development of high-permittivity dielectrics [41]. In energy harvesting systems, PVDF films can simultaneously act as mechanical energy converters and dielectric layers in flexible capacitors. This dual functionality makes PVDF a strategic material in wearable electronics and next-generation self-powered systems.

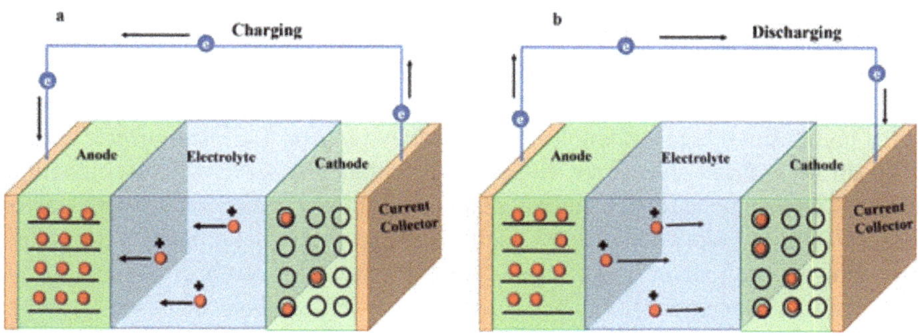

Fig. 7 Mechanism for charge/discharge of a lithium-ion battery [39]

2.7 Future Perspectives

Ongoing research aims to further enhance the performance and multifunctionality of PVDF-based flexible electronics. Key directions include the integration of PVDF with advanced nanomaterials for improved sensitivity, the development of fully biodegradable and recyclable PVDF composites [20], and the incorporation of artificial intelligence for real-time data analysis and adaptive response in wearable systems. These advancements are expected to drive the next wave of innovations in personalized healthcare, soft robotics, and smart textiles.

3 Biocompatibility and Durability Aspects

PVDF is a highly versatile piezoelectric polymer that has gained significant attention in the field of flexible and wearable electronics. Its unique combination of mechanical flexibility, chemical stability, and strong piezoelectric, pyroelectric, and triboelectric properties makes it particularly suitable for next-generation smart devices. PVDF can efficiently convert mechanical energy into electrical signals, which is a critical feature for applications such as motion sensors, pressure sensors, and energy harvesting systems. Its lightweight nature and processability into thin films, fibers, and membranes allow seamless integration into smart textiles, implantable devices, and self-powered wearable technologies. Moreover, PVDF's excellent thermal stability and resistance to harsh environmental conditions enhance its long-term performance and reliability in biomedical and industrial environments. These features make PVDF a key material in the development of multifunctional, self-powered, and flexible electronic systems [5].

3.1 Surface Compatibility and Bio-safety

PVDF is known for its excellent surface compatibility with biological tissues, making it highly suitable for wearable and implantable sensors [42]. Its intrinsic chemical stability, hydrophobic nature, and ability to be functionalized with biocompatible coatings contribute to reduced inflammatory responses when in contact with skin or internal tissues. Figure 8 highlights traditional approaches including auto transplantation (using the patient's own tissue), allotransplantation (from a human donor), and xenotransplantation (from animal sources) as well as advanced tissue engineering methods such as decellularized scaffolds, polymeric scaffolds, organ bioprinting, and the use of support systems [43]. These pathways collectively illustrate the diverse options and technological innovations available for clinical transplantation. Additionally, PVDF does not support microbial growth, providing a plaque-free interface, which is crucial for long-term biomedical applications such as epidermal

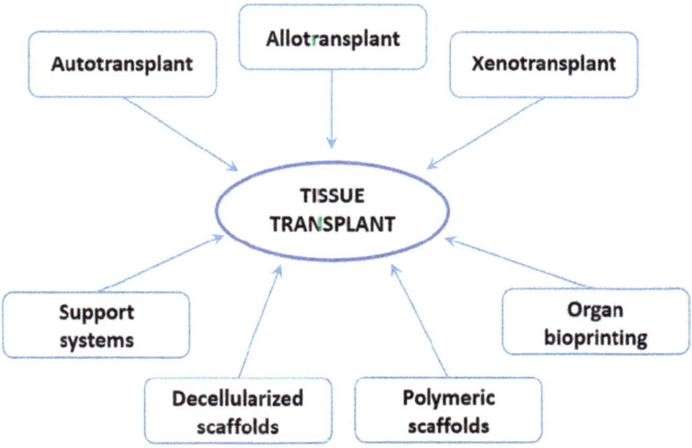

Fig. 8 Schematic representation of the principal strategies for obtaining grafts in tissue and organ transplantation [43]

electronics, wound monitoring, or implantable stimulators [44]. Studies have confirmed that PVDF and its copolymers exhibit low cytotoxicity and good hemocompatibility. Surface modifications like oxygen plasma treatment or protein immobilization further enhance its cellular adhesion without compromising piezoelectric performance. Recent advancements have explored nanostructuring and blending PVDF with materials like graphene oxide or hydroxyapatite to enhance its biocompatibility and cellular interactions. These modifications not only improve protein adsorption and cell proliferation but also help regulate inflammatory signaling at the bio-interface. Overall, the combination of mechanical adaptability, chemical inertness, and tunable surface properties positions PVDF as a promising material for next-generation bioelectronic devices and implantable therapeutic systems [5].

3.2 Environmental and Mechanical Stability

PVDF-based piezoelectric systems are integral to a wide range of flexible and wearable technologies, including electronic skin, pressure sensors, health monitors, and smart textiles. These applications, as shown in Fig. 9, require materials that can endure dynamic mechanical deformation, humidity, sweat, UV exposure, and temperature fluctuations. PVDF exhibits remarkable mechanical fatigue resistance, retaining its piezoelectric response over more than 10^5 bending cycles. Its strong C–F bonds and semicrystalline structure provide stability under harsh conditions, making it ideal for long-term, real-world deployment in personalized healthcare, human–computer interaction, and flexible electronics [46]. This environmental durability is crucial to ensure consistent performance in continuously worn or mobile sensing systems. In wearable electronics, components are frequently

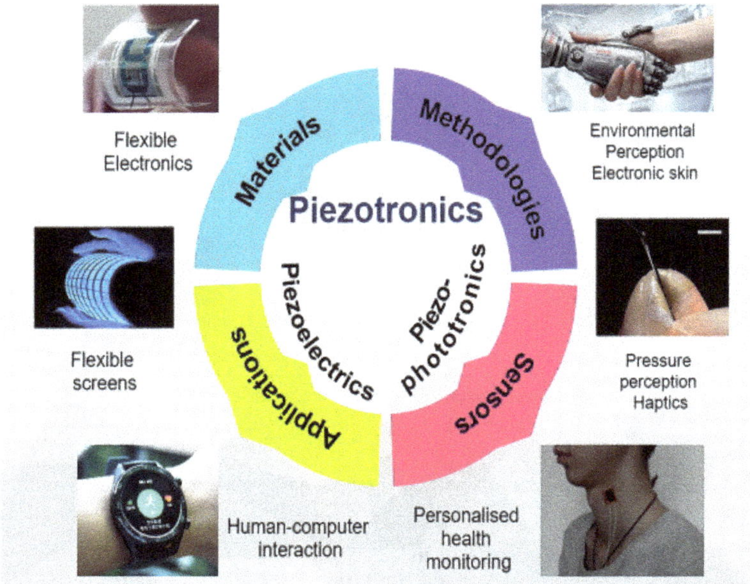

Fig. 9 Applications of piezotronics in flexible electronics, sensing, health monitoring, and human–machine interaction [45]

subjected to repetitive strain, such as bending, stretching, and twisting, during daily activities. PVDF maintains its structural and functional integrity under these conditions due to its high flexibility and toughness [26]. Its piezoelectric properties remain stable under continuous cyclic loading, making it suitable for applications such as electronic skin and pressure perception devices seen in Fig. 9. These properties are particularly vital in haptic feedback systems and robotic sensing, where materials must be both sensitive and resilient under fluctuating environmental stress. Altogether, PVDF's outstanding mechanical durability, chemical inertness, and thermal and UV stability make it an exceptional material for next generation piezotronic systems. The image encapsulates the breadth of its application areas, emphasizing PVDF's role in enabling durable, efficient, and multifunctional flexible electronic platforms.

3.3 Breathable Electronic Skin (E-Skin)

PVDF-based electronic skin (e-skin) is a flexible, breathable, and biocompatible material that closely mimics the tactile sensing functions of human skin while offering enhanced comfort and responsiveness for wearable electronics [47]. One of its key advantages lies in its porous nanofiber architecture, typically fabricated via electrospinning, which allows for continuous airflow through the structure. This breathability reduces sweat accumulation, skin irritation, and discomfort during prolonged use, making it highly suitable for continuous skin contact applications such as wearable health monitors and prosthetics. In addition to comfort, PVDF e-skin offers multimodal sensing capabilities. Its intrinsic piezoelectric properties enable the of mechanical stimuli such as pressure, strain, and motion. It can generate real-time electrical signals in response to physical deformation, allowing it to sense anything from gentle touches to joint movements. When integrated with thermo-responsive components, it can also detect temperature changes, thus functioning as a comprehensive smart skin system. The combination of lightweight design, stretchability, and high sensitivity makes PVDF e-skin ideal for advanced applications in robotics, where it can provide tactile feedback for robotic limbs; in prosthetics, where it helps restore sensory perception for amputees; and in human–machine interaction (HMI) systems, where it acts as an interface for gesture control or motion-triggered commands [48] (Fig. 10). Furthermore, its self-powered functionality thanks to its piezoelectric nature eliminates the need for external power sources, enhancing its utility in portable and energy-efficient systems. Studies have demonstrated that PVDF e-skin exhibits high durability over thousands of bending

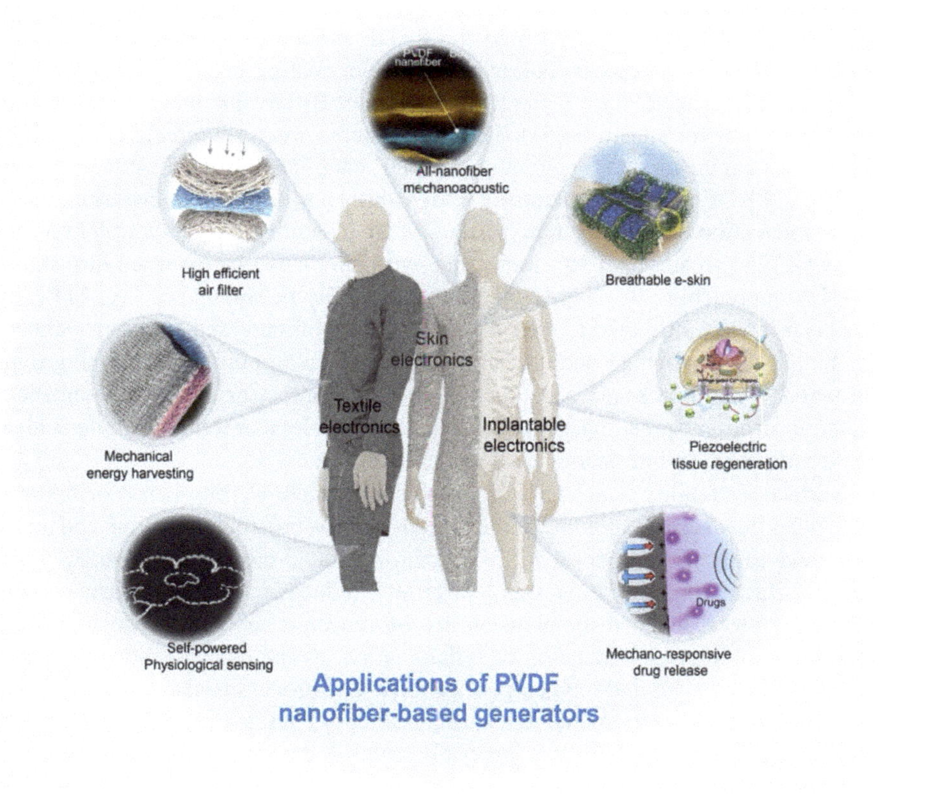

Fig. 10 Applications of PVDF nanofiber-based generators in textile, skin, and implantable electronics, including sensing, energy harvesting, air filtration, e-skin, tissue regeneration, and drug release [49]

cycles, excellent signal stability, and more than 9% air permeability, making it a leading candidate for next-generation wearable and implantable electronics.

3.4 Surface Wettability Modification of PVDF Films

PVDF is a widely used polymer in sensing, biomedical, and wearable electronics due to its excellent piezoelectric, mechanical, and chemical properties. However, one of the inherent limitations of PVDF is its naturally hydrophobic surface [50], which limits its interaction with aqueous environments and reduces its effectiveness in applications such as biosensing, tissue interfacing, or hydrophilic surface coatings. Therefore, surface wettability modification of PVDF films has become an essential strategy to enhance their compatibility and functionality for such advanced applications. To improve the surface properties, various physical and chemical techniques have been employed to transition PVDF from a hydrophobic to a superhydrophilic state. Among these, the combined use of rub-coating and spin-coating is a practical and scalable method. In rub-coating, an ink or polymeric solution is manually applied and gently rubbed onto the PVDF surface to promote adhesion and initiate surface modification. Following this, spin-coating helps to uniformly distribute a composite solution typically consisting of PVDF, ink, and polyvinyl alcohol (PVA) over the substrate [51]. This dual process approach effectively modifies the surface chemistry and topography of PVDF, enabling enhanced wettability. The resulting PVDF-ink-PVA composite film exhibits a dramatic transformation in water interaction behavior. While pristine PVDF generally displays a water contact angle (WCA) greater than 90°, indicating hydrophobicity, the modified film shows a WCA approaching 0°, signifying superhydrophilic behavior (see Fig. 11) [52]. This is primarily attributed to the synergistic effects of increased surface roughness (from physical rubbing) and the introduction of hydrophilic functional groups through PVA and other ink constituents. The hydrophilic surface greatly enhances water absorption, spreading, and interaction with biological fluids, making it ideal for sensor and skin-interfacing applications.

Surface wettability is not only crucial for immediate wetting behavior but also influences downstream biological interactions such as protein adsorption, cell adhesion, and antifouling properties [53]. In bio-integrated electronics, a superhydrophilic surface promotes better skin conformity, reduces interfacial resistance, and minimizes sweat-induced signal instability. Moreover, it improves biocompatibility by reducing the risk of irritation or inflammation, as the modified surface allows better fluid exchange and oxygen permeability. The process is also highly tunable. By adjusting parameters such as the concentration of PVA, spin speed, or number of coating cycles, researchers can precisely control the degree of hydrophilicity. Characterization techniques like contact angle goniometry, scanning electron microscopy (SEM), and atomic force microscopy (AFM) are typically used to analyze surface morphology and wettability before and after modification. These

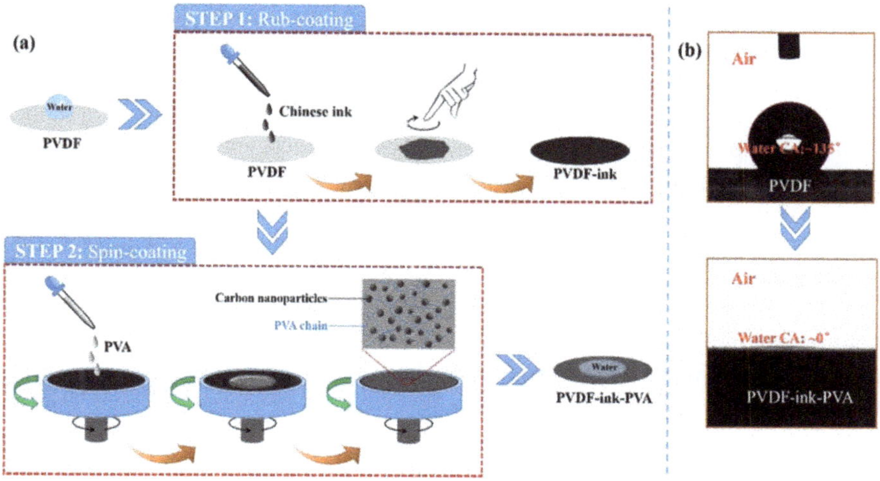

Fig. 11 (**a**) Schematic illustration of the sequential conversion of hydrophobic PVDF to superhydrophilic PVDF-ink-PVA via rub-coating with Chinese ink followed by spin-coating with PVA, highlighting the incorporation of carbon nanoparticles and PVA chains, and (**b**) water contact angle measurements demonstrate a transition from hydrophobic PVDF (contact angle ~135°) to superhydrophilic PVDF-ink-PVA (contact angle ~0°), confirming the dramatic enhancement in surface wettability [52]

insights help optimize the surface engineering process based on the intended application, whether for flexible biosensors, e-skin, or implantable devices.

4 Advanced Fabrication Techniques

The versatility of PVDF in flexible and wearable electronics is not solely attributed to its intrinsic material properties, but also to the advancement of fabrication techniques that allow the polymer to be processed into diverse architectures such as thin films, nanofibers, and composite structures. These techniques are critical for optimizing the piezoelectric, ferroelectric, and mechanical performance of PVDF-based devices and for enabling scalable, reliable integration into wearable platforms.

4.1 Thin Film Processing Methods

PVDF thin films are widely utilized in sensors, actuators, and energy-harvesting devices due to their flexibility, ease of handling, and tunable electroactive properties. Several methods are available for the preparation of high-quality PVDF thin films, including solution casting, spin-coating, and Langmuir–Blodgett (LB) deposition.

Solution casting is a straightforward and cost-effective technique where PVDF is dissolved in solvents such as dimethylformamide (DMF) or *N,N*-dimethylacetamide (DMAc) and cast onto substrates, followed by solvent evaporation [54, 55]. While simple, the control of film thickness, crystallinity, and surface morphology requires optimization of solution concentration, casting conditions, and annealing protocols [56]. Figure 12 details the sequential steps: dissolution of PVDF and benzoyl peroxide (BPO) in DMF, incorporation of surface modified barium titanate (APTS@ BTO) nanoparticles and cellulose nanocrystals (CNC), thorough mixing and sonication to ensure uniform dispersion, casting of the homogeneous mixture onto a substrate to form a film, and subsequent thermal treatment to remove solvent and induce cross-linking. This process yields flexible composite films with enhanced mechanical strength and tailored properties for advanced electronic applications [57].

Spin-coating offers precise control over film thickness and uniformity, making it ideal for microfabrication and integrated devices (Fig. 13). By adjusting the spin speed, solution viscosity, and annealing parameters, PVDF thin films with controlled phase composition and improved β-phase content can be achieved [59]. Spin-coated PVDF films have been successfully employed in piezoelectric sensors and capacitors for wearable applications.

Langmuir–Blodgett (LB) deposition enables the fabrication of highly ordered PVDF monolayers or multilayers with controlled molecular orientation, which enhances piezoelectric and ferroelectric responses [60]. Although this technique is more complex and typically limited to research-scale production, it provides valuable insights into nanoscale control of PVDF properties. Figure 14 illustrates the alignment of PVDF molecules at the air–water interface and their subsequent deposition onto a substrate to create a Langmuir–Blodgett (LB) film.

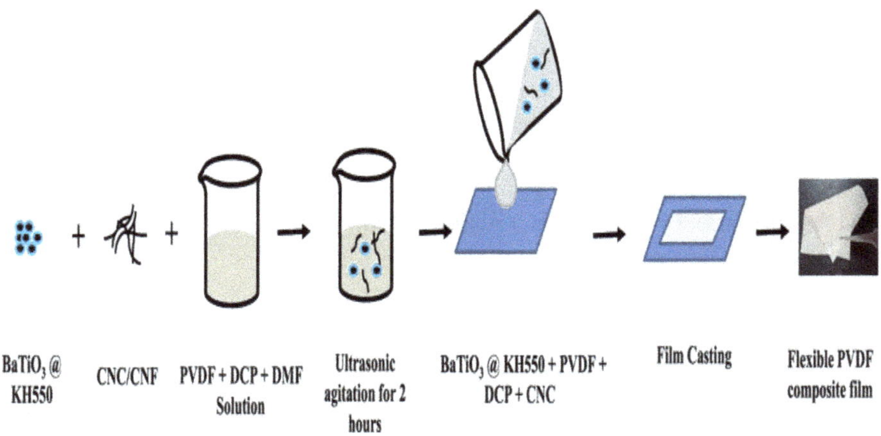

Fig. 12 Schematic representation of the synthesis process for PVDF/CNC/BTO composite films [57]

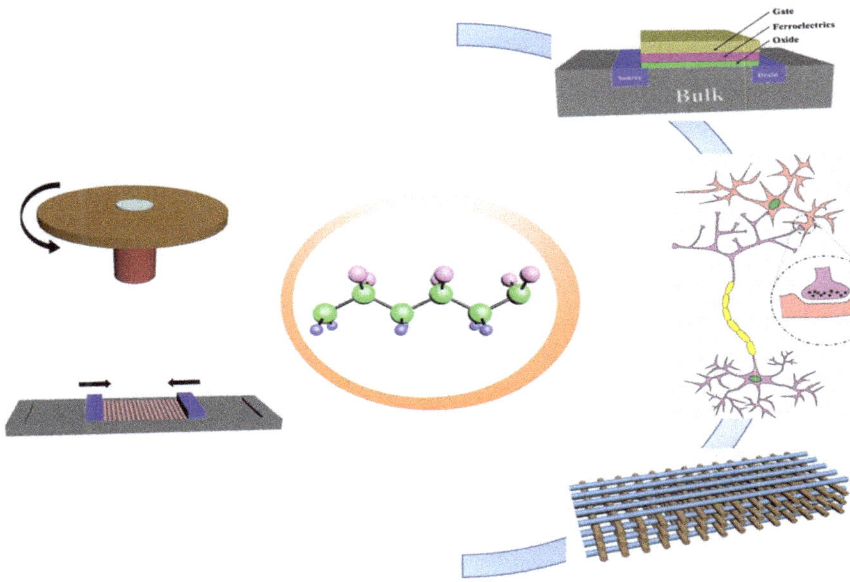

Fig. 13 Spin coating of PVDF for different electrical applications [58]

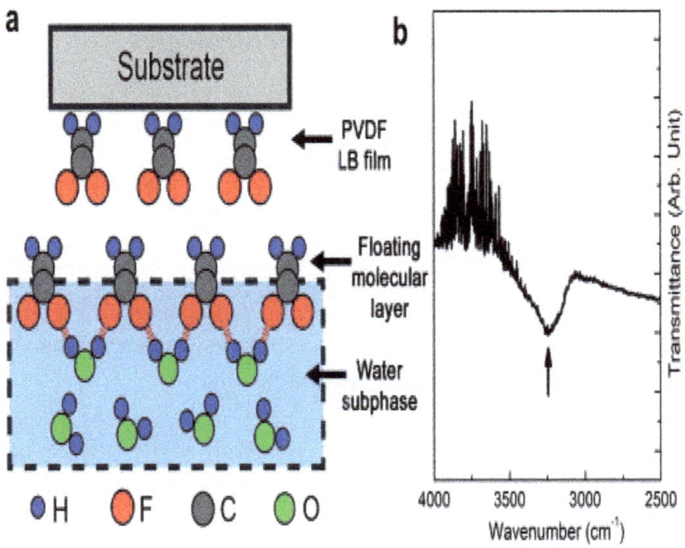

Fig. 14 (**a**) Schematic representation of PVDF molecule orientation at the water surface and the subsequent transfer to a substrate to form a Langmuir–Blodgett (LB) film. The PVDF chains are depicted as standing perpendicular to the plane of the paper; (**b**) FTIR spectrum of the PVDF LB film, where the indicated peak corresponds to the absorption band of O–H groups interacting with PVDF molecules via hydrogen bonding [61]

4.2 Printing and Scalable Manufacturing

To meet the growing demand for large-area, cost-effective, and scalable fabrication of PVDF-based devices, several printing and manufacturing techniques have been developed, including screen printing, inkjet printing, and roll-to-roll processing.

Screen printing is a mature technique that allows the deposition of PVDF-based inks onto flexible substrates in well-defined patterns [62]. The process offers simplicity, high throughput, and compatibility with various substrates, making it suitable for mass production of wearable sensors, energy harvesters, and antennas [63].

Inkjet printing enables high-resolution, mask-free patterning of PVDF films with minimal material waste [64]. The development of stable, printable PVDF inks has facilitated the fabrication of flexible piezoelectric devices on substrates such as polyethylene terephthalate (PET) and textiles, supporting the integration of PVDF into conformal and stretchable electronics.

Roll-to-roll (R2R) processing represents a scalable and continuous manufacturing approach ideal for producing large-area PVDF-based devices [17]. R2R techniques, combined with printing or coating processes, allow for the efficient fabrication of flexible sensors, energy harvesting films, and smart textiles with consistent performance over large areas.

4.3 Emerging Techniques and Hybrid Strategies

In addition to conventional methods, emerging fabrication approaches such as 3D printing, laser-assisted deposition, and additive manufacturing are being explored to further expand the capabilities of PVDF in wearable electronics. Hybrid fabrication strategies, where PVDF is combined with nanomaterials like graphene, carbon nanotubes, MXenes, or metallic nanoparticles, have been shown to significantly enhance mechanical strength, electrical conductivity, and functional performance [62]. These hybrid systems leverage the synergistic properties of PVDF and nanomaterials to develop next-generation sensors, actuators, and energy harvesters with improved sensitivity, durability, and multifunctionality.

4.4 Fabrication of Thin-Film Composite (TFC) Polyamide Membranes on PVDF Substrates

Thin-film composite (TFC) membranes, widely used in separation processes such as reverse osmosis, nanofiltration, and forward osmosis, are valued for their high permeability and selectivity. These membranes typically consist of a porous support layer often made from PVDF due to its excellent thermal stability, chemical resistance, and mechanical strength coated with an ultra-thin polyamide selective layer.

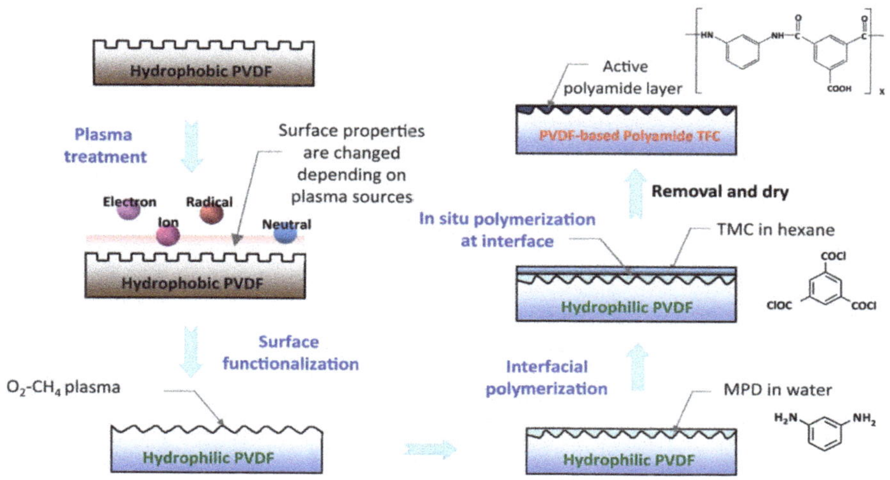

Fig. 15 Stepwise process for fabricating PVDF-based polyamide TFC membranes. Plasma treatment converts hydrophobic PVDF to hydrophilic, followed by interfacial polymerization of MPD and TMC to form the polyamide selective layer [50]

The polyamide layer is commonly fabricated via interfacial polymerization, where a diamine solution is first applied to the PVDF support, followed by an acyl chloride solution, resulting in a rapid reaction at the interface that forms a dense, selective film. This process often illustrated schematically (Fig. 15) by sequential deposition and reaction steps, produces membranes that combine the mechanical robustness and porosity of PVDF with the high selectivity of polyamide. As a result, PVDF-based TFC membranes are highly effective for applications such as desalination, wastewater treatment, and pharmaceutical filtration, with their performance tunable through careful control of fabrication parameters.

4.5 Fabrication Process of Porous PVDF Monoliths

The fabrication of porous PVDF monoliths via the TINIPS method involves first preparing a homogeneous, transparent PVDF solution by magnetic stirring and heating.

Once the solution cools to 25 °C, methanol is introduced as a non-solvent under vigorous stirring, which induces phase separation and results in solution turbidity as PVDF chains begin to aggregate due to the reduced solubility in DMF. To optimize the formation of a well-developed porous structure, the mixture is maintained at 20 °C for 24 h, further promoting phase separation. During this stage, the system divides into polymer-rich and solvent-rich regions, which evolve into the robust PVDF skeleton and interconnected microcells that define the final pore network. The organic solvents (DMF and methanol) are then exchanged with deionized

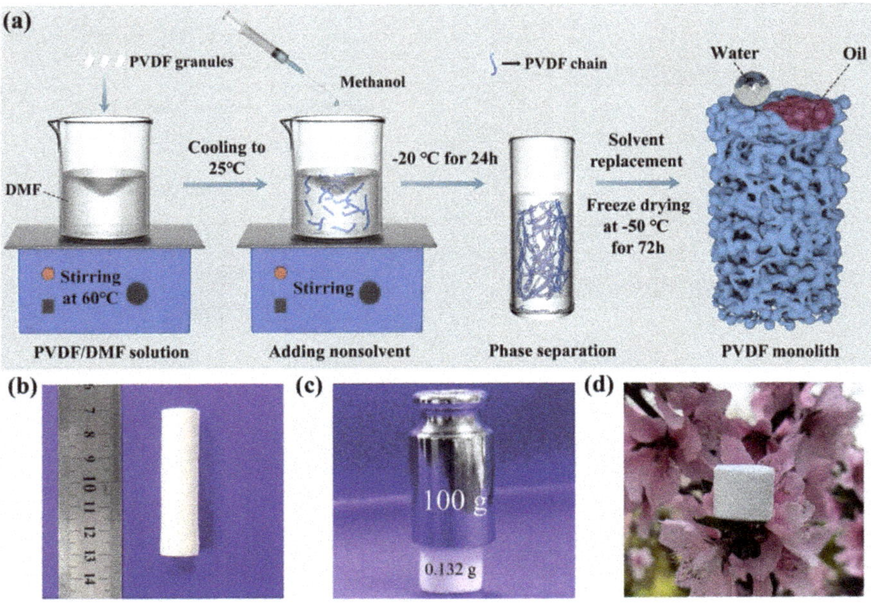

Fig. 16 (**a**) Schematic of preparation process for PVDF monolith; digital photographs of PVDF-13% monolith: (**b**) sample appearance, (**c**) supporting a weight of 100 g, and (**d**) standing on a small leaf [65]

water, and the remaining water is removed by freeze-drying, resulting in a three-dimensional, interconnected porous PVDF monolith (Fig. 16). Notably, the resulting PVDF-13% monolith demonstrates remarkable mechanical integrity, being able to support 757 times its own weight and stand stably on a small leaf, which highlights its strong framework and ultralow density (0.116 g/cm^3).

5 Application Case Studies

The translation of PVDF's unique material properties and advanced fabrication techniques into real-world devices is best illustrated through detailed application case studies. These examples highlight how PVDF's flexibility, piezoelectricity, and processability enable the creation of high-performance, wearable, and self-powered systems for diverse sectors such as health monitoring, energy harvesting, and smart textiles.

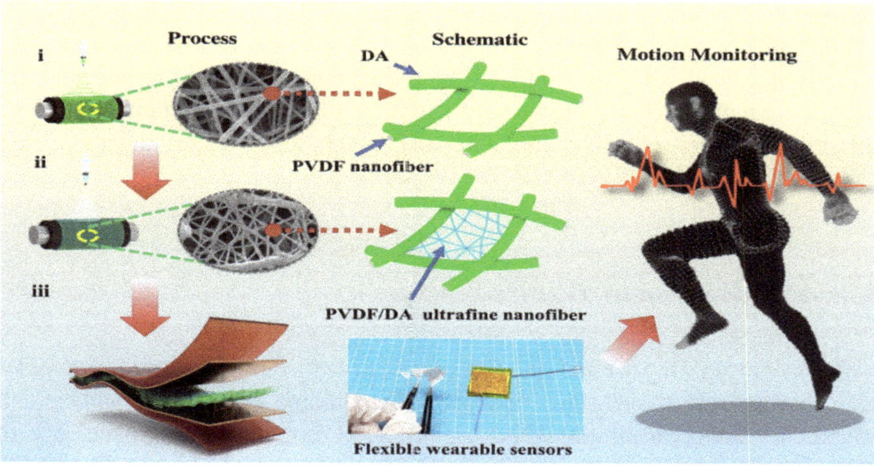

Fig. 17 Schematic illustration of the fabrication process for a composite nanofiber piezoelectric sensor featuring a hierarchical structure. The diagram highlights the sequential steps involved in creating the composite nanofiber mat and its integration into a flexible sensor device [66]

5.1 Flexible PVDF Sensors for Health and Motion Monitoring

PVDF's strong piezoelectric effect forms the basis of highly sensitive, flexible sensors engineered as thin films, nanofiber mats, or composite coatings. These can be integrated into wearable formats, including patches, bands, or even woven directly into garments. For instance, PVDF/graphene composite coatings on fabric have enabled wearable sensors capable of detecting subtle biomechanical signals, such as pulse, respiration, and joint movement. Such sensors exhibit high voltage sensitivity and can distinguish between different types of motion, making them suitable for health tracking and medical diagnostics. Integration into textiles ensures comfort and unobtrusive long-term monitoring, which is crucial for real-world adoption (Fig. 17).

5.2 Self-Powered PVDF Nanogenerators for Energy Harvesting

PVDF's piezoelectric and triboelectric properties have enabled the creation of nanogenerators that convert mechanical energy such as body movement or ambient vibrations into electrical power. These self-powered devices are especially valuable for wearable electronics, where battery replacement or charging is impractical. Recent advancements include PVDF-based piezoelectric nanogenerators (PENGs) and triboelectric nanogenerators (TENGs) that can power LEDs, wireless sensors, or small electronic modules directly from human motion. For example, PVDF films enhanced with cobalt-based metal–organic frameworks (Co-MOFs) have been used to fabricate.

TENGs capable of generating high output voltages, sufficient to operate GPS modules or mobile devices during daily activities. These devices demonstrate stable performance under repeated deformation and are robust enough for integration into clothing or accessories.

5.3 PVDF in Smart Textiles

PVDF's compatibility with textile substrates has led to the development of smart garments and electronic skins (e-skins) that combine comfort, durability, and advanced sensing capabilities. Electrospun PVDF nanofibers, often blended with other polymers or nanomaterials, are woven or coated onto fabrics to create self-powered tactile sensors.

5.4 Wireless and Underwater Sensing Applications

PVDF-based sensors have also been adapted for wireless and underwater applications. Flexible PVDF sensors embedded in footwear can wirelessly transmit gait and posture data, enabling real-time biomechanical analysis for rehabilitation or sports science. In underwater environments, funnel-type PVDF energy harvesters generate electricity from flow-induced vibrations, providing a reliable power source for underwater wireless sensor networks [67].

5.5 Performance Optimization and Device Design

The performance of PVDF-based devices is closely tied to their geometry and structural configuration. Studies have shown that parallel bimorph sensor layouts maximize energy output, while careful tuning of film thickness, surface area, and loading frequency further enhances efficiency [68]. These insights guide the rational design of PVDF devices tailored for specific applications, from high-sensitivity medical sensors to robust industrial harvesters (Table 2).

Table 2 Representative PVDF application case studies

Application area	Device/approach	Key features
Health monitoring	PVDF/graphene textile sensor	High sensitivity, comfort, real-time signals
Activity sensing	PVDF-nanographite in sensor	Wireless gait monitoring in socks
Energy harvesting	PVDF-based PENG/TENG	Self-powered, stable output
Underwater harvesting	Funnel-type PVDF harvester	Robust, aquatic environments
Electronic skin (E-skin)	Electrospun PVDF nanofiber e-skin	Tactile, strain, temperature sensing

5.6 *Comparative Study: PVDF Vs. Polypropylene in Hernia Repair*

PVDF is increasingly recognized for its excellent biocompatibility and chemical stability, making it a strong candidate for permanent surgical implants such as hernia meshes. Compared to polypropylene (PP), PVDF triggers a much lower inflammatory response when implanted, which reduces tissue irritation and the likelihood of postoperative complications. Its inert nature leads to minimal immune reaction, supporting better healing and improved patient comfort. PVDF also exhibits superior resistance to hydrolysis and oxidative degradation, ensuring long-term durability inside the body. In hernia repair, material choice is crucial for surgical success and minimizing complications. While polypropylene meshes (as shown in Fig. 18) have long been favored for their strength and affordability, they are associated with higher rates of inflammation, fibrosis, and chronic pain due to strong tissue reactions. PVDF, in contrast, offers greater flexibility and softness, which improves handling during minimally invasive procedures and allows better adaptation to surrounding tissues, reducing discomfort and foreign body sensation. Additionally, PVDF meshes are less likely to cause adhesions when placed near internal organs, lowering the risk of complications such as bowel obstruction. Overall, although polypropylene remains widely used, PVDF is emerging as a promising alternative for hernia repair because of its favorable tissue compatibility, mechanical properties, and lower surgical meshes where long-term patient outcomes are a priority.

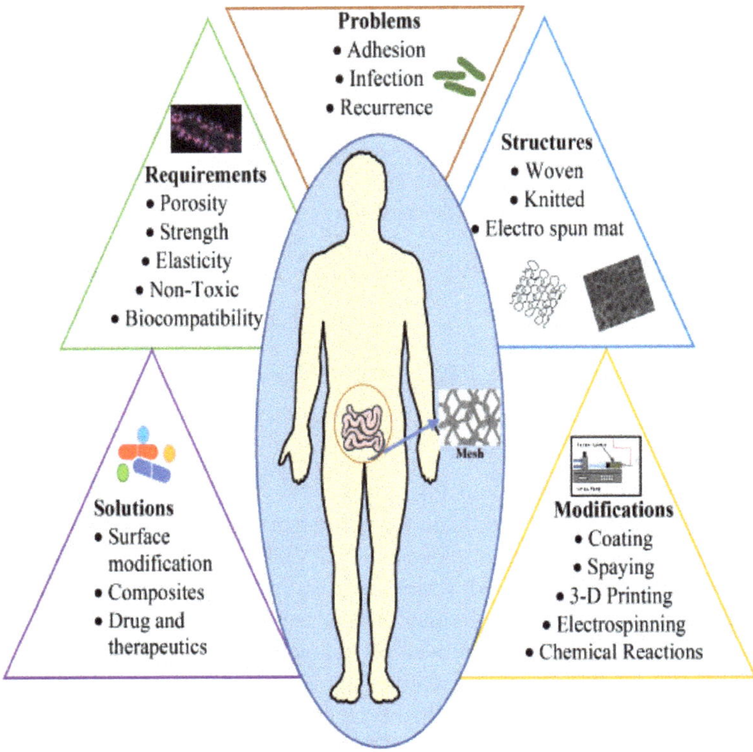

Fig. 18 Polypropylene-based implant for hernia correction [69]

6 Challenges and Future Directions

Despite the significant progress in the development of PVDF-based flexible and wearable electronics, several scientific and engineering challenges persist, limiting large-scale commercialization and long-term practical deployment. Addressing these challenges is essential for advancing PVDF from laboratory prototypes to robust, market-ready products that meet the stringent requirements of real-world applications.

6.1 Material and Structural Limitations

While PVDF exhibits remarkable piezoelectric, ferroelectric, and mechanical properties, its relatively low stretchability compared to elastomers restricts its integration into highly deformable devices. Efforts to enhance the stretchability of PVDF through blending with soft polymers (e.g., polyurethane, PDMS) or by creating nanocomposites have shown promise, but achieving an optimal balance between

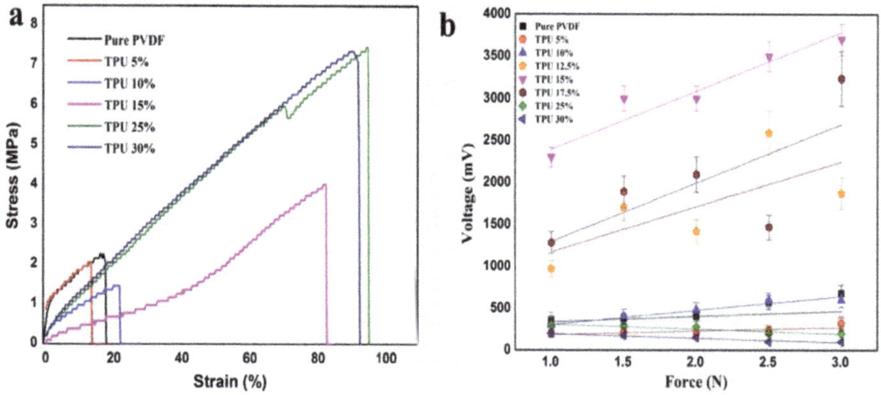

Fig. 19 (**a**) Stress–strain curves comparing pure PVDF nanofibers and PVDF/TPU composite nanofibers, demonstrating the enhanced stretchability and mechanical robustness achieved by blending PVDF with TPU; (**b**) relationship between peak-to-peak output voltage and applied force at a frequency of 8 Hz, illustrating the electromechanical performance of the composite under dynamic loading [70]

mechanical flexibility and electroactive performance remains an ongoing challenge. Furthermore, the fabrication of PVDF in complex, miniaturized structures for next-generation wearable electronics, such as flexible micro-sensors and implantable devices, requires precise control over material morphology, phase composition, and surface properties. While techniques like electrospinning and micro-patterning offer potential solutions, scalability and reproducibility are still under development (Fig. 19) [71].

6.2 *Device Integration and System-Level Challenges*

For PVDF-based sensors, energy harvesters, and actuators to function effectively in wearable systems, seamless integration with other device components, including electrodes, interconnects, and data processing units, is crucial. However, challenges related to adhesion, mechanical mismatch, and long-term electrical stability between PVDF and other materials persist. In particular, the development of stretchable, transparent, and conductive electrodes compatible with PVDF remains a bottleneck for fully flexible and conformal devices. Advanced electrode materials, such as silver nanowires, carbon-based composites, and conductive polymers, show potential but require further optimization for durability, biocompatibility, and large-area manufacturing. Moreover, system-level challenges, including power management for self-powered devices, wireless data transmission, and integration with Internet-of-Things (IoT) platforms, need to be addressed to realize truly autonomous, intelligent wearable systems.

6.3 Biocompatibility, Safety, and Environmental Concerns

Although PVDF is generally considered biocompatible, long-term studies on its performance and safety in direct contact with human skin or implanted environments are still limited. Comprehensive evaluations of cytotoxicity, allergenic responses, and degradation behavior under physiological conditions are necessary, especially for medical-grade devices. Additionally, the environmental impact of PVDF production, processing, and end-of-life disposal raises sustainability concerns. The fluoropolymer backbone of PVDF contributes to its durability but also to its persistence in the environment. Future research should focus on developing eco-friendly synthesis routes, recycling strategies, and biodegradable alternatives to conventional PVDF-based materials.

6.4 Future Directions

To overcome existing challenges and unlock the full potential of PVDF in flexible and wearable electronics, several research and development directions should be pursued:

- **Advanced Material Engineering**: Exploration of novel PVDF composites incorporating nanomaterials (e.g., MXenes, graphene, CNTs) to improve stretchability, mechanical robustness, and multifunctionality.
- **Scalable Fabrication**: Development of cost-effective, scalable manufacturing techniques, such as roll-to-roll printing and additive manufacturing, for large-area, high-performance PVDF-based devices.
- **Bio-Integrated Systems**: Enhanced focus on biocompatibility studies, implantable device design, and biodegradable PVDF-based materials for healthcare and biomedical applications.
- **Intelligent and Autonomous Devices**: Integration of PVDF sensors with flexible electronics, wireless communication modules, and AI-driven data processing for real-time monitoring and smart health management.
- **Sustainability**: Development of green synthesis approaches for PVDF and investigation of environmentally friendly disposal and recycling methods.

Addressing these challenges through interdisciplinary research and collaboration among material scientists, engineers, and healthcare experts will be critical for advancing PVDF-based wearable technologies and realizing their potential in next-generation flexible electronics.

7 Conclusion

PVDF has established itself as a cornerstone material in the rapidly evolving field of flexible and wearable electronics. Its unique combination of mechanical flexibility, chemical stability, and pronounced electroactive properties has enabled the development of highly sensitive, self-powered sensors, energy harvesters, and advanced biomedical devices. Through continuous progress in material processing ranging from thin films and nanofibers to hybrid composites PVDF based systems have achieved remarkable performance in health monitoring, human–machine interfaces, and smart textiles. The integration of PVDF with novel nanomaterials and scalable fabrication techniques have further broadened their application landscape, supporting multifunctional devices that are lightweight, durable, and adaptable to complex environments. Its inherent biocompatibility and environmental resilience make PVDF particularly suitable for long-term deployment in real-world and clinical settings. Despite these advances, several challenges remain, including enhancing stretchability, improving device integration, and advancing sustainable manufacturing practices. Addressing these issues will require interdisciplinary collaboration and continued innovation in material science, device engineering, and system integration. Looking ahead, PVDF is poised to play a pivotal role in the next generation of intelligent, autonomous, and sustainable electronic systems. As research continues to overcome existing limitations and unlock new functionalities, PVDF-based technologies are expected to drive significant progress in personalized healthcare, soft robotics, and energy-autonomous platforms, firmly positioning PVDF at the forefront of future flexible electronics.

References

1. Khan, B., Khalid, R.T., Masrur, M.H., Awais, M., Imdad, N., Hilal, M.E., Khan, W.U., Riaz, Z., Ahmed, S.B., Khan, B., et al.: Advances in polyvinylidene fluoride (PVDF) for self-powered wearable physiological monitoring and energy harvesting applications. Nano Energy. **143**, 111296 (2025)
2. Mondal, R., Hasan, M.A.M., Baik, J.M., Yang, Y.: Advanced pyroelectric materials for energy harvesting and sensing applications. Mater. Today. **66**, 273–301 (2023)
3. Pan, X., Wu, Y., Wang, Y., Zhou, G., Cai, H.: Mechanical energy harvesting based on the piezoelectric materials: recent advances and future perspectives. Chem. Eng. J. **497**, 154249 (2024)
4. Martins, P., Lopes, A.C., Lanceros-Mendez, S.: Electroactive phases of poly (vinylidene fluoride): determination, processing and applications. Prog. Polym. Sci. **39**(4), 683–706 (2014)
5. Costa, C.M., Cardoso, V.F., Martins, P., Correia, D.M., Gonçalves, R., Costa, P., Correia, V., Ribeiro, C., Fernandes, M.M., Martins, P.M., et al.: Smart and multifunctional materials based on electroactive poly (vinylidene fluoride): recent advances and opportunities in sensors, actuators, energy, environmental, and biomedical applications. Chem. Rev. **123**(19), 11392–11487 (2023)
6. Saxena, P., Shukla, P.: A comprehensive review on fundamental properties and applications of poly (vinylidene fluoride) (PVDF). Adv. Compos. Hybrid Mater. **4**(1), 8–26 (2021)

7. Vallem, V., Sargolzaeiaval, Y., Ozturk, M., Lai, Y.-C., Dickey, M.D.: Energy harvesting and storage with soft and stretchable materials. Adv. Mater. **33**(19), 2004832 (2021)
8. He, Z., Rault, F., Lewandowski, M., Mohsenzadeh, E., Salaün, F.: Electrospun PVDF nanofibers for piezoelectric applications: a review of the influence of electrospinning parameters on the β phase and crystallinity enhancement. Polymers. **13**(2), 174 (2021)
9. Zhang, M., Liu, C., Li, B., Shen, Y., Wang, H., Ji, K., Mao, X., Wei, L., Sun, R., Zhou, F.: Electrospun PVDF-based piezoelectric nanofibers: materials, structures, and applications. Nanoscale Adv. **5**(4), 1043–1059 (2023)
10. Ranjan, H., Ranjan, P., Sahu, T.K., Sharma, R.K., Kumar, P.: Reduced graphene oxide electrode-coating as anti-corrosive/anti-oxidative laminate for al/cu liquid-phase batteries. J. Mater. Res. **38**(7), 1792–1802 (2023)
11. Yu, Y., Sun, H., Orbay, H., Chen, F., England, C.G., Cai, W., Wang, X.: Biocompatibility and in vivo operation of implantable mesoporous PVDF-based nanogenerators. Nano Energy. **27**, 275–281 (2016)
12. Sun, W., Guo, Z., Yang, Z., Wu, Y., Lan, W., Liao, Y., Wu, X., Liu, Y.: A review of recent advances in vital signals monitoring of sports and health via flexible wearable sensors. Sensors. **22**(20), 7784 (2022)
13. Ranjan, H., Singh, C.P., Singh, V.P., Pandey, S.K.: Self-rectifying and forming-free resistive switching with Cu/BN/SiO_2/Pt bilayer device. Mater. Sci. Semicond. Process. **183**, 108744 (2024)
14. Ranjan, H., Singh, C.P., Singh, V.P., Pandey, S.K.: Efficient and robust resistive switching behaviour of MoS2 based memristor. IEEE Trans. Mater. Electron Devices. (2025)
15. Wang, C., He, T., Zhou, H., Zhang, Z., Lee, C.: Artificial intelligence enhanced sensors-enabling technologies to next-generation healthcare and biomedical platform. Bioelectron. Med. **9**(1), 17 (2023)
16. Rashid, A., Zubair, U., Ashraf, M., Javid, A., Abid, H.A., Akram, S.: Flexible piezoelectric coatings on textiles for energy harvesting and autonomous sensing applications: a review. J. Coat. Technol. Res. **20**(1), 141–172 (2023)
17. Stadlober, B., Zirkl, M., Irimia-Vladu, M.: Route towards sustainable smart sensors: ferroelectric polyvinylidene fluoride-based materials and their integration in flexible electronics. Chem. Soc. Rev. **48**(6), 1787–1825 (2019)
18. Ramasubramanian, B., Sundarrajan, S., Rao, R.P., Reddy, M.V., Chellappan, V., Ramakrishna, S.: Novel low-carbon energy solutions for powering emerging wearables, smart textiles, and medical devices. Energy Environ. Sci. **15**(12), 4928–4981 (2022)
19. Vijayakanth, T., Shankar, S., Finkelstein-Zuta, G., Rencus-Lazar, S., Gilead, S., Gazit, E.: Perspectives on recent advancements in energy harvesting, sensing and bio-medical applications of piezoelectric gels. Chem. Soc. Rev. **52**(17), 6191–6220 (2023)
20. Kujawa, J., Boncel, S., Al-Gharabli, S., Koter, S., Kaczmarek-Kędziera, A., Korczeniewski, E., Terzyk, A.P.: Current and future applications of PVDF-carbon nanomaterials in energy and sensing. Chem. Eng. J. **492**, 151856 (2024)
21. Zhou, Y., He, J., Wang, H., Qi, K., Nan, N., You, X., Shao, W., Wang, L., Ding, B., Cui, S.: Highly sensitive, self-powered and wearable electronic skin based on pressure-sensitive nanofiber woven fabric sensor. Sci. Rep. **7**(1), 12949 (2017)
22. do Nascimento, L.M.S., Bonfati, L.V., Freitas, M.L.B., Junior, J.J.A.M., Siqueira, H.V., Stevan Jr., S.L.: Sensors and systems for physical rehabilitation and health monitoring—a review. Sensors. **20**(15), 4063 (2020)
23. Farzin, M.A., Naghib, S.M., Rabiee, N.: Advancements in bio-inspired self-powered wireless sensors: materials, mechanisms, and biomedical applications. ACS Biomater. Sci. Eng. **10**(3), 1262–1301 (2024)
24. Dong, W., Li, M., Chen, C., Xie, K., Hong, J., Yang, L.: Flexible hybrid self-powered piezo-triboelectric nanogenerator based on BTO-PVDF/PDMS nanocomposites for human machine interaction. Sci. Rep. **15**(1), 1–11 (2025)

25. Arjun, A.M., Garlapati, K.K., Rasheed, P.A.: Recent advancements in MOF-based nanogenerators for bioelectronics. In: Bioelectronics, pp. 221–235. CRC Press (2022)
26. Keum, K., Kim, J.W., Hong, S.Y., Son, J.G., Lee, S.-S., Ha, J.S.: Flexible/stretchable supercapacitors with novel functionality for wearable electronics. Adv. Mater. **32**(51), 2002180 (2020)
27. Purushothaman, S.M., Tronco, M.F., Kottathodi, B., Royaud, I., Ponçot, M., Kalarikkal, N., Thomas, S., Rouxel, D.: A review on electrospun PVDF-based nanocomposites: recent trends and developments in energy harvesting and sensing applications. Polymer. **283**, 126179 (2023)
28. Pachkawade, V.: Bioinspired micro-electromechanical systems nano-electromechanical systems: an overview, applications, and perspective. In: Nature-Derived Sensors, pp. 267–293 (2025)
29. Guan, X., Xu, B., Wu, M., Jing, T., Yang, Y., Gao, Y.: Breathable, washable and wearable woven-structured triboelectric nanogenerators utilizing electrospun nanofibers for biomechanical energy harvesting and self-powered sensing. Nano Energy. **80**, 105549 (2021)
30. Papamatthaiou, S., Menelaou, P., Oussallam, B.E.A., Moschou, D.: Recent advances in bio-microsystem integration and lab-on-PCB technology. Microsyst. Nanoeng. **11**(1), 1–31 (2025)
31. Hossain, N., Al Mahmud, M.Z., Hossain, A., Rahman, M.K., Islam, M.S., Tasnim, R., Mobarak, M.H.: Advances of materials science in mems applications: a review. Results Eng. **22**, 102115 (2024)
32. Aubry, G., Lee, H.J., Lu, H.: Advances in microfluidics: technical innovations and applications in diagnostics and therapeutics. Anal. Chem. **95**(1), 444–467 (2023)
33. Singh, M., Apata, I.E., Samant, S., Wu, W., Tawade, B.V., Pradhan, N., Raghavan, D., Karim, A.: Nanoscale strategies to enhance the energy storage capacity of polymeric dielectric capacitors: review of recent advances. Polym. Rev. **62**(2), 211–260 (2022)
34. Xia, W., Zhang, Z.: PVDF-based dielectric polymers and their applications in electronic materials. IET Nanodielectr. **1**(1), 17–31 (2018)
35. Persano, L., Camposeo, A., Matino, F., Wang, R., Natarajan, T., Li, Q., Pan, M., Su, Y., Kar-Narayan, S., Auricchio, F., et al.: Advanced materials for energy harvesting and soft robotics: emerging frontiers to enhance piezoelectric performance and functionality. Adv. Mater. **36**(45), 2405363 (2024)
36. Riquelme, S.A., Ramam, K.: Dielectric and piezoelectric properties of lead free BZT-BCT/PVDF flexible composites for electronic applications. Mater. Res. Express. **6**(11), 116331 (2019)
37. He, Q., Sun, K., Shi, Z., Liu, Y., Fan, R.: Polymer dielectrics for capacitive energy storage: from theories, materials to industrial capacitors. Mater. Today. **68**, 298–333 (2023)
38. Rajeevan, S., John, S., George, S.C.: Polyvinylidene fluoride: a multifunctional polymer in supercapacitor applications. J. Power Sources. **504**, 230037 (2021)
39. Bicy, K., Gueye, A.B., Rouxel, D., Kalarikkal, N., Thomas, S.: Lithium-ion battery separators based on electrospun PVDF: a review. Surf. Interfaces. **31**, 101977 (2022)
40. Lee, S., Koo, H., Kang, H.S., Oh, K.-H., Nam, K.W.: Advances in polymer binder materials for lithium-ion battery electrodes and separators. Polymers. **15**(23), 4477 (2023)
41. Wang, X.W., Fan, J.T., Manikandan, M., Zhang, B.H., Guo, J.N., Chen, J.Y., Yang, F., Zheng, M.M., Zhang, H.X., Hou, M.Z., et al.: Recent advances in composite films of lead-free ferroelectric ceramics and poly (vinylidene fluoride) (PVDF) for energy storage capacitor: a review. J. Mater. Sci. **58**(1), 124–143 (2023)
42. Houis, S., Engelhardt, E.M., Wurm, F., Gries, T.: Application of polyvinylidene fluoride (PVDF) as a biomaterial in medical textiles. In: Medical and Healthcare Textiles, pp. 342–352. Elsevier (2010)
43. Massaro, M.S., Palek, R., Rosendorf, J., Červenková, L., Liška, V., Moulisova, V.: Decellularized xenogeneic scaffolds in transplantation and tissue engineering: immunogenicity versus positive cell stimulation. Mater. Sci. Eng. C. **127**, 112203 (2021)
44. Zhao, G., Chen, W.N.: Biofouling formation and structure on original and modified PVDF membranes: role of microbial species and membrane properties. RSC Adv. **7**(60), 37990–38000 (2017)

45. Wu, Y., Ma, Y., Zheng, H., Ramakrishna, S.: Piezoelectric materials for flexible and wearable electronics: a review. Mater. Des. **211**, 110164 (2021)
46. Wu, L., Jin, Z., Liu, Y., Ning, H., Liu, X., Alamusi, Hu, N.: Recent advances in the preparation of PVDF-based piezoelectric materials. Nanotechnol. Rev. **11**(1), 1386–1407 (2022)
47. Lee, Y., Park, J., Choe, A., Cho, S., Kim, J., Ko, H.: Mimicking human and biological skins for multifunctional skin electronics. Adv. Funct. Mater. **30**(20), 1904523 (2020)
48. Zarei, M., Lee, G., Lee, S.G., Cho, K.: Advances in biodegradable electronic skin: material progress and recent applications in sensing, robotics, and human–machine interfaces. Adv. Mater. **35**(4), 2203193 (2023)
49. Gong, W., Yang, W., Fan, F., Liu, X., Wang, J., Wang, X.-Q., Hou, C., Wang, Z., Dickey, M.D., Wang, H.: PVDF nanofibers for body-area triboelectric generators. Nano Energy. **131**, 110277 (2024)
50. Kang, G.-d., Cao, Y.-m.: Application and modification of poly (vinylidene fluoride) (PVDF) membranes—a review. J. Membr. Sci. **463**, 145–165 (2014)
51. Ribeiro, C., Costa, C.M., Correia, D.M., Nunes-Pereira, J., Oliveira, J., Martins, P., Gonçalves, R., Cardoso, V.F., Lanceros-Méndez, S.: Electroactive poly (vinylidene fluoride)-based structures for advanced applications. Nat. Protoc. **13**(4), 681–704 (2018)
52. Wang, M., Xu, Z., Hou, Y., Li, P., Sun, H., Niu, Q.J.: Fabrication of a superhydrophilic PVDF membrane with excellent chemical and mechanical stability for highly efficient emulsion separation. Sep. Purif. Technol. **251**, 117408 (2020)
53. Afonso, E., Martínez-Gómez, A., Tiemblo, P., García, N.: Exploring chemical and structural features to tailor wetting properties of PVDF and PVDF/PMMA surfaces. Polymer. **262**, 125441 (2022)
54. Sengupta, A., Das, S., Dasgupta, S., Sengupta, P., Datta, P.: Flexible nanogenerator from electrospun PVDF–polycarbazole nanofiber membranes for human motion energy-harvesting device applications. ACS Biomater. Sci. Eng. **7**(4), 1673–1685 (2021)
55. Dallaev, R., Pisarenko, T., Sobola, D., Orudzhev, F., Ramazanov, S., Trčka, T.: Brief review of PVDF properties and applications potential. Polymers. **14**(22), 4793 (2022)
56. Amani, H., Arzaghi, H., Bayandori, M., Dezfuli, A.S., Pazoki-Toroudi, H., Shafiee, A., Moradi, L.: Controlling cell behavior through the design of biomaterial surfaces: a focus on surface modification techniques. Adv. Mater. Interfaces. **6**(13), 1900572 (2019)
57. Ram, F., Kaviraj, P., Pramanik, R., Krishnan, A., Shanmuganathan, K., Arockiarajan, A.: PVDF/$BaTiO_3$ films with nanocellulose impregnation: investigation of structural, morphological and mechanical properties. J. Alloys Compd. **823**, 153701 (2020)
58. Yin, Z., Tian, B., Zhu, Q., Duan, C.: Characterization and application of PVDF and its copolymer films prepared by spin-coating and Langmuir–Blodgett method. Polymers. **11**(12), 2033 (2019)
59. Ma, Z., Lu, X., Wu, C., Gao, Q., Zhao, L., Zhang, H., Liu, Z.: Functional surface modification of PVDF membrane for chemical pulse cleaning. J. Membr. Sci. **524**, 389–399 (2017)
60. Gorbachev, I.A., Smirnov, A.V., Shamsutdinova, E.S., Kashin, V.V., Yudin, S.G., Anisimkin, V.I., Kolesov, V.V., Kuznetsova, I.E.: Studying the structural and piezoelectric properties of PVDF films obtained using Langmuir–Blodgett technology. Bull. Russ. Acad. Sci. Phys. **85**, 603–607 (2021)
61. Chen, S., Li, X., Yao, K., Tay, F.E.H., Kumar, A., Zeng, K.: Self-polarized ferroelectric PVDF homopolymer ultra-thin films derived from Langmuir–Blodgett deposition. Polymer. **53**(6), 1404–1408 (2012)
62. Maji, A., Kuila, C., Murmu, N.C., Kuila, T.: Facile fabrication of covalently grafted functionalized graphene-PVDF composites as strain sensor with tunable sensitivity: application to the structural health and human motion monitoring. Sens. Actuators A Phys. **369**, 115081 (2024)
63. Rathod, V.T., Roy Mahapatra, D., Jain, A., Gayathri, A.: Characterization of a large-area PVDF thin film for electro-mechanical and ultrasonic sensing applications. Sens. Actuators A Phys. **163**(1), 164–171 (2010)

64. Abdolmaleki, H., Haugen, A.B., Merhi, Y., Nygaard, J.V., Agarwala, S.: Inkjet-printed flexible piezoelectric sensor for self-powered biomedical monitoring. Mater. Today Electron. **5**, 100056 (2023)
65. Wang, B., Wang, B., Zhang, Y., Ma, S., Yang, X., Feng, Y., Liu, C., Shen, C.: Superhydrophobic porous polyvinylidene fluoride monolith with outstanding environmental suitability for highefficient continuous oil/water separation under harsh conditions. J. Environ. Chem. Eng. **10**(3), 107480 (2022)
66. Xiong, J., Wang, L., Liang, F., Li, M., Yabuta, Y., Iqbal, M.A., Mayakrishnan, G., Shi, J., Kim, I.S.: Flexible piezoelectric sensor based on two-dimensional topological network of PVDF/DA composite nanofiber membrane. Adv. Fiber Mater. **6**(4), 1212–1228 (2024)
67. Lee, J., Ahn, J., Jin, H., Lee, C.H., Jeong, Y., Lee, K., Seo, H.-S., Cho, Y.: A funnel type PVDF underwater energy harvester with spiral structure mounted on the harvester support. Micromachines. **13**(4), 579 (2022)
68. Uddin, M., Alford, S., Aziz, S.M.: Evaluating energy generation capacity of PVDF sensors: effects of sensor geometry and loading. Materials. **14**(8), 1895 (2021)
69. Saha, T., Wang, X., Padhye, R., Houshyar, S.: A review of recent developments of polypropylene surgical mesh for hernia repair. OpenNano. **7**, 100046 (2022)
70. Shehata, N., Nair, R., Boualayan, R., Kandas, I., Masrani, A., Elnabawy, E., Omran, N., Gamal, M., Hassanin, A.H.: Stretchable nanofibers of polyvinylidenefluoride (PVDF)/thermoplastic polyurethane (TPU) nanocomposite to support piezoelectric response via mechanical elasticity. Sci. Rep. **12**(1), 8335 (2022)
71. Nurmakanov, Y., Kalimuldina, G., Nauryzbayev, G., Adair, D., Bakenov, Z.: Structural and chemical modifications towards high-performance of triboelectric nanogenerators. Nanoscale Res. Lett. **16**(1), 122 (2021)

PVDF Nanocomposites for Advanced Functional Applications

Bharat Bhushan Bhatt and Dipti Gupta

Abstract Poly(vinylidene fluoride) (PVDF) nanocomposites represent a transformative class of functional materials that synergistically integrate high-performance polymers with nanoscale additives to meet the evolving demands of modern technologies. Incorporation of 2D materials such as graphene and functionalized carbon nanotubes (CNTs) has been shown to act as effective nucleating agents, increasing β-phase formation up to 100% and piezoelectric response–d_{33} values exceeding 50 pC/N against 20–30 pC/N of bulk PVDF. Electrical conductivity reaches 1.2×10^3 $cm^2/V{\cdot}s$ in annealed graphene/PVDF systems, alongside dielectric constant enhancements of over 300%. Hybrid PVDF–perovskite composite, enabled by defect engineering and interfacial coupling, demonstrated modulation of its optoelectronic behavior through mechanical stimulus.

These nanocomposites are also widely utilized in self-powered wearable electronics, electromagnetic interference (EMI) shielding, and piezoelectric nanogenerators (PENGs), with record performances such as approximately 2.5 mW/cm^2 power density and over 24% efficiency in perovskite-based solar cells, ultrasensitive pressure detection capabilities (<10 Pa) by graphene-embedded wearable sensing membranes. To support industrial translation, scalable processing techniques—including electrospinning, solution casting, and in situ perovskite crystallization—are critically evaluated for their compatibility with roll-to-roll manufacturing and 3D printing, further broadening the commercial viability and application landscape of PVDF nanocomposites. The integration of PVDF nanocomposites into flexible electronics and Internet-of-Things (IoT) platforms holds promise for revolutionizing energy autonomy and human–machine interfaces.

This chapter synthesizes foundational insights regarding how nanofillers modify PVDF's electroactive β-phase and its associative properties and examines the mechanisms by which these nanocomposites are used for certain prominent functional applications. Further, it outlines a strategic roadmap for commercialization, posi-

B. B. Bhatt · D. Gupta (✉)
Department of Metallurgical Engineering and Materials Science, Indian Institute of Technology Bombay, Mumbai, India
e-mail: diptig@iitb.ac.in

D. Punetha (ed.), *Polyvinylidene Fluoride (PVDF)*, Engineering Materials,
https://doi.org/10.1007/978-981-95-6718-8_9

tioning PVDF nanocomposites as cornerstone materials for sustainable, multifunctional technologies.

Keywords PVDF nanocomposites · Graphene · CNTs · Perovskites · Industrial scalability · Piezoelectricity · Flexible electronics · Energy harvesting · Hybrid composites

1 Introduction and Fundamentals of PVDF

Poly(vinylidene fluoride) (PVDF) is a semi-crystalline polymer that exhibits unique structural and electroactive properties, making it superior to many other polymers for applications such as sensing, actuation, energy harvesting, and flexible electronics. Its molecular structure, composed of alternating $-CH_2-$ and $-CF_2-$ units, allows multiple crystalline phases (α, β, γ, δ, ε), with the electroactive β-phase being the most important due to its all-trans conformation and large dipole moment, which imparts strong piezoelectric, pyroelectric, and ferroelectric properties [1, 2]. PVDF's ability to form various polymorphs is critical for tailoring its functional properties (Fig. 1).

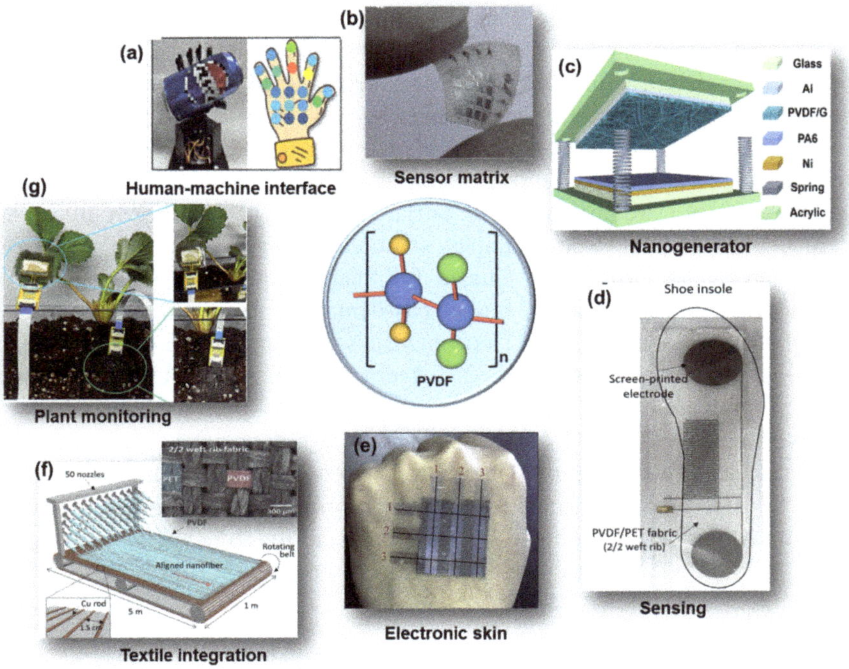

Fig. 1 Various applications of PVDF nanocomposites: (**a**) human–machine interface [3]; (**b**) sensor matrix for strain measurement [4]; (**c**) triboelectric nanogenerator [5]; (**d**) wearable sensor [6]; (**e**) electronic skin [7]; (**f**) textile integration [6]; (**g**) plant monitoring [8]

1.1 PVDF's Unique Structural Advantages

PVDF's flexibility and high piezoelectric response enable sensitive detection of mechanical and acoustic stimuli. For example, electrospun PVDF nanofibers have been used to fabricate flexible acoustic frequency sensors that exhibit broad frequency detection with high sensitivity and mechanical robustness, suitable for wearable health monitoring [9]. The polymer's conformability allows intimate contact with irregular surfaces, enhancing sensor performance, whereas its relative inertness against corrosive environments makes it attractive as a base functional material for several sensing applications.

Conversely, the electrostrictive and piezoelectric properties of PVDF allow it to convert electrical signals into mechanical deformation efficiently. Lightweight and flexible PVDF actuators have been used to fabricate artificial muscles and micro-actuators in MEMS devices, outperforming brittle ceramic piezoelectrics [10]. The fast response and durability of PVDF films make them ideal for biomedical and robotics applications.

The widest use of the piezoelectric properties of PVDF has been for energy harvesting such as piezoelectric nanogenerators (PENGs) that convert mechanical energy into electrical energy. Recent studies show that PVDF combined with graphene or perovskite nanocrystals significantly enhances output voltage and current by increasing β-phase content and conductivity [11], enabling self-powered wearable electronics [12].

PVDF's mechanical flexibility, chemical stability, and ease of processing allow its integration into flexible electronic devices such as bendable displays and electronic skins. Its electroactive properties enable sensing and actuation functions in flexible formats [2].

1.2 Fabrication Techniques for PVDF Devices

Common fabrication methods for PVDF-based devices include: (a) solution casting and spin coating for thin films with controlled thickness; (b) electrospinning for producing aligned nanofibers with enhanced β-phase due to mechanical stretching during fiber formation [9]; (c) melt processing for large-scale production in which thermal control is used to induce β-phase; (d) electrical poling that aligns dipoles post-fabrication to maximize piezoelectricity; (e) in situ nanofiller incorporation for embedding nanomaterials during polymer processing to enhance properties [11, 12].

1.3 Limitations of Bulk PVDF

Despite these advantages, pristine PVDF has notable drawbacks: (a) piezoelectric output lower than ceramic counterparts, which limits its energy harvesting efficiency [11, 13]; (b) phase control difficulty that prevents uniform high β-phase content in bulk films; (c) mechanical brittleness of bulk films under cyclic stress, reducing durability; (d) low electrical conductivity that limits charge transport and sensor sensitivity; (e) processing constraints optimized for bulk PVDF may not yield optimal microstructures for electroactivity [2].

1.4 Overcoming Limitations via Nanocomposites

These limitations are effectively addressed by forming PVDF nanocomposites with conductive and functional nanomaterials such as graphene, carbon nanotubes (CNTs), and perovskites. As discussed in detail in Sect. 2.1, nanofillers act as nucleating agents for the β-phase, improve mechanical flexibility and toughness, and enhance electrical conductivity and interfacial polarization. For example, the PVDF nanocomposites with expanded graphite and hexagonal boron nitride exhibited significantly improved piezoelectric coefficients and sensor output voltages compared to bulk PVDF [11, 12].

2 Role of Nanostructures in Enhancing Properties

The integration of nanostructures into PVDF matrices has revolutionized the performance of PVDF-based materials. Nanofillers such as graphene, CNTs, h-BN, metal oxides, and perovskite nanocrystals induce significant changes in PVDF's crystalline structure, phase content, and interfacial interactions [1, 9, 12]. These changes are primarily attributed to the following mechanisms:

2.1 Nucleation of Electroactive β-Phase

The nanofillers induce local strain and dipole alignment, facilitating the all-trans conformation of PVDF chains and increasing β-phase content without requiring mechanical stretching or electrical poling. For example, Adaval et al. [12] demonstrated that 2D nanomaterials like hexagonal boron nitride and expanded graphite increase β-phase fraction and improving piezoelectric coefficients [12]. The PVDF/h-BN nanocomposite showed a polar fraction of ~57% at 0.50 wt% of h-BN concentration with an increased β-phase content, whereas the PVDF/EG nanocomposite showed a higher content of γ-phase as compared to β-phase. Sensor devices were fabricated based on PVDF/EG and PVDF/h-BN nanocomposite films, which showed a maximum peak-to-peak voltage of ~24 V (at 0.17 wt% EG) and ~14 V (at 0.16 wt% h-BN), respectively (see Fig. 2). Similarly, ZnO nanoparticles incorporated into PVDF have been shown to increase β-phase content significantly, enhancing piezoelectric response without mechanical stretching or electrical poling

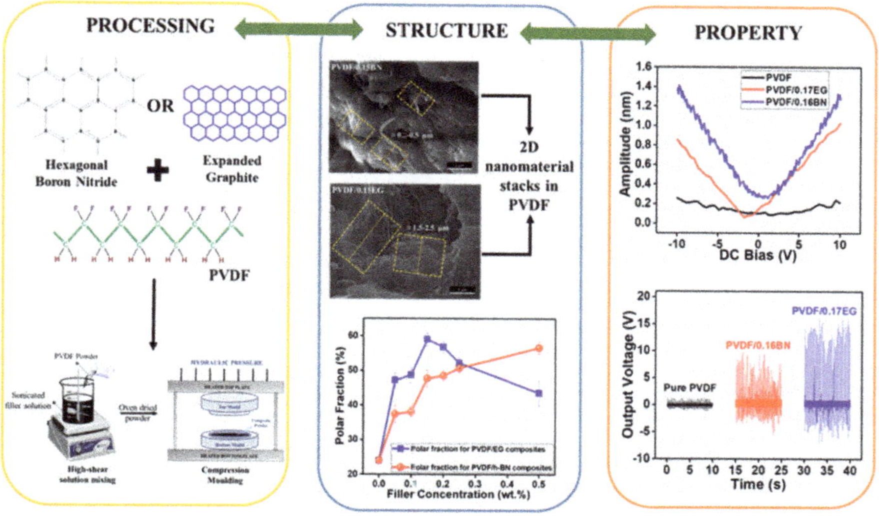

Fig. 2 2D Nanomaterials (hexagonal boron nitride and expanded graphite) incorporated PVDF nanocomposites: processing, morphology, and piezoelectric properties [12]

[14]. Peña-Oliveras et al. [15] studied the effect of iron oxide, magnesium oxide, and zinc oxide mixing with PVDF [15]. Zhang et al. [16] enhanced the dielectric properties of PVDF by making its composites with nano iron oxide-deposited barium titanate hybrid particles to increase dielectric permittivity, reduce dielectric loss, coupled with low conductivity for the electronics industry [16].

2.2 *Interfacial Polarization and Dielectric Enhancement*

Nanostructures create heterogeneous interfaces within the PVDF matrix, leading to Maxwell–Wagner–Sillars interfacial polarization, which enhances dielectric permittivity and energy storage capability [17]. For instance, Zhang et al. [16] reported that Fe_3O_4 nanoparticles and barium titanate hybrids embedded in PVDF significantly increase the dielectric constant due to interfacial polarization effects under an external magnetic field [16]. Similarly, findings of Adaval et al. [12] showed that incorporation of 2D nanomaterials in PVDF improves dielectric properties through enhanced interfacial polarization [12]. Jian et al. [18] utilized 2D nanohybrid $BaTiO_3$ nanoparticle@TiO_2 nanosheet as fillers in PVDF and exploited the core-shell interface as 2D electron scatterer for improving the energy density in high-power capacitors [18] (see Fig. 3). PVDF nanocomposites with noble metal-coated multiwalled carbon nanotubes (nAu–fMWCNTs) exhibit significantly improved dielectric permittivity due to interfacial polarization effects, β-phase formation, and restrained conductive pathways, resulting in low dielectric loss [19].

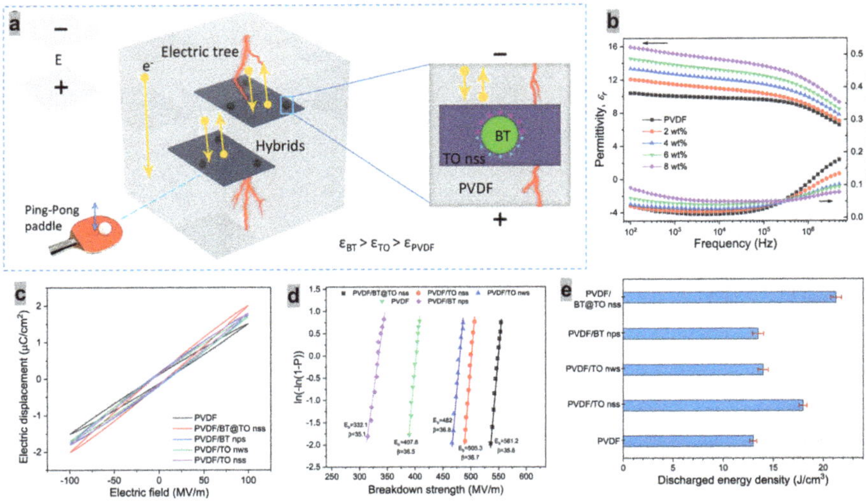

Fig. 3 Demonstration of the mechanisms in the PVDF/BT@TO ns composites: (**a**) schematic showing the role of the filler for electron scattering and interface engineering in the PVDF/BT@TO ns composites; (**b**) frequency-dependent ε_r and tanδ of the PVDF/BT@TO ns composites at room temperature; (**c**) D–E hysteresis loops of PVDF and the various composites (PVDF/TO nss, PVDF/TO nws, PVDF/BT nps, and PVDF/BT@TO nss) with a 4 wt% filler loading at 100 MV/m; (**d**) Weibull distribution of PVDF and the various composites (PVDF/TO nss, PVDF/TO nws, PVDF/BT nps, and PVDF/BT@TO nss) at 4 wt% filler loading; (**e**) U_d of PVDF and the various composites at 4 wt% filler loading [18]

Similarly, $BaTiO_3$ nanowires in PVDF-HFP matrices enhance dielectric properties through superior interfacial polarization and high aspect ratios, as demonstrated by stress-induced alignment under external fields [20]. Additionally, core-shell structured $BaTiO_3$@SiO_2 nanoparticles in PVDF composites mitigate electric field concentration at filler interfaces, balancing high permittivity and breakdown strength [20]. Padurariu et al. [19] combined 3D FEM simulations with experimental data to study the effect of microstructures on the dielectric properties in PVDF-Perovskite nanocomposites [19] (see Fig. 4).

2.3 *Mechanical Reinforcement*

Nanostructures significantly enhance the mechanical properties of PVDF by reinforcing effective load transfer, mechanical interlocking, and restricting polymer chain mobility. Carbon-based nanofillers such as carbon nanotubes (CNTs), graphene, carbon black (CB), and silicon dioxide (SiO_2) nanoparticles have been extensively studied for this purpose. The mechanical enhancement depends on the type of PVDF polymer matrix, the filler type, and their content. For example, carbon nanotubes incorporated into various PVDF copolymers (PVDF-HFP, PVDF 6010, PVDF 5130) increase stiffness and tensile strength, with PVDF-HFP composites showing strain values up to 300% (see Fig. 5), which is about ten times higher than neat

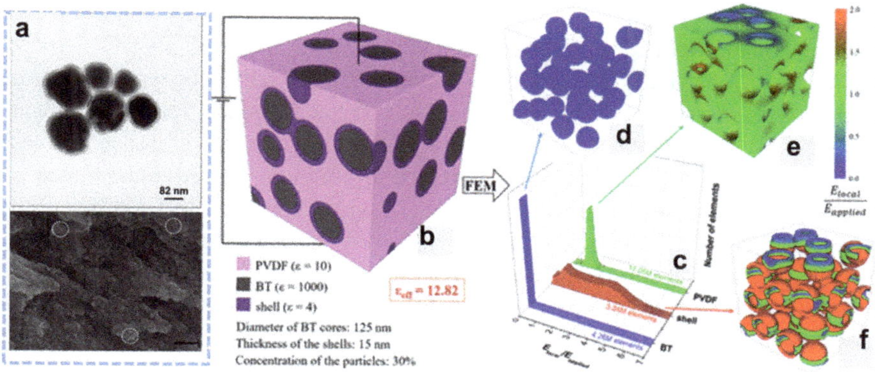

Fig. 4 Role of microstructures in the dielectric properties of PVDF-based nanocomposites containing high-permittivity fillers for energy storage: (**a**) morphology of $BaTiO_3@SiO_2$; (**b**) microstructure of a PVDF composite with $BaTiO_3@SiO_2$ inclusions (30 vol%); (**c**) local field distribution expressed on a number of elements basis and the local field maps in (**d**) the BT inclusions; (**e**) polymer matrix; and (**f**) silica shell, simulated by FEM. Shell thickness is 15 nm [19]

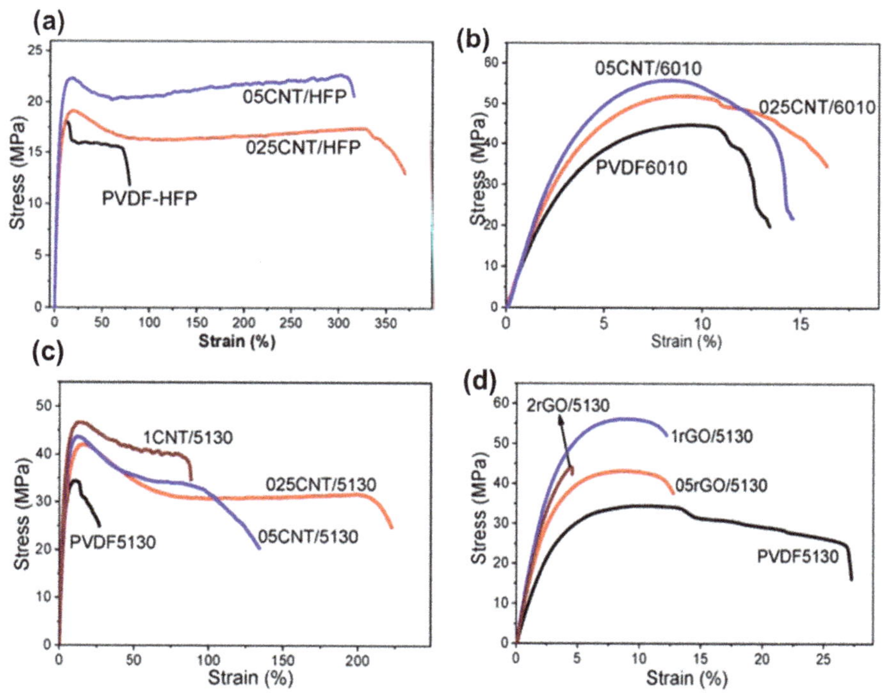

Fig. 5 Stress–strain response for PVDF and the corresponding composites for (**a**) PVDF-HPF, (**b**) PVDF6010 and (**c**) PVDF5010 for different CNT contents. (**d**) PVDF5130 reinforced with rGO [21]

PVDF [21]. This effect is ascribed to the heterogeneity of the composites caused by filler agglomeration [22] that limits the mechanical strain for all PVDF matrices, PVDF-HFP, 6010, and 5130. However, excessive CNT loading can lead to agglomeration, causing brittleness and reduced strain at break. Similarly, reduced graphene oxide (rGO) composites tend to decrease maximum strain but improve stiffness [21]. This behavior highlights the importance of optimizing filler content and dispersion for mechanical performance. Furthermore, graphene-related materials (GRMs) and CNTs reinforce PVDF by transferring stress efficiently from the polymer matrix to the nanofillers. Raman spectroscopy studies confirm effective stress transfer at the nanoscale, which correlates with improved mechanical properties [23]. This nanoscale reinforcement mechanism is crucial for applications requiring both flexibility and mechanical robustness. Recent advances also include PVDF nanofibers doped with ceria nanoparticles, where electrostatic interactions and hydrogen bonding between PVDF chains and ceria enhance mechanical integrity alongside piezoelectric performance (see Fig. 6) [24].

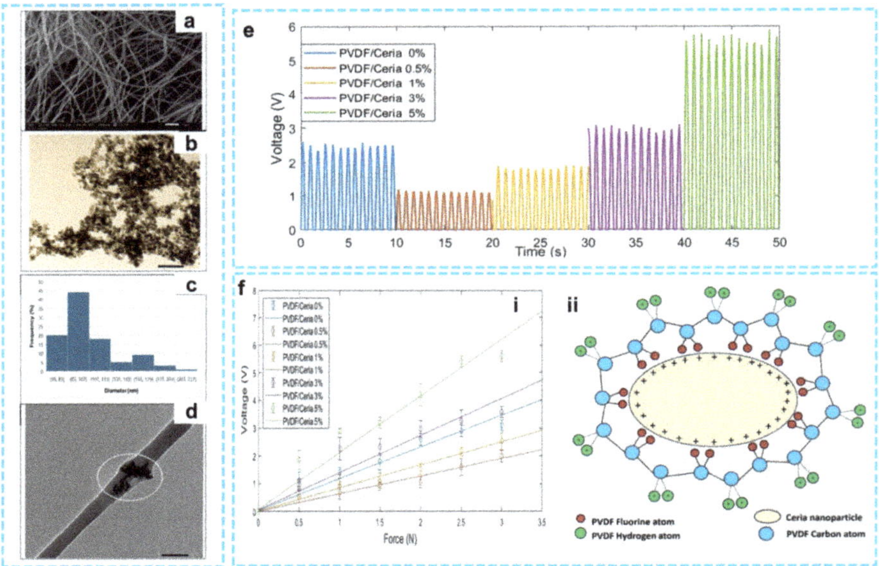

Fig. 6 Nanoscale reinforcement of PVDF fibers with ceria nanoparticles to enhance mechanical integrity and piezoelectric performance: (**a**) SEM image of ceria embedded PVDF electrospun nanofibers in 2 μm scale; (**b**) TEM image of ceria NPs in 50 nm scale; (**c**) nanofibers diameter distribution curve of PVDF/ceria with ratio 95:5 wt%, respectively; (**d**) TEM image of 5% PVDF/Ceria; (**e**) AC sinusoidal voltage developed at 3 N force for different PVDF/ceria concentrations; (**f**) (**i**) The peak-to-peak output voltage generated by the piezomats for different applied forces/compressed loads. (**ii**) Schematic representation of the dipole alignment of PVDF according to nucleating agent of ceria NPs [24]

2.4 Improved Electrical Conductivity and Charge Transport

Conductive nanostructures such as multiwalled carbon nanotubes (MWCNTs), graphene, and hybrid fillers enhance PVDF's electrical conductivity by forming percolation networks and optimizing charge transport pathways. For example, thermally treated MWCNTs in PVDF composites exhibit a dramatic increase in electrical conductivity from 2×10^{-12} S/cm to 4×10^{-8} S/cm due to defunctionalization of MWCNT surfaces, which restores their intrinsic conductivity while maintaining mechanical stability [25]. This thermal retro-Diels–Alder process effectively reduces the percolation threshold, enabling efficient charge transport at low filler loadings.

Hybrid fillers, such as Co_3O_4-functionalized MWCNTs, further enhance conductivity by leveraging synergistic interactions between the polymer matrix and nanofillers. PVDF composites with 0.5 wt% Co_3O_4 and 0.3 wt% MWCNTs exhibit significant improvements in dielectric properties and electrical conductivity, attributed to strong interfacial interactions and reduced agglomeration [26]. Similarly, $BaTiO_3$ nanofillers in PVDF increase AC conductivity by three orders of magnitude compared to pure PVDF, with charge transport governed by ionic hopping and Schottky emission at higher electric fields [27].

Annealing graphene/PVDF composites at 150 °C optimizes electrical conductivity by redistributing graphene sheets and reducing defects, achieving a Hall mobility of 1.2×10^3 cm²/V/s [28] (see Fig. 7).

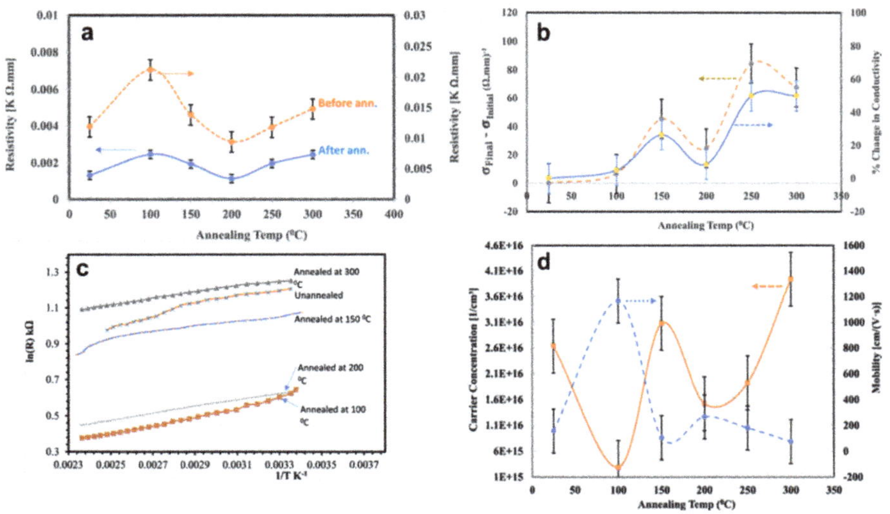

Fig. 7 Tuning the electrical properties of graphene/polyvinylidene fluoride (Gr/PVDF) nanocomposite films by thermal annealing: (**a**) Resistivity of Gr/PVDF vs temperature before and after annealing. (**b**) Change of conductivity in Gr/PVDF films with different annealing temperatures (right side gives $(\sigma_{annealed} - \sigma_{unannealed})/\sigma_{unannealed}) \times 100\%$, and the left side gives $(\sigma_{annealed} - \sigma_{unannealed})$. (**c**) Arrhenius plots of resistance (R) versus reciprocal of temperature (1/K) for Gr/PVDF composites annealed at different temperatures. (**d**) Carrier concentration and Hall mobility of Gr/PVDF with annealing temperature [28]

On of the key parameter in loading PVDF with a conductive filler such as graphene is the desired filler concentration. To optimize dielectric property, one requires little amount of filler whereas to establish conductive percolation network, much higher amount is needed. For instance, Hung et al. [29] studied very high loading of PVDF with graphene (Fig. 8) [29]. They did not get any conductivity till 25% loading but β-phase appeared. Conductivity appeared at 50% loading (see Fig. 8b). Applied voltage reversibly expands membrane thickness (Fig. 8e), adjusting

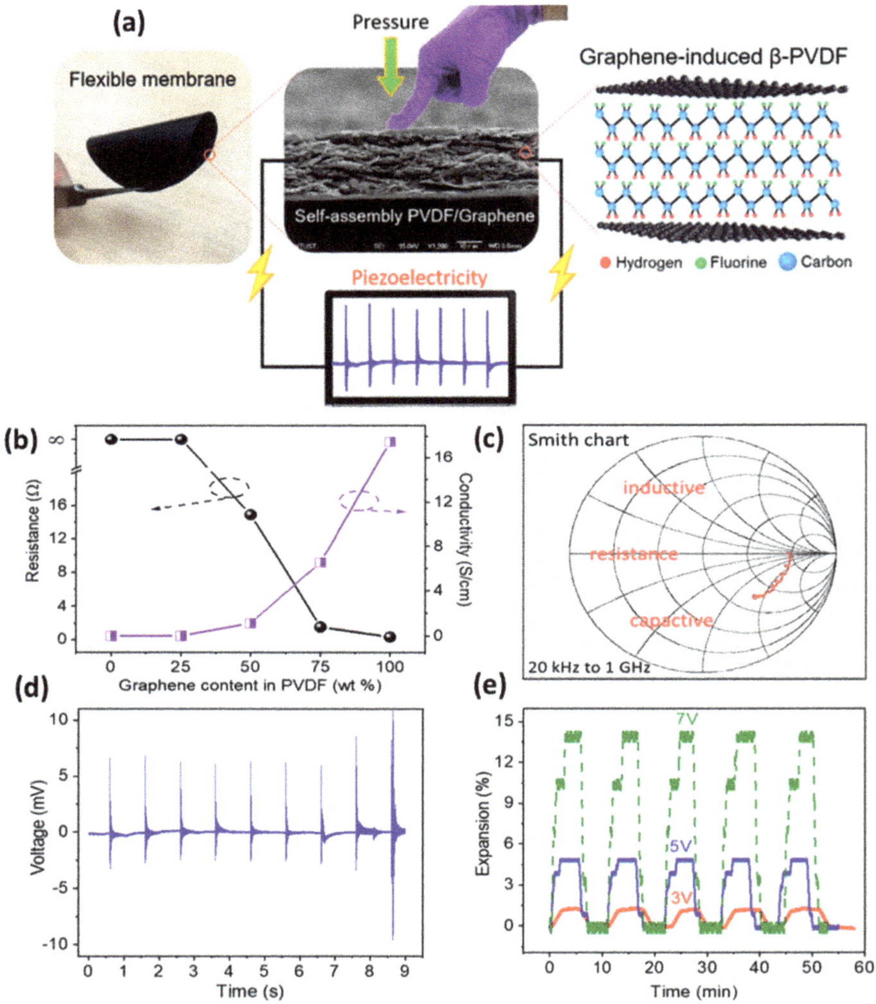

Fig. 8 Electrical tunable PVDF/graphene membrane for controlled molecule separation (**a**) scheme for piezoelectricity and molecular structure of flexible self-assembled PVDF/graphene membrane. Electrical properties of the PVDF/graphene composite membrane: (**b**) four-point probe, (**c**) Smith chart, (**d**) piezoelectric response, and (**e**) motorized microscope of the composite membrane with different applied voltages [29]

nanochannels via electroactive PVDF deformation. Graphene's conductivity is essential for transmitting this stimulus. Conductivity-enabled voltage control tailors free volume, creating selective water channels. Higher voltages enhance hydrophilicity (contact angle reduction) and molecular sieving.

3 Hybrid Composites with Graphene, Carbon Nanotubes (CNTs), and Perovskites

3.1 Graphene-Based PVDF Nanocomposites

Graphene and graphene-related additives such as rGO nanosheets in PVDF transform the crystalline structure from the non-polar α-phase to the polar β-phase, enhancing piezoelectricity and pyroelectricity significantly [30, 31]. This phase transformation is attributed to strong interfacial interactions, including hydrogen bonding and π–π stacking between PVDF chains and graphene derivatives, which align the polymer chains into the all-trans conformation characteristic of the β-phase [32].

Graphene and rGO provide conductive pathways within the polymer matrix, reducing the percolation threshold and enabling electron transport at low filler loadings [33]. For example, PVDF/rGO composites show a substantial increase in dielectric constant (from ~6 to over 20 at low rGO content) with minimal dielectric loss, making them ideal for capacitive applications [30]. The electrical conductivity can be further tuned by thermal annealing, which improves filler dispersion and interfacial bonding, as demonstrated in graphene/PVDF films annealed at 150 °C [28].

The high aspect ratio and exceptional mechanical properties of graphene nanosheets reinforce the polymer matrix, increasing tensile strength and Young's modulus [34]. Additionally, the thermal decomposition temperature of PVDF increases by approximately 15 °C upon rGO incorporation, indicating improved thermal stability [30]. This enhancement is due to the barrier effect of graphene sheets that hinder thermal degradation and improve heat dissipation.

The dispersion state of graphene derivatives within PVDF critically affects the composite properties. GO, with abundant oxygen-containing groups, disperses better in polar solvents and PVDF, improving interfacial adhesion and uniformity. Strategies such as functionalization of graphene or combining GO with rGO help optimize dispersion and synergistically improve the composite's functional properties [35].

3.2 Applications of Graphene-Based Nanocomposites

Graphene and graphene-like materials offer a wide range of material properties to exploit in cost-effectively fabricating different kinds of devices. An example of the diverse applicability of tailor-made rGO/PVDF nanocomposites is given below in Fig. 9 and Fig. 10 by Yuanfei et al. [36] in which they showed how rGO's versatility (conductivity, adsorption, photoresponse) complements PVDF's mechanical and piezoelectric properties, enabling integration of sensors and energy units into a single compact, wearable E-skin system. The E-skin incorporates micro-supercapacitors (MSCs), a pressure sensor, a photodetector, and a gas sensor, all fabricated from rGO-on-PVDF-NFs, simplifying design and manufacturing [36]. The construction and properties of MSCs are shown in Fig. 9. For MSCs, rGO's high electrical conductivity creates percolation pathways within the PVDF matrix, enabling efficient charge transport. This boosts the specific capacitance (595.4 F/g at 5 A/g) by facilitating ion adsorption and electron transfer at the electrode–electrolyte interface. The nanofiber morphology of PVDF increases surface area, enhancing ion accessibility, while rGO prevents structural degradation during charge-discharge cycles (97.1% capacitance retention after 3000 cycles).

Further, the oxygen-containing groups in rGO improve wettability with the PVA-KOH gel electrolyte, promoting ion diffusion and interfacial charge storage. For pressure sensing, rGO-coated PVDF-NFs deform under pressure, altering inter-fiber contact points. This modulates the conductive network, translating mechanical stress into measurable resistance changes. The high sensitivity (0.13 μA per 5 kPa) arises from rGO's high surface area and piezoresistive response. PVDF's elasticity ensures structural resilience, while rGO's adhesion to the nanofibers maintains conductivity during repeated deformation. This enables detection of subtle forces, such as a 35 mg mung bean or physiological pulses. For gas sensing, rGO's high surface area and residual oxygen groups provide active sites for gas adsorption (e.g., acetone), see Fig. 10j. Gas molecules alter rGO's electron density, changing conductivity. PVDF-NFs act as a deformable scaffold, maintaining rGO's integrity during bending (Fig. 10d, f, i). To fabricate a photodetector, rGO's broad optical absorption (UV to visible) generates electron–hole pairs upon illumination (Fig. 11). PVDF's piezoelectric properties may assist in charge separation under strain, enhancing photocurrent generation. Whereas, rGO's conductive network accelerates carrier mobility, reducing recombination. The photo-responsivity (up to 0.71 at 233 nm UV light) and stability under bending stem from rGO's optoelectronic properties and PVDF's mechanical flexibility.

This ensures consistent performance (response/recovery times of 5.5 s/30 s) even under mechanical stress. Usually, piezoelectric devices based on PVDF, as well as other materials, are in the shape of one-dimensional nanofibers [37] and two-dimensional nanofilms [38]. The possible deformation and contact area of these low-dimensional devices are small and, consequently, the conversion efficiency is also small and is suitable for miniature devices (Fig. 11).

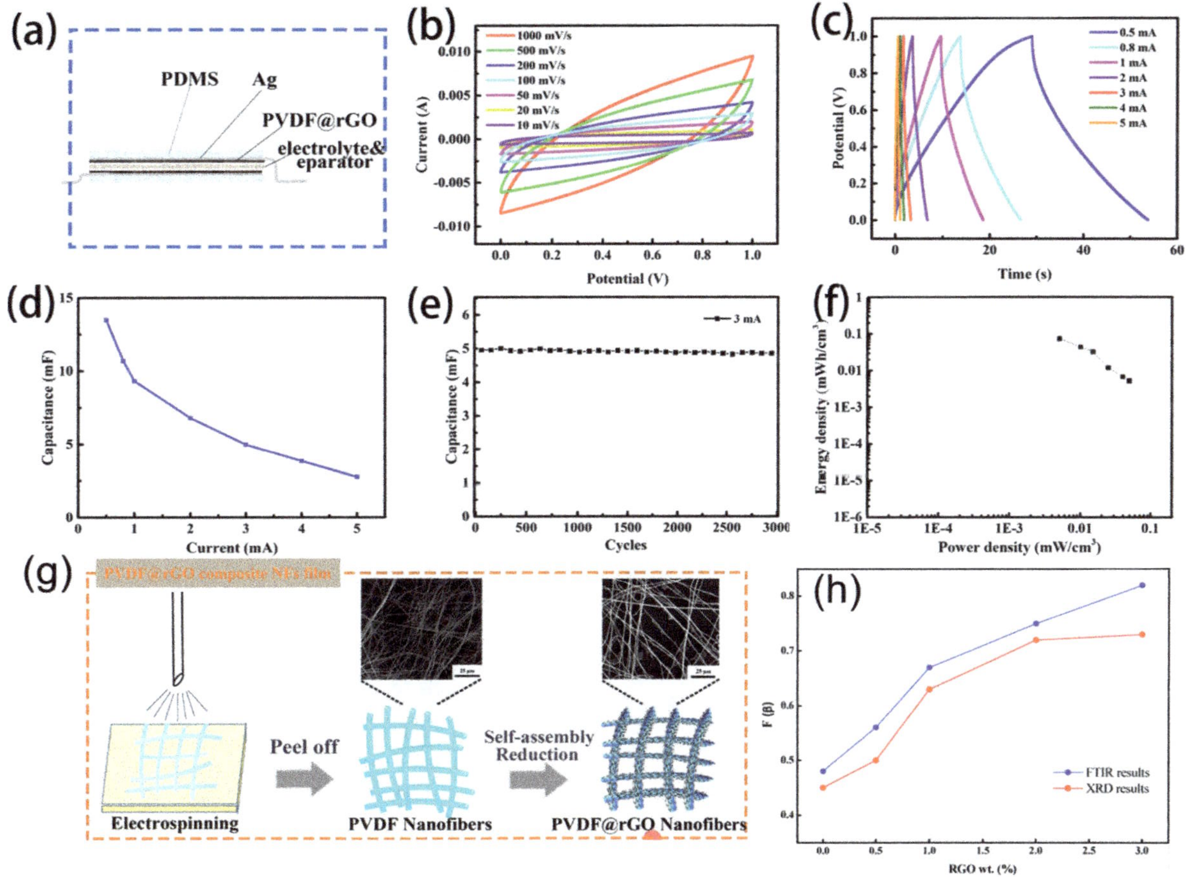

Fig. 9 Micro-supercapacitors with the rGO-on-PVDF-NFs as the electrodes: (**a**) device structures; (**b**) CV curves of the MSCs; (**c**) galvanostatic CD curves of the MSCs; (**d**) capacitances of the MSCs at varied galvanostatic CD current densities; (**e**) cycling performance of the MSCs at the current density of 3 mA; (**f**) energy and power densities of the MSCs; (**g**) schematic illustration of the synthesis of rGO-on-PVDF-NFs [36]; and (**h**) the dependence of β-phase fraction on RGO content [31]

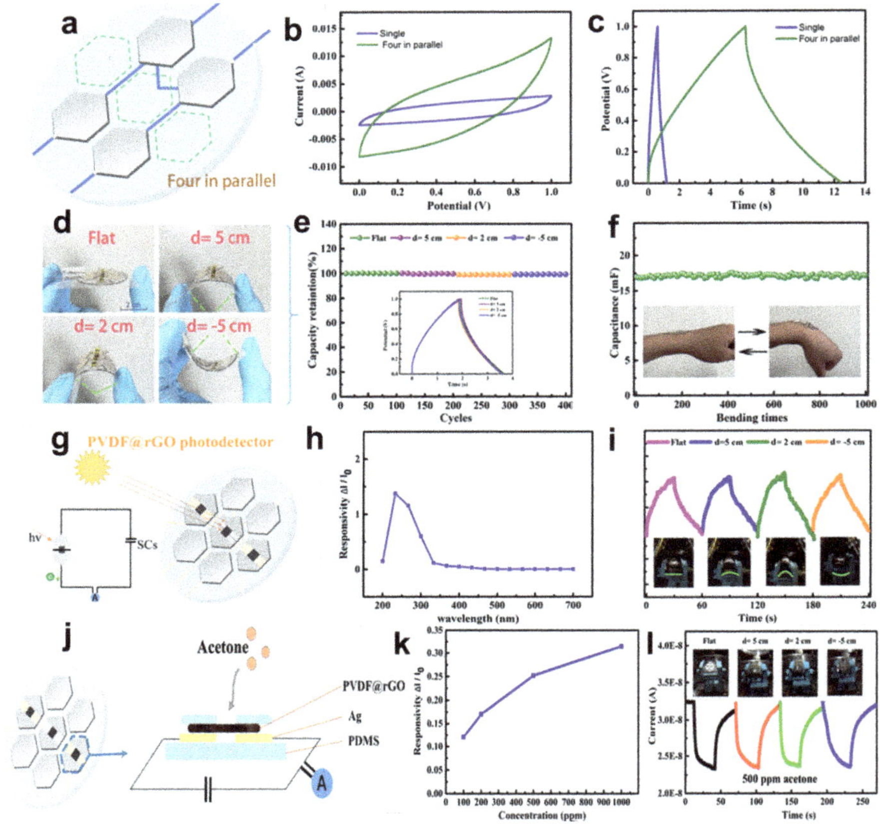

Fig. 10 All rGO-on-PVDF-nanofibers based self-powered electronic skins: (**a**) four micro-supercapacitor (MSCs) in parallel on the multifunctional E-skins (**b**) CV and (**c**) galvanostatic CD curves of one MSC and four MSCs in parallel at the scan rate of 100 mV/s and the current of 5 mA, respectively; (**d**) flexibility performance of the four MSCs in parallel; (**e**, **f**) the capacitance stability of the MSCs during repeatedly bending–restoring cycles at galvanostatic current of 10 mA; (**g**) sensing mechanism of the photodetector powered by integrated MSCs; (**h**) the current on/off ratios of the device upon different wavelength light; (**i**) sensing of the device under different bending state exposed to 20 mW/cm^2 white light; (**j**) gas sensor (e.g. acetone) on the multifunctional E-skins; (**k**) the responsivity of the sensor under different acetone concentrations; (**l**) sensing stability of the device under different bending state in 500 ppm acetone [36]

To remove this drawback, Song et al. [40] fabricated 3D piezoelectric devices of higher stress-to-electricity conversion efficiency using a laser-based 3D printing technique, called selective laser sintering (SLS). Here again, they found graphene/PVDF nanocomposite as suitable for SLS 3D printing due to the synergistic interaction between PVDF chains and graphene nanoparticles that provided an excellent workable material for laser-based SLS with enhanced stress-to-electricity conversion efficiency [40].

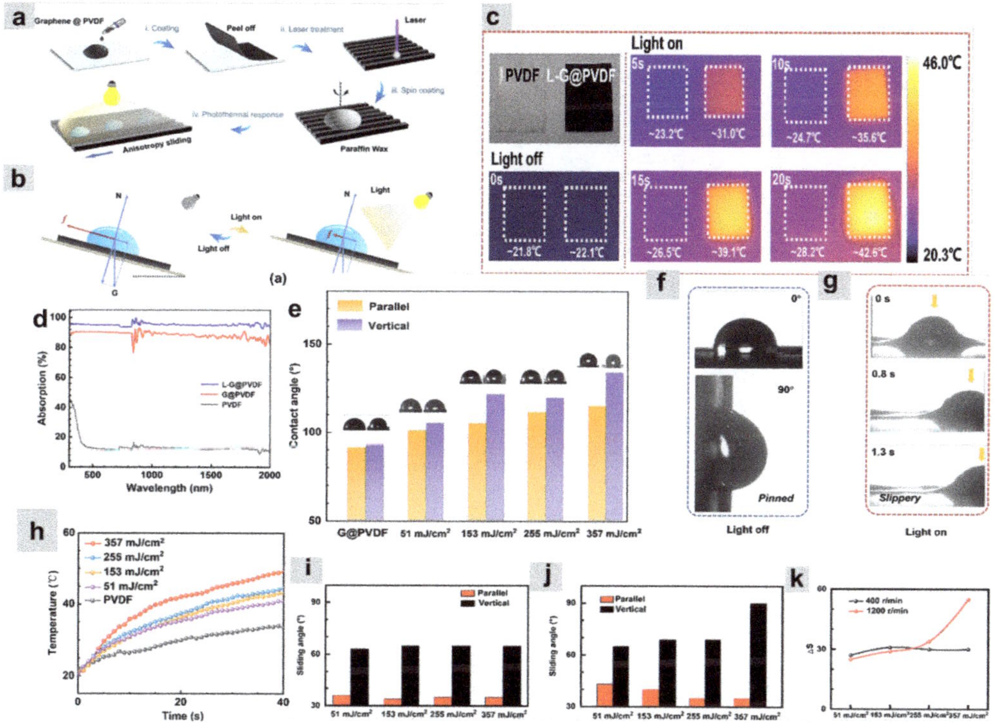

Fig. 11 Photothermal responsive slippery surfaces based on laser-structured graphene@PVDF composites: (**a**) schematic diagrams for preparing and (**b**) responsive mechanism of photothermal responsive slippery surfaces based on laser-structured G@PVDF composites; (**c**) photographs and infrared thermal images of PVDF and L-G@PVDF; (**d**) absorption spectra of PVDF, G@PVDF, and L-G@PVDF films; (**e**) wettability: water contact angles of G@PVDF and L-G@PVDF under different laser fluence; (**f**) droplet pins to the paraffin-infused L-G@PVDF surface without light irradiation; (**g**) a water droplet slides on the surface with light irradiation; (**h**) surface temperature varies of PVDF and L-G@PVDF when exposed to 191 mW/cm^2 light irradiation for 40 s; sliding angles of paraffin-infused L-G@PVDF surface at the spin-coating speed of (**i**) 400 r/min and (**j**) 1200 r/min. (**k**) ΔS variation under different laser fluences and spin-coating speeds [39]

3.3 CNTs-Based Nanocomposites with PVDF

The presence of CNTs significantly enhances the electrical conductivity of PVDF nanocomposites by forming conductive networks within the polymer matrix. This reduces the percolation threshold to as low as 1–2 wt% CNT loading, enabling a transition from insulating to conductive behavior with conductivities reaching up to 0.05 S/cm at higher CNT contents [26, 41]. Functionalization of CNTs, such as silanization with 3-aminopropyl triethoxysilane (APTES), further improves dispersion and interfacial bonding, resulting in a substantial increase in the β-phase fraction (up to 103%) and enhanced dielectric, ferroelectric, and piezoelectric properties [42]. For example, PVDF/MWCNT-APTES nanocomposites exhibited a piezoelectric response with an amplitude signal of 464 mV and a power density of ~456 μW/cm^2 in nanogenerator applications [42].

Mechanically, CNTs reinforce the PVDF matrix due to their high aspect ratio and exceptional tensile strength, leading to improvements in Young's modulus and thermal stability. The thermal decomposition temperature of PVDF increases by tens of degrees celsius upon CNT incorporation, attributed to the barrier effect of CNTs that hinder thermal degradation and enhance heat dissipation [26, 43]. Thermal analysis (TGA, DSC) confirms that the addition of CNTs elevates the onset temperature of degradation and improves the overall thermal stability of the nanocomposites [26, 43].

The dispersion state of CNTs within PVDF is critical for optimizing composite properties. Pristine CNTs tend to agglomerate due to strong van der Waals interactions, which can deteriorate mechanical and electrical performance. Surface functionalization (e.g., silanization) or combination with other nanofillers (e.g., Co_3O_4) improves CNT dispersion, interfacial adhesion, and synergistically enhances the β-phase formation and composite properties [26, 42]. Low-temperature processing techniques combined with functionalized CNTs have been shown to promote self-organization of dipoles in PVDF, further boosting the electroactive β-phase and related properties [42].

3.4 Applications of CNTs–PVDF Nanocomposites

3.4.1 Piezoelectric Energy Harvesting and Nanogenerators

Carbon nanotubes (CNTs) contribute to PVDF's piezoelectric energy harvesting capability in two ways: (a) by working as a nucleating agent for increasing the available electroactive β-phase of PVDF and (b) by forming conductive networks that facilitate charge transport and enhance device efficiency.

Bairagi et al. [44] demonstrated that incorporating 0.1 wt% CNTs into a PVDF/KNN (potassium sodium niobate) composite boosted the nanogenerator's output voltage to ~23.24 V and power density to 52.29 μW/cm^2 under mechanical stress, doubling the performance compared to CNT-free composites [44]. Chandran et al. [42] used surface-functionalized MWCNTs (multi-walled CNTs) in PVDF to increase the β-phase fraction by over 100%, resulting in a piezoelectric response of 464 mV and power density of ~456 μW/cm^2, suitable for wearable energy harvesting devices [42]. PVDF/ZnO/CNT composites have been fabricated for human-motion-based energy harvesting, demonstrating effective conversion of biomechanical energy during walking or running [45].

Badatya et al. [46] reported the fabrication of a lightweight, flexible PVDF–CNT foam nanogenerator with a high β-phase fraction (~65%) induced by CNTs (see Fig. 12). The foam exhibits self-poling behavior due to local electric fields from CNTs, achieving a piezoelectric charge coefficient (d_{33}) of 9.4 pC/N. The nanogenerator generates an output voltage of ~12 V and a current density of 30 nA/cm^2 under small compressive pressure without electrical poling. It also demonstrates excellent hydrophobicity and stable performance under humid conditions, highlighting its potential for wearable and self-powered nanosensors [46]. The synergy between CNTs and PVDF enhances the mechanical flexibility and durability of nanogenerators, making them ideal for portable and wearable electronics [41]. Thermal and mechanical stability improvements due to CNTs ensure long-term reliability of piezoelectric energy harvesters [42].

3.4.2 Flexible and Wearable Sensors

CNT-PVDF composites exhibit high gauge factors and excellent durability under cyclic bending, making them suitable for wearable strain and pressure sensors. Yang et al. [7] developed a wearable capacitive E-skin sensor using electrospun

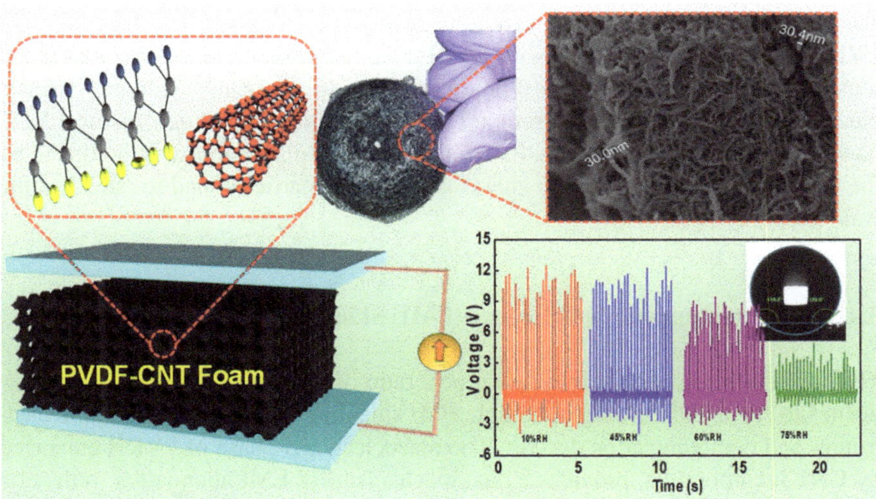

Fig. 12 Hydrophobic PVDF–CNTs foam based piezoelectric nanogenerator [46]

PVDF-CNTs composite (see Fig. 13) [7]. Dispersed CNTs (below percolation threshold) created micro-interfaces within PVDF fibers, amplifying interfacial polarization under electric fields. Applied pressure reduces electrode distance ($d \rightarrow d'$), increasing electric field strength ($E = U/d$). This intensifies polarization, boosting permittivity (ε) further [7].

Yang et al (2019) showed how pristine and functionalized CNTs improve strain sensing properties of pure PVDF [7]. Pristine CNTs (ufCNT) exhibit weak interfacial adhesion due to van der Waals forces and high interfacial tension. This creates microvoids at the CNT–PVDF interface, amplifying contact resistance changes under strain, which increase initial resistance and enhance piezoresistive response (Fig. 13e). Whereas mild acid-treated CNTs (fCNT) have polar groups (e.g., –COOH, –OH) on CNT surfaces, enabling dipole–dipole interactions with the polar γ-phase of PVDF. This strengthens interfacial adhesion, improving stress transfer and reducing CNT network disruption under strain. Strong interfaces in fCNT composites resist irreversible network damage during cyclic loading.

For effects of processing methods, low-shear dispersion in solvent in the solution casting method preserves longer CNT lengths (aspect ratio), facilitating percolation at low CNT content (0.35–0.75 wt% ufCNT). High-shear processing of melt mixing shortens CNTs (0.15 wt% fCNT) but improves dispersion via mechanical entanglement (Fig. 13). Longer CNTs (solution casting) enable higher sensitivity in tensile mode (e.g., ufCNT-S: $F = 11.65$ vs. ufCNT-M: $F \approx 8$) (Fig. 13f). This is attributed to the piezoresistance of intact CNTs and tunneling effects. Shorter CNTs (melt mixing) favor the bending mode due to tunneling-dominated resistance changes. It is noteworthy that how the sensing performance changes with CNT wt% (Fig. 13i, j, k).

Apart from enhancing the piezoelectric properties of PVDF, CNTs—owing to its structural versatility—can also work as a bridge between the organic polymer chain of PVDF and the inorganic additives like ZnO that are otherwise difficult to integrate with it. To overcome the difficulty of organic–inorganic interface between PVDF and ZnO, CNTs attach with PVDF chains through hydrogen bonding and thus facilitate numerous sites for β-phase creation [46, 47]. ZnO, itself, has intrinsic piezoelectric property which promotes the polarization generation in the PVDF nanofibers when dispersed in the PVDF matrix. Their dispersibility is facilitated by the CNTs attached to the PVDF chains as ZnO nanoparticles tend to 'stick' to the walls of CNTs [48].

3.4.3 Electromagnetic Interference (EMI) Shielding

The high electrical conductivity and aspect ratio of CNTs enable PVDF nanocomposites to serve as lightweight, flexible EMI shielding materials for wearable electronics and communication devices. The piezoelectric β-phase of PVDF, enhanced by CNT incorporation, provides a dual mechanism for EMI attenuation: reflection from conductive CNT networks and absorption via dipole polarization. CNT–PVDF films (<1 mm thickness) achieve >20 dB shielding effectiveness (SE) in the X-band

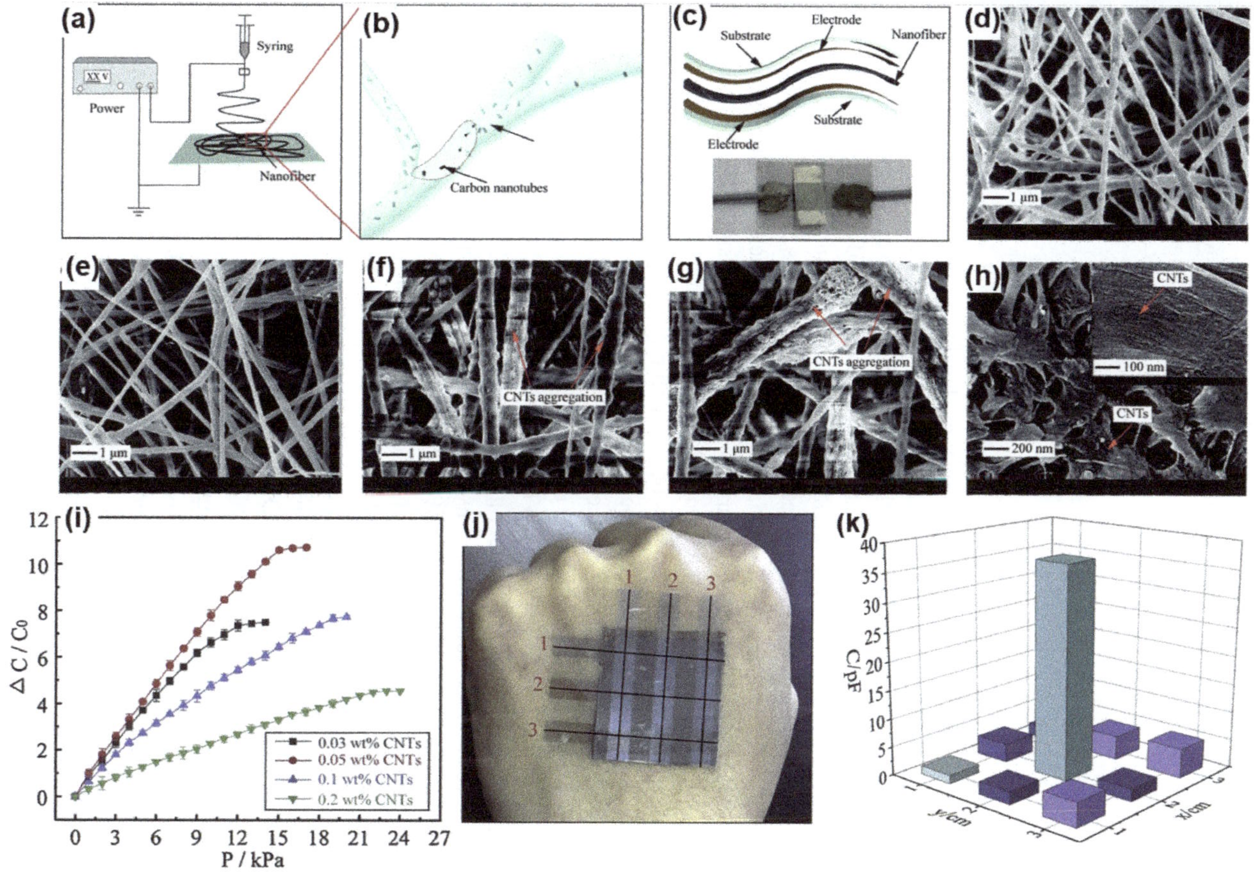

Fig. 13 Wearable capacitive sensor based on PVDF–CNTs nanocomposite: (**a**) schematic of the fabrication of the composite nanofiber. (**b**) Schematic diagram of the composite nanofiber. (**c**) Diagram of the sensor. (**d–h**) The microstructures of the composite nanofiber membrane with 0.03, 0.05, 0.1, 0.2 wt% carbon nanotube additions, respectively. (**i**) The relative change in capacitance of the sensor with different weight ratio CNTs addition under pressure applied; (**j, k**) The diagram of sensor network attached on the hand [7]

(8–12 GHz), ideal for wearable devices and foldable displays [49]. Pourzahedi et al. [50] showed that compared to conventional aluminium shields, CNT–PVDF composites reduce shielding mass by >50% while maintaining comparable shielding effectiveness (SE ~75 dB). This mass reduction decreases satellite launch fuel consumption by 6%, yielding a 30% reduction in cumulative energy demand (CED) over aluminum. The analysis highlights CNT–PVDF's potential for lightweight aerospace applications [50] (Fig. 14).

3.4.4 Dielectric Capacitors

Incorporating CNTs into PVDF enhances dielectric properties by improving charge polarization, electrical conductivity, and energy density while maintaining low dielectric loss. The addition of CNTs facilitates the formation of the electroactive β-phase in PVDF through interfacial interactions between the π-conjugated CNTs and the $-CH_2/-CF_2$ dipoles of PVDF. This results in improved dielectric constant and energy storage density. The percolation threshold of CNTs in PVDF is critical as excessive loading leads to high conductivity and dielectric loss.

For PVDF to be employed as functional materials in fields such as high charge-storage capacitors, electrostriction for artificial muscles, and "smart skins", it is required to enhance its dielectric constant substantially. Addition of CNTs shows that the dielectric property of composites depends on physical properties, preparation method, and interface interaction between the fillers and the polymer, especially the size and shape of CNTs. Wang et al. [51] studied the effect of percolation threshold of CNTs in PVDF matrix on its dielectric properties [51]. They found that the percolation threshold for MWNT/PVDF composites was exceptionally low at 1.61 vol%, attributed to the high aspect ratio and conductivity of MWNTs. Near the percolation threshold, the dielectric constant of the composite reached ~200, 20 times higher than pure PVDF. At 2.0 vol% MWNT loading, the dielectric constant peaked at 300. This enhancement was achieved without chemical functionalization of MWNTs. The dielectric loss remained below 0.4 across all frequencies, making the composites suitable for practical applications. Li et al. [52] increased the dielectric constant of CNTs using functionalization with carboxylic and ester groups. The carboxylic-functionalized MWCNT/PVDF composite achieved a record dielectric constant of 3600 at 1 kHz (8 vol% MWCNT), attributed to interfacial effects and mini-capacitor formation. Ester-functionalized composites reached 2400. Pristine MWCNT composites showed a sharp insulator–conductor transition, while functionalized composites exhibited gradual conductivity increases due to hindered tunneling from surface groups. Dielectric loss rose post-percolation, linked to CNT-induced energy dissipation. The mini-capacitor model explained the high dielectric constants, where functionalized MWCNTs created more isolated conductive networks [52].

The energy density of PVDF–CNT nanocomposites was further increased using innovative methods such as using additives and novel processing protocols. For example, Uyor [53] used a lower concentration of CNTs in the PVDF matrix but

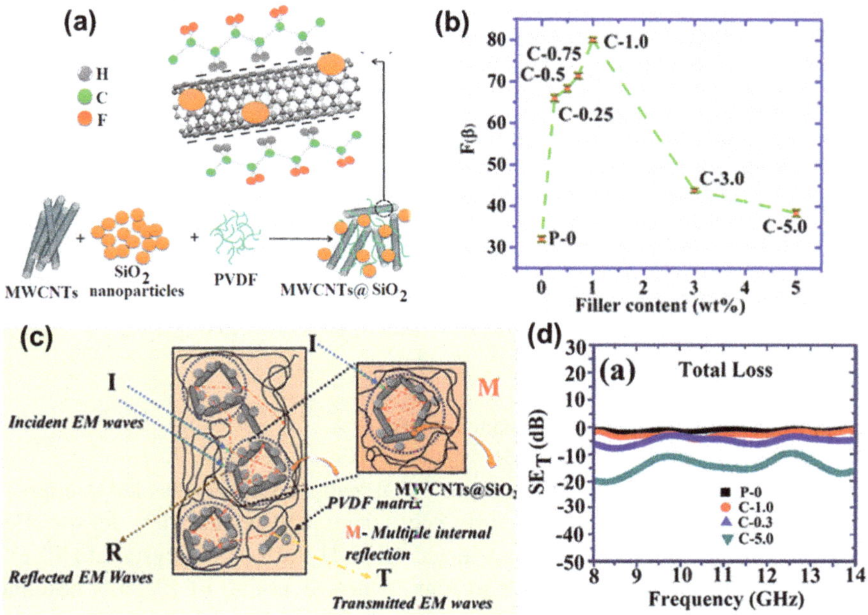

Fig. 14 CNTs@SiO_2 heterogeneous nanofiller-based polymer composites for EMI shield (**a**) schematic representation of the electrostatic interaction between the negative surface charges of the MWCNT@SiO_2 nanoparticles and the positive $-CH_2$ bond of PVDF matrix; (**b**) variation of electroactive β-phase fraction [F(β)] with filler concentration; (**c**) schematic of EMI shielding mechanism for the prepared PVDF composite; (**d**) variation of total EM loss (SE_T) with frequency for the samples C-1.0, C-3.0, and C-5.0 within the frequency range of 8–14 GHz [49]

added another additive, boron nitride (BN), 5–10 wt% for nano reinforcement while preventing the agglomeration of CNTs. As a result, the dielectric constant increased from 9.1 for PVDF to 42.8 for the PVDF–CNT–BN nanocomposite, with a minimum loss of 0.1 registered for PVDF-0.1 wt% CNTs-10 wt% BN at 100 Hz. Apart from improvement in dielectric properties, it showed good thermal stability improvement of 55 °C [53].

3.5 *PVDF-Perovskites Based Nanocomposites*

Perovskite materials, characterized by the ABX_3 crystal structure (A = organic/inorganic cation, B = metal cation, X = halide), have transformed the field of optoelectronics due to their exceptional photophysical properties. Hybrid organic–inorganic perovskites (HOIPs), such as methylammonium lead iodide ($MAPbI_3$), exhibit high absorption coefficients (>10^4/cm), long charge carrier diffusion lengths (>1 μm), and tunable direct bandgaps (1.2–3.0 eV) through compositional engineering of halides or cations.

The dipole–dipole interactions between the polar surface of perovskites and PVDF chains align polymer dipoles, stabilizing the β-phase without post-processing. For instance, Si et al. [54] prepared $FAPbBr_3$/PVDF composite aerogel films *via* the freeze–drying process for energy harvesting and photodetector applications. The uniformly distributed $FAPbBr_3$ NPs in the PVDF matrix stabilized the γ-phase (97%) effectively at an 8 wt% $FAPbBr_3$ NP loading. Thus, it emerges as a prominent material for piezoelectric nanogenerator applications with a piezoelectric coefficient (d_{33}) of ~32 pC/N [54].

3.5.1 Enhanced Properties of PVDF Nanocomposites Through Perovskite Integration

Piezoelectric and Ferroelectric Enhancements

The integration of halide perovskites into PVDF nanocomposites has led to unprecedented improvements in piezoelectric coefficients and polarization density. For instance, adding just 0.94% self-nucleated $CH_3NH_3PbBr_3$ nanocrystals to PVDF resulted in a polarization of ~19.6 μC/cm^2, a near-doubling of PVDF's baseline performance [55]. This enhancement arises from field-induced Frenkel defect pairs in the perovskite, stabilized by the polymer's poled ferroelectric domains through strong interfacial coupling [55]. Similarly, $CsPbI_3$@PVDF composites exhibit a piezoelectric charge coefficient (d_{33}) of 28.4 pm/V, doubling the output of pristine PVDF [56]. The perovskite's inherent ionic displacement under stress synergizes with PVDF's β-phase dipolar alignment, amplifying charge separation efficiency [56, 57].

Stability and Environmental Resistance

Perovskite additives also address PVDF's susceptibility to environmental degradation. A macromolecular intermediate phase formed by cross-linking PVDF with perovskite grains via long-chain polymers reduces grain-boundary defects, enhancing moisture and thermal stability. For example, PVDF-terpolymer composites with 0.1 wt% $(PEA)_2PbI_4$ additive show a 25% increase in dielectric breakdown strength due to reduced carrier density [58].

Dielectric and Optical Functionality

The introduction of lead-free double perovskites like $Cs_2KInCl_6:Sb^{3+}/Mn^{2+}$ into PVDF matrices extends functionality beyond mechanical properties. These composites achieve a photoluminescence quantum yield of 86.98% via energy transfer between self-trapped excitons and Mn^{2+} dopants, enabling applications in white-light-emitting diodes [59]. Dielectric constant enhancements are similarly notable;

Zr-doped $Cs_2Ag_{0.3}Na_{0.7}InCl_6$ double perovskites grown *in situ* within PVDF yield a 44.4 V open-circuit voltage in piezoelectric nanogenerators, a sixfold improvement over conventional blends [57]. The perovskite's wide bandgap and high permittivity augment PVDF's charge storage capacity while maintaining low dielectric loss [57, 58].

Mechanical Flexibility and Scalability

Critically, perovskite-PVDF nanocomposites retain the polymer's inherent flexibility despite inorganic filler addition. $CH_3NH_3PbBr_3$-PVDF films achieve 19.6 μC/cm^2 polarization with >300% tensile strain tolerance, outperforming rigid ceramics like PZT [55]. *In situ* synthesis methods further ensure uniform dispersion of perovskite nanocrystals, preventing agglomeration-induced brittleness [57]. Scalable fabrication is demonstrated through solution-processed $CsPbI_3$@PVDF films stable for 60 days under ambient conditions, showcasing industrial viability [56].

3.5.2 Applications of PVDF-Perovskite Nanocomposites

Solar Cells and Photovoltaics

Solar cells based on PVDF-Perovskite composites leverage polarization-enhanced mechanisms to stabilize internal electric fields, offering a practical pathway to boost efficiency through controlled polarization states. Key mechanisms include the piezo-phototronic effect, intrinsic spontaneous polarization, and external electric field alignment processes [60].

Ahn et al. [61] showed how PVDF–HFP–Perovskite composite-based solar cells have better reliability in the harsh environment of space (Fig. 15). The fluorine-rich PVDF formed hydrogen bonds with organic cations (e.g., FA^+/MA^+) and coordinated with Pb^{2+} ions, passivating surface and bulk defects. This led to a power conversion efficiency (PCE) of 22.54% for PVDF-HFP PSCs, surpassing the 19.85% of pristine devices. Under simulated space conditions, PVDF–HFP–Perovskite solar cells retained 70% of their initial PCE after 600 h in a high vacuum (2×10^3 torr), compared to only 32% for pristine perovskite cells. The PVDF-HFP mitigated cation evaporation (e.g., MA^+) by forming hydrophobic barriers and stabilizing the perovskite lattice. Additionally, PVDF–HFP–Perovskite exhibited resilience to gamma-ray irradiation (2.3 Mrad dose), with efficiency losses primarily attributed to substrate darkening rather than perovskite degradation. In contrast, pristine cells suffered from trap-state formation and recombination losses under radiation (Fig. 15) [61].

Triboelectric Nanogenerators (TENG)

Karimy et al. [3] enhanced the triboelectric nanogenerator (TENG) performance of PVDF by integrating polymorphic perovskite nanofillers into polyvinylidene fluoride (PVDF) matrix through 3D direct-ink writing (DIW) [3]. By embedding δ-$FAPbI_3$ into PVDF, the composite film achieves an 83% electroactive β-phase content (vs. 45% in pristine PVDF), crucial for dipole alignment and piezoelectricity (Fig. 16).

The electronegative PbI_3^- framework in $FAPbI_3$ interacts with $-CH_2$ dipoles in PVDF, inducing realignment of PVDF chains into the polar β-phase conformation (TTTT). Further annealing converts δ-$FAPbI_3$ to α-$FAPbI_3$, creating a mesoporous structure that enhances surface charge density and reduces electron–hole recombination. Pristine PVDF is tribo-negative (work function ≈6.67 eV). $FAPbI_3$ lowers PVDF's work function (PVDF-$FAPbI_3$(α): 5.20 eV), making it tribo-positive. When paired with tribo-negative PDMS (work function ≈7.59 eV), the larger work function difference drives stronger electron transfer during contact electrification. The optimized PVDF-$FAPbI_3$(α) composite exhibits exceptional triboelectric performance, generating a peak-to-peak voltage of 392 V, a current of 61 μA, and a power density of 2587 μW/cm^2—fivefold improvements over pristine PVDF. Demonstrations include powering 200 LEDs, charging capacitors, and harvesting biomechanical energy from human motion (e.g., walking, running). The TENG also shows promise in tactile sensing for robotics, with pressure-sensitive voltage mapping. Similar results and application were reported by Jiao et al. [62].

Piezoelectric Nanogenerators (PENG)

Chen et al. [63] presented a high-performance piezoelectric nanogenerator (PENG) based on all-inorganic $CsPbBr_3$ perovskite nanocrystals embedded in polyvinylidene fluoride (PVDF) fibers [63]. The authors developed an in situ growth method during electrospinning to synthesize $CsPbBr_3$@PVDF composite fibers, addressing the instability issues of lead halide perovskites while leveraging their superior piezoelectric properties. By encapsulating $CsPbBr_3$ within PVDF, the composite fibers achieved uniform nanocrystal distribution (≈20 nm diameter) and enhanced protection against thermal, moisture, and chemical degradation.

The optimized PENG demonstrated exceptional output performance, with an open-circuit voltage (Voc) of 103 V and a short-circuit current density (Isc) of 170 μA/cm^2, surpassing previous perovskite-based PENGs [63].

Sultana et al. [64] reported a photoactive PENG and photodetector using methylammonium lead iodide ($CH_3NH_3PbI_3$) and PVDF [64]. Electrostatic interaction between MAPI's negatively charged PbI_3^- framework and PVDF's $-CH_2$ dipoles drives β-phase nucleation. $CH_3NH_3PbI_3$ decomposition during film formation creates homogeneous pores (average diameter ~540 nm), enhancing piezoelectric strain response.

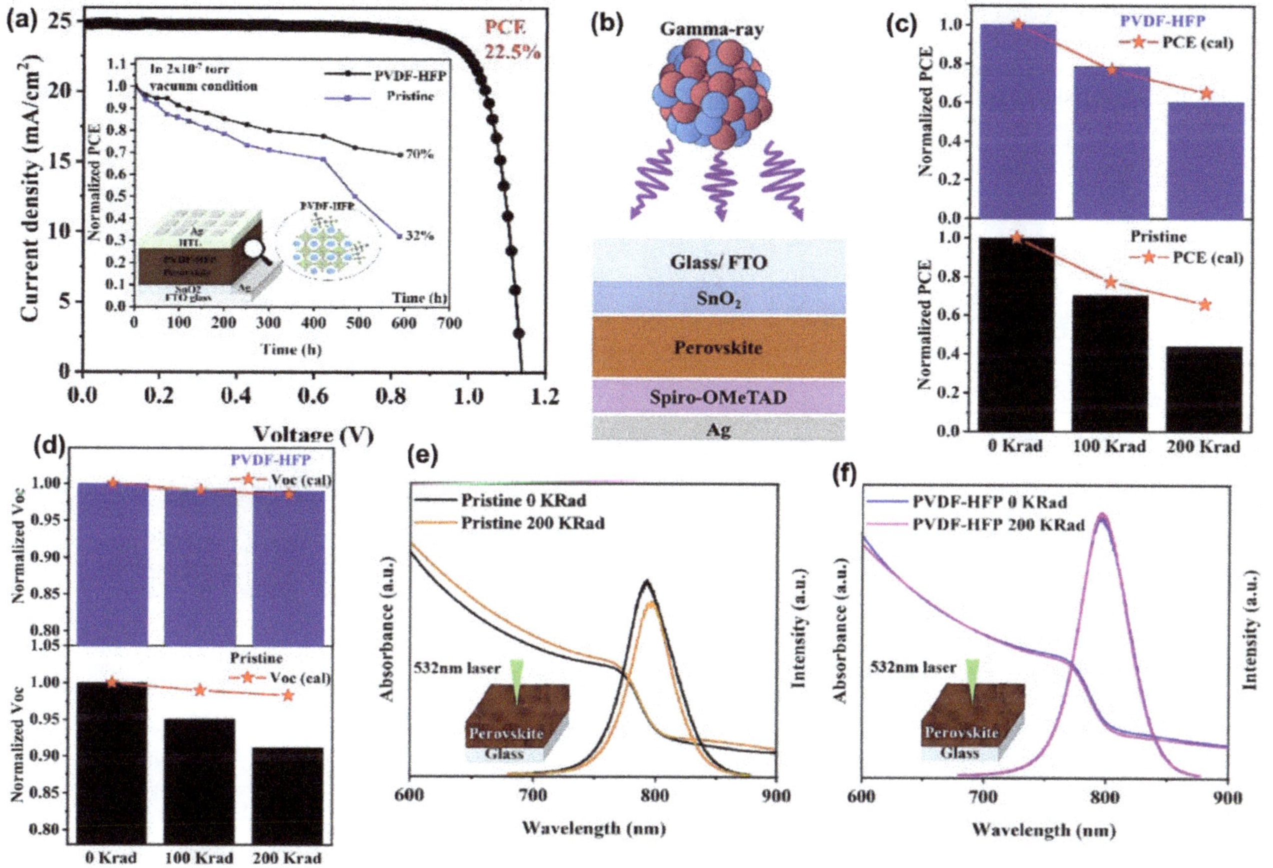

Fig. 15 PVDF-HFP in perovskite solar cells for improved space mission reliability: (**a**) remarkable PCE of 22.5%, improved PCE due to PVDF-HFP; (**b**) schematic diagram of PSC exposed to gamma ray; (**c**) PCE and (**d**) Voc evolution of PSCs under gamma ray radiation. Red lines show the expected behavior in case gamma rays induce only substrate darkening without damaging the PV stack. (**e**) UV–Vis absorption and (**f**) PL spectra of pristine and PVDF-HFP perovskite films under gamma ray radiation [61]

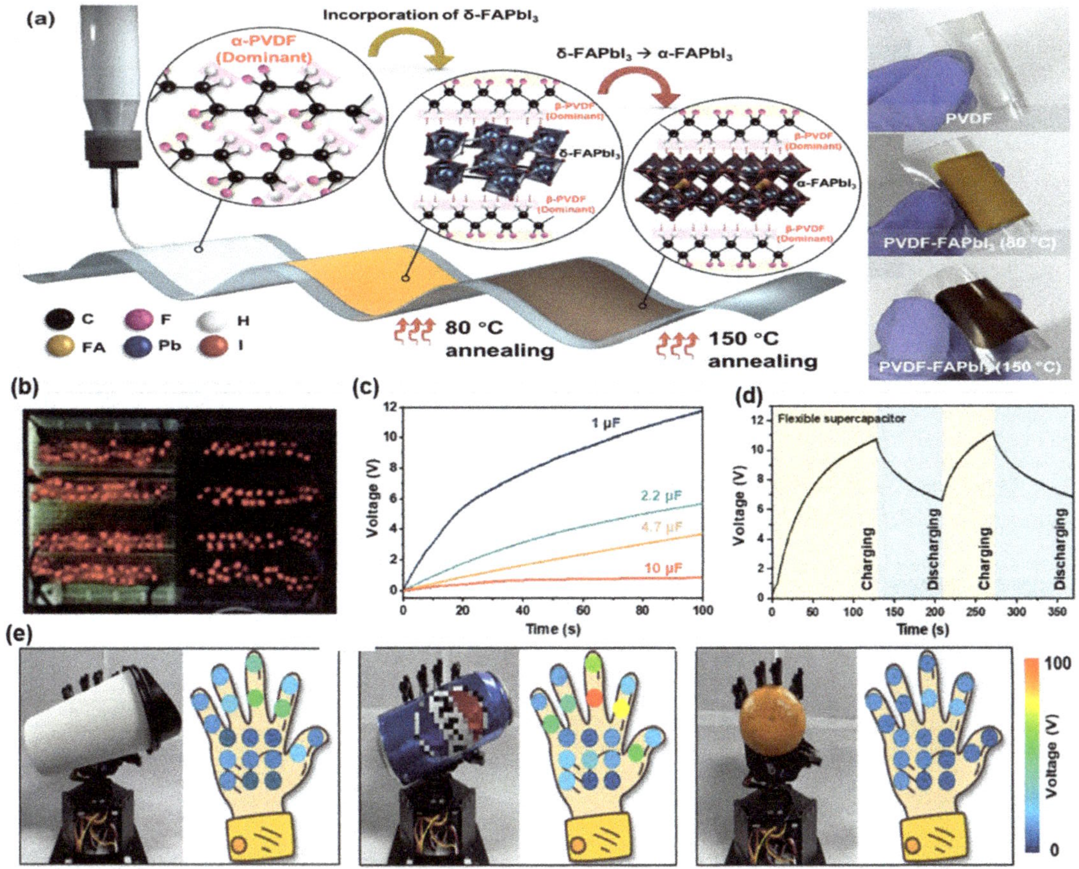

Fig. 16 (**a**) Schematic illustration of 3D direct-ink writing process and crystal phase changes (insets) of PVDF-FAPbI$_3$ composite films prepared at different annealing temperatures; (**b**) digital photograph showing the as-prepared TENG device successfully powering up 200 commercial red LEDs; (**c**) charging voltage curves of various commercial capacitors as charged by TENGs at 3 Hz frequency. (**d**) Charging and discharging cyclic characteristics of a self-designed flexible supercapacitor; (**e**) digital images showing the robotic hand grasping different items: a paper cup, a metal can, and an orange, along with its corresponding spatially mapping pressure distributions through voltage signals outputs from the contact area [3].

Wang et al. [8] in Fig. 17, electrospun a novel lead-free perovskite/polymer nanocomposite ($Cs_3Bi_2Br_9$/PVDF-HFP), designed as a flexible wearable multimodal sensor for long-term in situ monitoring of agricultural microenvironments [8]. This structure combines breathability, stretchability (800% tensile strain), and hydrophobicity (contact angle ~102°), preventing water ingress and electrode degradation. The perovskite acts as an electron-capturing layer (surface potential < −184.9 mV), enhancing charge transfer efficiency and sensor response. The use of lead-free perovskite eliminates toxicity concerns in soil. Deployed on potted strawberries for 17 days, the sensors monitored root pH (5.95–6.37) and leaf humidity (22.5–52.18% RH) with high accuracy (RMSE: 0.05 pH, 2.14% RH), matching commercial probes [8].

Another interesting application of PVDF-perovskite nanocomposites is in lasing action. Halide perovskites, while promising for low-threshold lasers, suffer from instability in polar environments and air. Nie et al. [65] address this by synthesizing $CsPbBr_3$@PVDF composites through a two-step process: spin-coating a precursor solution and thermally inducing crystallization within the PVDF matrix [65]. This encapsulation preserves the optical properties of $CsPbBr_3$ and provides exceptional chemical resistance. The PVDF matrix enables precise tuning of microcrystal dimensions (20–68 μm in length) by adjusting growth temperatures (40–80 °C), directly influencing lasing characteristics. Remarkably, the encapsulated microcrystals retain their optical performance in water, 5% HCl, and 5% NaOH for 35 days, far surpassing previous stability benchmarks. The work demonstrates robust polarization-dependent lasing and two-photon absorption, confirming its practical viability under harsh conditions [65].

3.6 *Hybrid Nanocomposites*

When more than one of these nanofillers is combined in a PVDF matrix, novel synergistic effects emerge, leading to multifunctional nanocomposites with superior and often unprecedented performance.

3.6.1 Synergistic Enhancement of Functional Properties

As described in Fig. 18, Yang et al. [66] developed an EMI shielding device using a hybrid PVDF-CNTs-graphene-NiCo nanocomposite [66]. Herein, 1D CNTs connect 2D graphene nanoplates and NiCo chains, forming continuous conductive pathways. High surface area of graphene provides primary conductivity (912 S/cm with 8 wt% NiCo) whereas metallic NiCo chains boost conductivity and serve as "micro-antennas" to scatter microwaves and generate magnetic hysteresis and eddy currents, converting EM energy into heat. Saturation magnetization (Ms) scales with NiCo content (up to 2.49 emu/g). EM energy is further dissipated by the charge accumulation at graphene/CNT/NiCo interfaces. The β-crystalline phase of PVDF

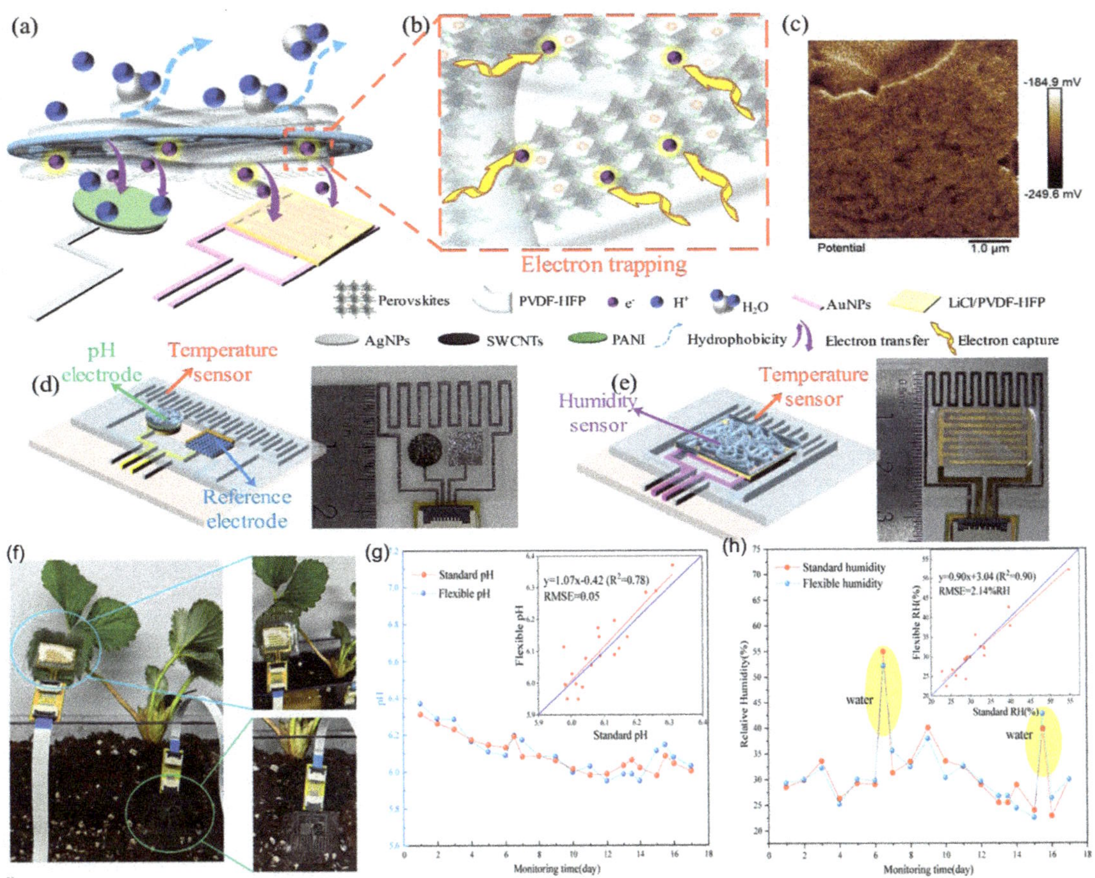

Fig. 17 Flexible multimodal sensors enhanced by electrospun lead-free perovskite and PVDF-HFP composite form-stable mesh membranes for in situ plant monitoring: (**a**) mechanism of LPPC action. (**b**) Electron capture of perovskite $Cs_3Bi_2Br_9$. (**c**) Surface potential of LPPC from Kelvin probe force microscopy (KPFM). The model and physical diagrams of (**d**) LPPC-pH/thermo sensor and (**e**) LPPC-humidity/thermo sensor; (**f**) in situ monitoring. (**g**) Root environment pH monitoring. (**h**) Leaf humidity monitoring [8]

enhances dipole polarization. Layered structure (graphene sheets + NiCo chains) creates micro-reflective surfaces which prolongs EM wave path and increases absorption (SE_A). Graphene boosts conductivity to nine times of that of NiCo alone and CNTs isolates NiCo/graphene, preventing aggregation and enhancing network connectivity. The composite shows excellent performance for EMI shielding in which PVDF-CNT-G-0.5NiCo (0.5 wt% NiCo) achieves $SE_T = 35.0$ dB, meeting commercial standards (>20 dB) with minimal filler. Besides these, simple solution mixing together with compression molding enables easy scale-up. Oriented NiCo chains formed via magnetic-field-assisted solvothermal synthesis. This work demonstrates a viable strategy for high-performance EMI shields in aerospace, electronics, and defense applications [66].

Similarly, He et al. [67] developed a solvent-free, industrially viable method to fabricate hierarchical PVDF/CNT/graphene nanocomposites for wearable EMI shielding in which a dual-phase electromagnetic synergy was engineered between 1D CNTs ("ballistic electron highways") and 2D GnPs ("polarization interfaces") into a 3D network to optimize conductivity and EM wave dissipation [67].

The combination of CNTs and graphene-based materials (such as graphene nanoribbons, GNRs) in PVDF leads to giant synergistic effects in dielectric properties. For example, ternary nanocomposites of multiwalled CNTs (MWCNTs), GNRs, and PVDF show much higher real permittivity and lower energy loss compared to binary systems. The MWCNTs form an efficient percolative network, while GNRs act as nanoelectrodes, creating effective nanocapacitor structures and enhancing dielectric response with reduced dissipation [68].

3.6.2 Mechanisms Behind the Synergy

- **Percolation Networks**: CNTs form conductive pathways, while graphene or GNRs serve as secondary fillers or nanoelectrodes, enhancing charge storage and transport.
- **Interfacial Interactions**: Functional groups on GO/rGO and the surface chemistry of perovskites or metal oxides create strong interfaces, improving dispersion and stress transfer.
- **Morphological Control**: Nanofillers influence PVDF crystallization, promoting the formation of desired crystalline phases and morphologies (Table 1).

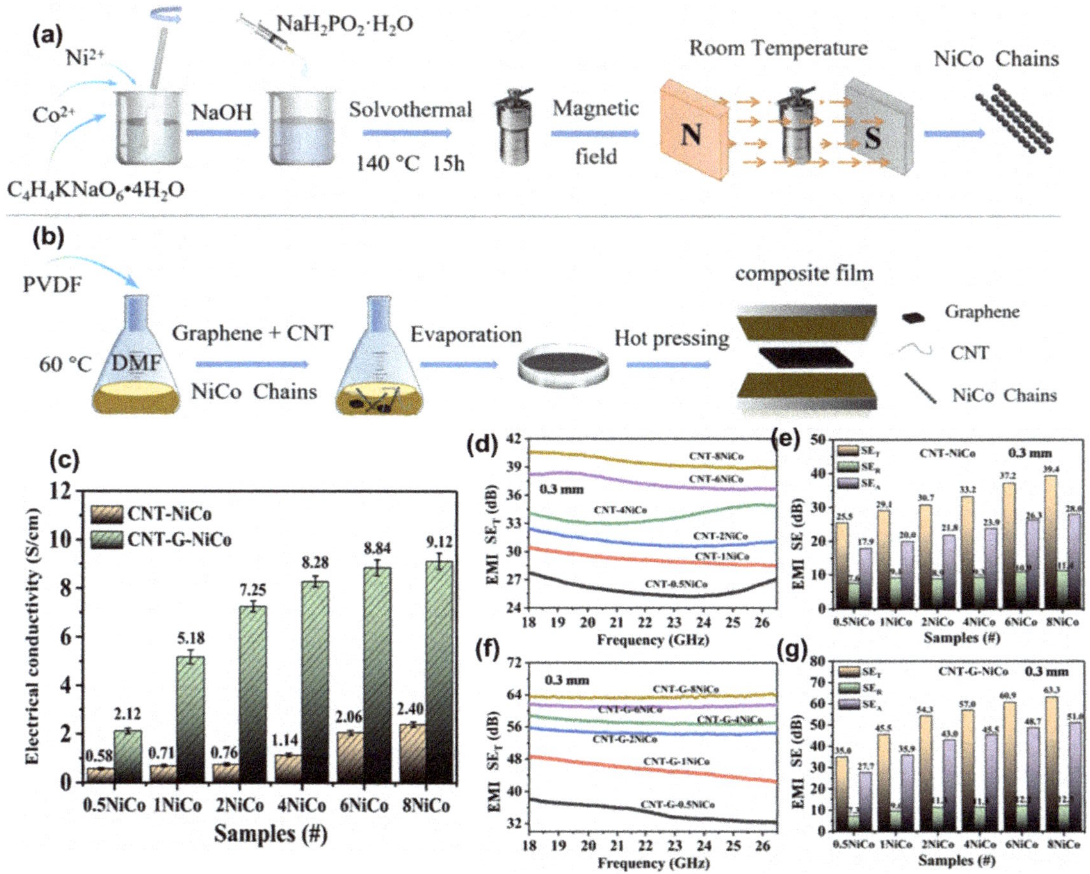

Fig. 18 PVDF-CNT-graphene-NiCo chains composite films with excellent electromagnetic interference shielding performance: preparation process of (**a**) NiCo chains and (**b**) the PVDF-CNT-G-NiCo composite films; (**c**) electrical conductivity of PVDF-CNT-NiCo and PVDF-CNT-G-NiCo composite films with different NiCo quality contents; (**d**) EMI SE and (**e**) average SE_T, SE_A, and SE_R of PVDF-CNT-NiCo with different NiCo contents; (**f**) EMI SE and (**g**) average SE_T, SE_A, and SE_R of PVDF-CNT-G-NiCo with different NiCo quality contents at the thickness of 0.3 mm [66]

Table 1 Comparison of properties and benefits of graphene, CNTs, and perovskite in PVDF

Nanofiller	Dielectric enhancement	Piezoelectric enhancement	Mechanical/thermal stability	Conductivity	Phase control (β-phase)	Unique benefits
Graphene	High (interfacial polarization, increased permittivity) [35]	Moderate (nucleation effect)	High (reinforcement, thermal conductivity) [35]	High	Promotes β-phase	Large surface area, flexibility, multifunctionality [35]
CNTs	High (percolation, strong networks) [41, 68]	High (strong nucleating agent, high d33) [41]	High (stiffness, strength) [41]	Very high	Strongly promotes β-phase	Superior percolation, flexibility, aspect ratio [41, 68]
Perovskite	Very high (high-k, interfacial polarization) [69]	Moderate (depends on type)	Moderate (thermal stability) [69]	Moderate	Can influence phase	Nonlinear optical properties, tunable band gap [69]
Graphene + CNTs	Giant synergistic effect (highest permittivity, lowest loss) [68]	High (additive nucleation)	Very high (combined reinforcement) [68]	Very high	Enhanced β-phase	Nanocapacitor effect, multifunctionality [68]
CNTs + perovskite	Enhanced dielectric and optical properties (tunable band gap, refractive index) [69]	High (synergy in nucleation)	High (improved interface) [69]	High	Enhanced β-phase	Tunable electronic/optical properties [69]
GO/ rGO + perovskite	High (synergistic interfacial polarization) [70]	Moderate	Very high (strong interface, stability) [70]	Moderate	Enhanced β-phase	Improved charge transport, energy storage [70]

4 Industrial Scalability and Commercialization

Polyvinylidene fluoride (PVDF) is a versatile piezoelectric polymer widely studied for energy harvesting, sensing, and electromagnetic shielding applications. Enhancing PVDF's intrinsic properties by incorporating nanomaterials such as graphene, carbon nanotubes CNTs, and perovskite nanocrystals has led to significant performance improvements. The focus now shifts towards translating these advances into scalable manufacturing and commercial products.

4.1 *Scalability of Fabrication Techniques*

Recent studies demonstrate scalable solution spinning and melt processing methods for fabricating PVDF nanocomposites with graphene and CNTs. For example, a facile solution spinning strategy enables the production of PVDF/CNT/GnP hierarchical composites with excellent electromagnetic interference EMI shielding performance and mechanical durability [67]. This method is environmentally friendly, avoiding toxic solvents and energy-intensive steps, making it suitable for industrial scale-up. Additionally, solid-state shear milling (S3M) technology has been employed to promote the formation of the electroactive β-phase in PVDF/graphene composites during conventional melt extrusion and injection molding [71]. Incorporating perovskite nanocrystals into PVDF via in situ growth techniques has shown promise in enhancing piezoelectric nanogenerator (PENG) performance [57]. The in situ method simplifies the fabrication process by combining nanocrystal synthesis and composite formation, reducing steps and potential contamination. Such streamlined processes are advantageous for scaling production.

4.2 *Enhanced Properties and Their Industrial Applications*

PVDF/CNT/GnP composites achieved EMI shielding effectiveness exceeding 23 dB in the X-band frequency range, outperforming many metal-based materials, which is currently the industry standard [67]. These composites also maintain performance under mechanical deformation and humidity exposure, critical for wearable electronics. Lead-free halide perovskite/PVDF composites have demonstrated promising energy harvesting capabilities with output voltages around 40 V and power densities suitable for portable electronics [72].

Flexible energy harvesters and sensors for wearable and portable devices benefit from the enhanced piezoelectricity and mechanical robustness of these composites. EMI shielding materials made from PVDF/graphene/CNT composites are lightweight, flexible, and environmentally stable, addressing growing demands in smart textiles and electronics packaging. Energy storage and sensing devices leverage the improved dielectric and ferroelectric properties imparted by nanofillers.

4.3 Commercialization Outlook

The demand for flexible, lightweight, and multifunctional materials in wearable electronics, energy harvesting, EMI shielding, and sensors is rapidly growing. PVDF nanocomposites with graphene, CNTs, and perovskite fillers align well with these trends, offering enhanced performance and flexibility. The compatibility of PVDF nanocomposites with conventional polymer processing techniques (e.g., melt extrusion, solution spinning) facilitates their adoption in existing industrial setups. Scalable, solvent-free, and low-energy processes further support commercial viability [67, 71]. However, high-quality graphene and CNTs remain relatively expensive and require controlled synthesis to ensure batch-to-batch consistency. Similarly, scalable production of defect-free perovskite nanocrystals with controlled doping is still evolving. Cost-effective raw material supply and processing will be key to commercialization.

5 Challenges and Future Perspectives

5.1 Challenges for PVDF-Graphene Nanocomposite

5.1.1 Dispersion and Interfacial Compatibility

One of the foremost challenges in graphene-PVDF nanocomposites is achieving uniform dispersion of graphene within the polymer matrix. Graphene's strong tendency to agglomerate due to van der Waals forces leads to heterogeneous microstructures that impair piezoelectric and dielectric properties. The weak interfacial interactions and presence of pores or voids can reduce dielectric permittivity and mechanical integrity. Although PVDF's fluorine atoms have a high affinity for carbon, the non-functionalized graphene surface often results in poor interfacial bonding, limiting load transfer and electrical connectivity. This challenge is compounded during melt or solution processing, where graphene restacking can occur, reducing effective surface area and nucleation sites for the electroactive β-phase of PVDF [28, 73].

5.1.2 Thermal and Mechanical Stability

While graphene enhances thermal conductivity and mechanical stiffness, the composite's overall thermal stability is sensitive to processing conditions such as annealing temperature. Annealing can induce cross-linking and intercalation effects, altering electrical conductivity and mechanical properties unpredictably. Balancing enhanced thermal stability with retention of flexibility is a delicate task, especially for wearable or flexible devices requiring repeated mechanical deformation [28, 73].

5.1.3 Percolation Threshold and Electrical Conductivity

Graphene loading must be carefully optimized. Below the percolation threshold, electrical conductivity is insufficient for many applications; above it, the piezoelectric output can drastically decrease due to leakage currents and short-circuiting pathways. Theoretical studies show that particle shape and size influence this threshold, but controlling these parameters at scale remains a challenge. Incorporating metal oxide nanoparticles on graphene surfaces has improved dispersibility and electrical performance, yet scalable, cost-effective synthesis routes are still under development [28].

5.2 Challenges for PVDF-CNTs Nanocomposite

5.2.1 Dispersion and Phase Control

Similar to graphene, CNTs tend to form bundles and aggregates within PVDF, complicating uniform dispersion. The electrospinning process, commonly used to fabricate CNTPVDF fibers, reveals that CNTs chemically bond with PVDF, altering the polymer's crystalline phase distribution. Notably, CNT addition can increase the α- and γ-phases at the expense of the desirable β-phase, which governs piezoelectricity. This phase alteration affects the material's electromechanical properties and requires precise control over CNT concentration and processing parameters [74].

5.2.2 Scalability of Functionalization

Functionalization of CNTs is essential for improving dispersion and interfacial bonding, but it introduces complexity and cost. Scaling these chemical modifications while maintaining consistent quality is a significant barrier. Moreover, the presence of CNTs protruding from fibers, as observed in SEM imaging, could affect surface roughness and device integration, requiring further optimization [74].

5.2.3 Electrical and Dielectric Optimization

CNTs enhance electrical conductivity and dielectric permittivity, but excessive loading can lead to percolation-induced conductivity that diminishes piezoelectric output. Hybrid fillers, such as gold nanoparticle-functionalized CNTs, have shown promise, yet their synthesis and integration at industrial scales are still nascent. The balance between conductivity and dielectric loss remains a critical design challenge [74].

5.3 Challenges in PVDF Nanocomposites with Perovskites

5.3.1 Stability and Environmental Sensitivity

Perovskite nanocrystals like $MAPbBr_3$ offer extraordinary enhancements in polarization and piezoelectric output at very low volume fractions of 1%. However, their inherent instability in ambient conditions—sensitivity to moisture, oxygen, and heat—poses a major obstacle. Degradation leads to loss of crystallinity and electroactive phase content, compromising device longevity. Although in situ synthesis within PVDF and polymer encapsulation strategies improve stability, long-term durability under real-world operating conditions remains insufficiently demonstrated [55].

5.3.2 Toxicity and Regulatory Concerns

Most high-performance perovskites contain lead, raising environmental and health concerns that could limit commercialization. Developing lead-free perovskite analogs with comparable ferroelectric and piezoelectric properties is an active research frontier but has yet to reach maturity. Regulatory frameworks for lead-containing materials are increasingly stringent, necessitating robust encapsulation or alternative compositions for market acceptance [55].

5.3.3 Interfacial Engineering and Mechanical Flexibility

The cohesive binding between perovskite nanocrystals and PVDF matrix is crucial to enhance electromechanical coupling while maintaining mechanical flexibility. Achieving uniform dispersion without agglomeration and controlling nanoparticle size and surface chemistry are complex [75]. The lattice contraction and Frenkel defect stabilization mechanisms that enhance polarization also introduce internal stresses that could affect mechanical integrity over repeated use [55].

5.4 Future Perspectives

5.4.1 Advanced Processing Techniques

Emerging fabrication methods such as solid-state shear milling (S^3M) for graphene-PVDF composites, electrospinning with controlled CNT functionalization, and in situ perovskite crystallization within polymer matrices offer promising routes to overcome dispersion and interfacial challenges. These scalable techniques must be further optimized to ensure reproducibility, cost-effectiveness, and compatibility with roll-to-roll manufacturing for flexible electronics [28, 55, 74].

5.4.2 Multifunctional and Hybrid Nanofillers

Combining graphene, CNTs, and perovskites in hybrid nanocomposites could synergistically leverage their complementary properties—mechanical strength, electrical conductivity, and ferroelectricity. Designing hierarchical architectures with controlled filler localization and orientation may enable tunable properties for specific applications, such as self-powered sensors or wearable health monitors. However, this introduces complexity in synthesis and characterization that requires advanced modeling and machine learning-guided optimization [28, 55, 74].

5.4.3 Lead-Free and Environmentally Friendly Alternatives

The development of lead-free perovskites or alternative ferroelectric nanomaterials with comparable performance is critical for sustainable commercialization. Efforts to synthesize double perovskites or doped variants with reduced toxicity are underway, but challenges in achieving stability and high piezoelectric response persist. Parallel advances in encapsulation materials and device architectures will be necessary to mitigate environmental risks [55].

5.4.4 Standardization and Reliability Testing

For industrial adoption, standardized testing protocols for mechanical fatigue, environmental aging, and electrical reliability must be established. Accelerated lifetime testing and real-world deployment studies will provide critical data to guide materials design and manufacturing processes. Integration with existing electronic platforms demands compatibility with flexible substrates, adhesives, and encapsulation layers [28, 55, 74].

6 Conclusion

Nanocomposites of PVDF have emerged as a transformative platform for next-generation functional materials, owing to their unique ability to synergistically combine the intrinsic electroactive properties of PVDF with the tailored functionalities of nanoscale additives. The strategic incorporation of nanostructures—including graphene, carbon nanotubes (CNTs), and perovskites—addresses the inherent limitations of bulk PVDF by substantially enhancing its electroactive β-phase content, piezoelectric response, dielectric permittivity, mechanical robustness, and electrical conductivity.

The chapter systematically demonstrates how nanofillers act as effective nucleating agents, promoting the formation of the electroactive β-phase and enabling superior piezoelectric and ferroelectric properties without the need for extensive mechanical or electrical processing. Interfacial polarization at the nanofiller–polymer interface further elevates dielectric properties and energy storage capabilities, while the formation of percolation networks by conductive nanostructures dramatically improves charge transport and sensor sensitivity. Mechanical reinforcement through nanofillers ensures that PVDF nanocomposites maintain flexibility and toughness, crucial for applications in flexible electronics, wearable devices, and energy harvesting systems.

The versatility of PVDF nanocomposites is underscored by their successful deployment in a wide spectrum of advanced applications, including self-powered sensors, piezoelectric nanogenerators, electromagnetic interference shielding, optoelectronic devices, and high-performance capacitors. Record-setting performances—such as enhanced power densities, increased efficiency in energy harvesting, and ultrasensitive pressure detection—highlight the commercial and technological potential of these materials.

Scalable fabrication techniques, such as electrospinning, solution casting, and in situ nanofiller incorporation, are critically evaluated for their compatibility with industrial processes like roll-to-roll manufacturing and 3D printing. Despite progress, challenges persist in nanofiller dispersion, interfacial compatibility, perovskite stability under environmental stressors, and lead toxicity concerns. Future efforts should prioritize advanced processing (e.g., solid-state shear milling), lead-free alternatives, hybrid nanofiller architectures, and standardized reliability testing.

In summary, PVDF nanocomposites represent a cornerstone material system for the future of multifunctional, sustainable, and high-performance technologies. Their continued development and commercialization are poised to revolutionize energy autonomy, flexible electronics, Internet-of-Things (IoT), human–machine interfaces, and smart device integration.

References

1. Shwetha, D., Shenoy, S.: A review on properties and applications of polyvinylidene fluoride (PVDF) and metal oxide nanocomposites thin films. IJFMR Int. J. Multidiscip. Res. **6**(5), 27068 (2024). https://doi.org/10.36948/IJFMR.2024.V06I05.27068
2. Costa, C.M., et al.: Smart and multifunctional materials based on electroactive poly(vinylidene fluoride): recent advances and opportunities in sensors, actuators, energy, environmental, and biomedical applications. Chem. Rev. **123**(19), 11392–11487 (2023). https://doi.org/10.1021/ACS.CHEMREV.3C00196/ASSET/IMAGES/LARGE/CR3C00196_0046.JPEG
3. Zain Karimy, N.H., et al.: Highly efficient 3D-printed PVDF-based triboelectric nanogenerators featuring polymorphic perovskite nanofillers. Adv. Funct. Mater. **35**, 2424271 (2025). https://doi.org/10.1002/ADFM.202424271
4. Lu, K., et al.: Ultra-sensitive strain sensor based on flexible poly(vinylidene fluoride) piezoelectric film. Nanoscale Res. Lett. **13**(1), 1–6 (2018). https://doi.org/10.1186/S11671-018-2492-7/FIGURES/6
5. Shi, L., et al.: High-performance triboelectric nanogenerator based on electrospun PVDF-graphene nanosheet composite nanofibers for energy harvesting. Nano Energy. **80**, 105599 (2021). https://doi.org/10.1016/J.NANOEN.2020.105599
6. Bin Kim, D., et al.: Weave-pattern-dependent fabric piezoelectric pressure sensors based on polyvinylidene fluoride nanofibers electrospun with 50 nozzles. npj Flex. Electron. **6**(1), 1–9 (2022). https://doi.org/10.1038/S41528-022-00203-6
7. Yang, X., Wang, Y., Qing, X.: A flexible capacitive sensor based on the electrospun PVDF nanofiber membrane with carbon nanotubes. Sens. Actuators A Phys. **299**, 111579 (2019). https://doi.org/10.1016/J.SNA.2019.111579
8. Wang, L., Wang, Q., Yao, C., Li, M., Liu, G., Zhang, M.: Flexible multimodal sensors enhanced by electrospun lead-free perovskite and PVDF-HFP composite form-stable mesh membranes for in situ plant monitoring. Anal. Chem. **96**(29), 11923–11931 (2024). https://doi.org/10.1021/ACS.ANALCHEM.4C01684/SUPPL_FILE/AC4C01684_SI_003.MP4
9. Purushothaman, S.M., et al.: A review on electrospun PVDF-based nanocomposites: recent trends and developments in energy harvesting and sensing applications. Polymer (Guildf). **283**, 126179 (2023). https://doi.org/10.1016/J.POLYMER.2023.126179
10. Sukumaran, S., Chatbouri, S., Rouxel, D., Tisserand, E., Thiebaud, F., Ben Zineb, T.: Recent advances in flexible PVDF based piezoelectric polymer devices for energy harvesting applications. J. Intell. Mater. Syst. Struct. **32**(7), 746–780 (2021). https://doi.org/10.1177/1045389X20966058
11. Wang, Y., Zhu, L., Du, C.: Progress in piezoelectric nanogenerators based on PVDF composite films. Micromachines. **12**(11), 1278 (2021). https://doi.org/10.3390/MI12111278
12. Adaval, A., et al.: 2D nanomaterials incorporated poly(vinylidene fluoride) nanocomposites: morphology, crystalline structure, and dielectric, ferroelectric, and piezoelectric properties. J. Phys. Chem. C. **127**(13), 6483–6502 (2023). https://doi.org/10.1021/ACS.JPCC.2C08689/ASSET/IMAGES/LARGE/JP2C08689_0016.JPEG
13. Khadka, A., et al.: Integrating PVDF-based piezoelectric nanogenerators with highly conductive carbon nanofibers for energy-harvesting applications. Nano Energy. **139**, 110991 (2025). https://doi.org/10.1016/J.NANOEN.2025.110991
14. Singh, H.H., Singh, S., Khare, N.: Enhanced β-phase in PVDF polymer nanocomposite and its application for nanogenerator. Polym. Adv. Technol. **29**(1), 143–150 (2018). https://doi.org/10.1002/PAT.4096
15. Peña-Oliveras, O., Javier-Boodhan, B., La Santa, A., Gonzalez-Sanchez, J.: Effect of nanostructures addition and enhancement of poly(vinylidene difluoride) (PVDF) energy harvesting. Mater. Sci. Appl. **15**, 228–244 (2024). https://doi.org/10.4236/msa.2024.157016
16. Zhang, C., et al.: Enhanced dielectric properties of poly(vinylidene fluoride) composites filled with nano iron oxide-deposited barium titanate hybrid particles. Sci. Rep. **6**(1), 1–9 (2016). https://doi.org/10.1038/SREP33508

17. Samet, M., Kallel, A., Serghei, A.: Maxwell–Wagner–Sillars interfacial polarization in dielectric spectra of composite materials: scaling laws and applications. J. Compos. Mater. **56**(20), 3197–3217 (2022). https://doi.org/10.1177/00219983221090629/SUPPL_FILE/SJ-PDF-1-JCM-10.1177_00219983221090629.PDF
18. Jian, G., et al.: High energy density of $BaTiO_3$@TiO_2 nanosheet/polymer composites via ping–pong-like electron area scattering and interface engineering. NPG Asia Mater. **14**(1), 1–11 (2022). https://doi.org/10.1038/s41427-022-00356-w
19. Padurariu, L., et al.: Role of microstructures in the dielectric properties of PVDF-based nanocomposites containing high-permittivity fillers for energy storage. ACS Appl. Mater. Interfaces. **15**(10), 13535–13544 (2023)
20. Feng, Y., et al.: Enhanced dielectric properties of PVDF-HFP/$BaTiO_3$-nanowire composites induced by interfacial polarization and wire-shape. J. Mater. Chem. C. **3**(6), 1250–1260 (2015). https://doi.org/10.1039/C4TC02183E
21. Vicente, J., Costa, P., Lanceros-Mendez, S., Abete, J.M., Iturrospe, A.: Electromechanical properties of PVDF-based polymers reinforced with nanocarbonaceous fillers for pressure sensing applications. Materials (Basel). **12**(21), 3545 (2019). https://doi.org/10.3390/MA12213545
22. Georgousis, G., et al.: Strain sensing in polymer/carbon nanotube composites by electrical resistance measurement. Compos. Part B Eng. **68**, 162–169 (2015). https://doi.org/10.1016/J.COMPOSITESB.2014.08.027
23. Papageorgiou, D.G., Li, Z., Liu, M., Kinloch, I.A., Young, R.J.: Mechanisms of mechanical reinforcement by graphene and carbon nanotubes in polymer nanocomposites. Nanoscale. **12**(4), 2228–2267 (2020). https://doi.org/10.1039/C9NR06952F
24. Shehata, N., et al.: Multifaceted enhancement of piezoelectricity and optical fluorescence in electrospun PVDF-ceria nanocomposite. Sci. Rep. **15**(1), 1–21 (2025). https://doi.org/10.1038/s41598-025-98048-2
25. Chang, C.M., Liu, Y.L.: Electrical conductivity enhancement of polymer/multiwalled carbon nanotube (MWCNT) composites by thermally-induced defunctionalization of MWCNTs. ACS Appl. Mater. Interfaces. **3**(7), 2204–2208 (2011). https://doi.org/10.1021/AM200558F/ASSET/IMAGES/LARGE/AM-2011-00558F_0003.JPEG
26. Rehman, Z.U., et al.: Thermal and electrical properties of PVDF modified Co_3O_4 functionalized MWCNTs. RSC Adv. **15**(11), 8740–8749 (2025). https://doi.org/10.1039/D4RA07239A
27. Kaur, G., Rana, D.S.: Investigation on dielectric performance and charge transport mechanism in synthesized PVDF films embedded with $BaTiO_3$ nanofillers for energy harvesting applications. Optoelectron. Adv. Mater. Rapid Commun. **15**, 613–623 (2021)
28. Samoei, V.K., Maharjan, S., Sano, K., Jayatissa, A.H.: Effect of annealing on graphene/PVDF nanocomposites. ACS Omega. **8**(15), 13876–13883 (2023). https://doi.org/10.1021/ACSOMEGA.3C00283/ASSET/IMAGES/LARGE/AO3C00283_0010.JPEG
29. Hung, W.S., et al.: Electrical tunable PVDF/graphene membrane for controlled molecule separation. Chem. Mater. **32**(13), 5750–5758 (2020). https://doi.org/10.1021/ACS.CHEMMATER.0C01547/ASSET/IMAGES/MEDIUM/CM0C01547_M001.GIF
30. Han, P., Fan, J., Zhu, L., Min, C., Shen, X., Pan, T.: Structure, thermal stability and electrical properties of reduced graphene/poly(vinylidene fluoride) nanocomposite films. J. Nanosci. Nanotechnol. **12**(9), 7290–7295 (2012). https://doi.org/10.1166/JNN.2012.6584
31. Ismail, A.M., Mohammed, M.I., Fouad, S.S.: Optical and structural properties of polyvinylidene fluoride (PVDF)/reduced graphene oxide (RGO) nanocomposites. J. Mol. Struct. **1170**, 51–59 (2018). https://doi.org/10.1016/J.MOLSTRUC.2018.05.083
32. Adaval, A., et al.: Poly(vinylidene fluoride)/graphene oxide nanocomposites for piezoelectric applications: processing, structure, dielectric and ferroelectric properties. Nano-Struct. Nano-Objects. **31**, 100899 (2022). https://doi.org/10.1016/J.NANOSO.2022.100899
33. Maharjan, S., Samoei, V.K., Amili, O., Sano, K., Honma, H., Jayatissa, A.H.: Design and fabrication of a graphene/polyvinylidene fluoride nanocomposite-based airflow sensor. ACS Omega. **7**(9), 7981–7988 (2022). https://doi.org/10.1021/ACSOMEGA.1C07101/ASSET/IMAGES/LARGE/AO1C07101_0012.JPEG

34. Dilip Kulkarni, N., Kumar, M., Kumari, P.: PVDF/RGO based piezoelectric nanocomposite films for enhanced mechanical and dielectric properties. Mater. Today Proc. **76**, 81–87 (2023). https://doi.org/10.1016/J.MATPR.2022.09.043
35. Pusty, M., Shirage, P.M.: Insights and perspectives on graphene-PVDF based nanocomposite materials for harvesting mechanical energy. J. Alloys Compd. **904**, 164060 (2022). https://doi.org/10.1016/J.JALLCOM.2022.164060
36. Ai, Y., et al.: All rGO-on-PVDF-nanofibers based self-powered electronic skins. Nano Energy. **35**, 121–127 (2017). https://doi.org/10.1016/J.NANOEN.2017.03.039
37. Zhang, X., et al.: Recent advances in functional fiber electronics. SusMat. **1**(1), 105–126 (2021). https://doi.org/10.1002/SUS2.1
38. Adhikary, P., Garain, S., Ram, S., Mandal, D.: Flexible hybrid eu^{3+} doped P(VDF-HFP) nanocomposite film possess hypersensitive electronic transitions and piezoelectric throughput. J. Polym. Sci. Part B Polym. Phys. **54**(22), 2335–2345 (2016). https://doi.org/10.1002/POLB.24144
39. Jiao, Z.Z., Zhou, H., Han, X.C., Han, D.D., Zhang, Y.L.: Photothermal responsive slippery surfaces based on laser-structured graphene@PVDF composites. J. Colloid Interface Sci. **629**, 582–592 (2023). https://doi.org/10.1016/J.JCIS.2022.08.153
40. Song, S., Li, Y., Wang, Q., Zhang, C.: Boosting piezoelectric performance with a new selective laser sintering 3D printable PVDF/graphene nanocomposite. Compos. Part A Appl. Sci. Manuf. **147**, 106452 (2021). https://doi.org/10.1016/J.COMPOSITESA.2021.106452
41. Cerrada, M.L., Arranz-Andrés, J., Caballero-González, A., Blázquez-Blázquez, E., Pérez, E.: The β form in PVDF nanocomposites with carbon nanotubes: structural features and properties. Polymer. **15**(6), 1491 (2023). https://doi.org/10.3390/POLYM15061491
42. Chandran, A.M., Mural, P.K.S.: Surface silanized MWCNTs doped PVDF nanocomposite with self-organized dipoles: an intrinsic study on the dielectric, piezoelectric, ferroelectric, and energy harvesting phenomenology. Sustain. Energy Fuels. **6**(6), 1641–1653 (2022). https://doi.org/10.1039/D1SE01256H
43. Farhan, et al.: Investigation on crystal-structure, thermal and electrical properties of PVDF nanocomposites with cobalt oxide and functionalized multi-wall-carbon-nanotubes. Nanomaterials. **12**(16), 2796 (2022). https://doi.org/10.3390/NANO12162796
44. Bairagi, S., Ali, S.W.: Investigating the role of carbon nanotubes (CNTs) in the piezoelectric performance of a PVDF/KNN-based electrospun nanogenerator. Soft Matter. **16**(20), 4876–4886 (2020). https://doi.org/10.1039/D0SM00438C
45. Kumar, A., et al.: Energy harvesting by piezoelectric polyvinylidene fluoride/zinc oxide/carbon nanotubes composite under cyclic uniaxial tensile deformation. Polym. Compos. **44**(8), 4746–4756 (2023). https://doi.org/10.1002/PC.27437;REQUESTEDJOURNAL:JOURNAL:15480569;WGROUP:STRING:PUBLICATION
46. Badatya, S., Bharti, D.K., Sathish, N., Srivastava, A.K., Gupta, M.K.: Humidity sustainable hydrophobic poly(vinylidene fluoride)-carbon nanotubes foam based piezoelectric nanogenerator. ACS Appl. Mater. Interfaces. **13**(23), 27245–27254 (2021). https://doi.org/10.1021/ACSAMI.1C02237/ASSET/IMAGES/LARGE/AM1C02237_0011.JPEG
47. Baji, A., Mai, Y.W., Abtahi, M., Wong, S.C., Liu, Y., Li, Q.: Microstructure development in electrospun carbon nanotube reinforced polyvinylidene fluoride fibers and its influence on tensile strength and dielectric permittivity. Compos. Sci. Technol. **88**, 1–8 (2013). https://doi.org/10.1016/J.COMPSCITECH.2013.08.021
48. Lin, G., Hu, H., Liu, J., Shang, S.: Highly sensitive PVDF/CNTs/ZnO flexible piezoelectric sensor for detecting human subtle motion. IEEE Sens. J. **25**(8), 12797–12804 (2025). https://doi.org/10.1109/JSEN.2025.3548556
49. Kar, E., Bose, N., Dutta, B., Mukherjee, N., Mukherjee, S.: MWCNT@SiO_2 heterogeneous nanofiller-based polymer composites: a single key to the high-performance piezoelectric nanogenerator and X-band microwave shield. ACS Appl. Nano Mater. **1**(8), 4005–4018 (2018). https://doi.org/10.1021/ACSANM.8B00770/ASSET/IMAGES/LARGE/AN-2018-007703_0011.JPEG

50. Pourzahedi, L., Zhai, P., Isaacs, J.A., Eckelman, M.J.: Life cycle energy benefits of carbon nanotubes for electromagnetic interference (EMI) shielding applications. J. Clean. Prod. **142**, 1971–1978 (2017). https://doi.org/10.1016/J.JCLEPRO.2016.11.087
51. Wang, L., Dang, Z.M.: Carbon nanotube composites with high dielectric constant at low percolation threshold. Appl. Phys. Lett. **87**(4), 042903 (2005). https://doi.org/10.1063/1.1996842/331004
52. Li, Q., Xue, Q., Hao, L., Gao, X., Zheng, Q.: Large dielectric constant of the chemically functionalized carbon nanotube/polymer composites. Compos. Sci. Technol. **68**(10–11), 2290–2296 (2008). https://doi.org/10.1016/J.COMPSCITECH.2008.04.019
53. Uyor, U.O., Popoola, A.P.I., Popoola, O.M.: Flexible dielectric polymer nanocomposites with improved thermal energy management for energy-power applications. Front. Energy Res. **11**, 1114512 (2023). https://doi.org/10.3389/FENRG.2023.1114512/BIBTEX
54. Si, S.K., et al.: In situ-grown organo-lead bromide perovskite-induced electroactive γ-phase in aerogel PVDF films: an efficient photoactive material for piezoelectric energy harvesting and photodetector applications. Nanoscale. **12**(13), 7214–7230 (2020). https://doi.org/10.1039/D0NR00090F
55. Wang, Y., et al.: Halide perovskite inducing anomalous nonvolatile polarization in poly(vinylidene fluoride)-based flexible nanocomposites. Nat. Commun. **15**(1), 1–9 (2024). https://doi.org/10.1038/s41467-024-48348-4
56. Zhu, W., et al.: Poly(vinylidene fluoride)-stabilized black γ-phase $CsPbI_3$ perovskite for high-performance piezoelectric nanogenerators. ACS Omega. **7**(12), 10559–10567 (2022). https://doi.org/10.1021/ACSOMEGA.2C00091/SUPPL_FILE/AO2C00091_SI_001.PDF
57. He, X., et al.: In situ grown $Cs_2Ag_{0.3}Na_{0.7}InCl_6$ double perovskites within a PVDF matrix for high-performance flexible piezoelectric nanogenerators. ACS Appl. Mater. Interfaces. **17**, 11144–11153 (2025). https://doi.org/10.1021/acsami.4c21765
58. Lu, Y., et al.: Remarkable dielectric breakdown strength enhancement of a PVDF terpolymer using a 2D hybrid organic inorganic perovskite as a functional additive. J. Mater. Chem. C. **7**(43), 13390–13395 (2019). https://doi.org/10.1039/C9TC03220G
59. Dong, Y., et al.: In situ preparation of highly luminescent Sb^{3+}/Mn^{2+} co-doped Cs_2KInCl_6 lead-free double perovskites in a PVDF matrix and application to white light emitting diodes and anti-counterfeiting. J. Mater. Chem. C. **12**(43), 17458–17468 (2024). https://doi.org/10.1039/D4TC03527E
60. Nie, J., Zhang, Y., Wang, J., Li, L., Zhang, Y.: Recent progress in polarization-enhanced PVDF-based perovskite solar cells. Sol. RRL. **7**(24), 2300786 (2023). https://doi.org/10.1002/SOLR.202300786;JOURNAL:JOURNAL:2367198X;WGROUP:STRING:PUBLICATION
61. Ahn, S., Chiu, W.H., Chu, W.C., Chen, P.Y., Lin, T.H., Lee, K.M.: A systematic investigation of PVDF-HFP in perovskite solar cells for improved space mission reliability. Chem. Eng. J. **496**, 153974 (2024). https://doi.org/10.1016/J.CEJ.2024.153974
62. Jiao, Y., et al.: Compositional engineering of hybrid organic–inorganic lead–halide perovskite and PVDF-graphene for high-performance triboelectric nanogenerators. ACS Appl. Mater. Interfaces. **16**(3), 3532–3541 (2024). https://doi.org/10.1021/ACSAMI.3C17203/SUPPL_FILE/AM3C17203_SI_003.MP4
63. Chen, H., et al.: Piezoelectric nanogenerator based on in situ growth all-inorganic $CsPbBr_3$ perovskite nanocrystals in PVDF fibers with long-term stability. Adv. Funct. Mater. **31**(19), 2011073 (2021). https://doi.org/10.1002/ADFM.202011073;JOURNAL:JOURNAL:10990712;PAGE:STRING:ARTICLE/CHAPTER
64. Sultana, A., Sadhukhan, P., Alam, M.M., Das, S., Middya, T.R., Mandal, D.: Organo-lead halide perovskite induced electroactive β-phase in porous PVDF films: an excellent material for photoactive piezoelectric energy harvester and photodetector. ACS Appl. Mater. Interfaces. **10**(4), 4121–4130 (2018). https://doi.org/10.1021/ACSAMI.7B17408/ASSET/IMAGES/LARGE/AM-2017-17408A_0001.JPEG

65. Nie, L., et al.: Highly stable and controllable lasing actions from PVDF encapsulated $CsPbBr_3$ perovskite microcrystals. J. Mater. Chem. C. **10**(43), 16301–16308 (2022). https://doi.org/10.1039/D2TC03148E
66. Yang, R., et al.: Promising PVDF-CNT-graphene-NiCo chains composite films with excellent electromagnetic interference shielding performance. J. Alloys Compd. **908**, 164538 (2022). https://doi.org/10.1016/J.JALLCOM.2022.164538
67. He, P., et al.: Synergistically engineered PVDF/CNT/GnP hierarchical nanocomposites via scalable solution spinning for ultradurable, superhydrophobic EMI shielding wearables. J. Mater. Chem. C. **13**(24), 12242–12253 (2025). https://doi.org/10.1039/D5TC01041A
68. Arjmand, M., Sadeghi, S., Khajehpour, M., Sundararaj, U.: Carbon nanotube/graphene nanoribbon/polyvinylidene fluoride hybrid nanocomposites: rheological and dielectric properties. J. Phys. Chem. C. **121**(1), 169–181 (2017). https://doi.org/10.1021/ACS.JPCC.6B10741/ASSET/IMAGES/MEDIUM/JP-2016-10741K_0009.GIF
69. Taha, T.A.M., Alanazi, S.S., El-Nasser, K.S., Alshammari, A.H., Ismael, A.: Structure–property relationships in PVDF/$SrTiO_3$/CNT nanocomposites for optoelectronic and solar cell applications. Polymers (Basel). **16**(6), 736 (2024). https://doi.org/10.3390/POLYM16060736
70. Palomba, M., Longo, A.: Special Issue: Advanced nanocomposite materials based on graphene oxide/reduced graphene oxide: potential applications and perspectives. Materials (Basel). **15**(24), 8983 (2022). https://doi.org/10.3390/MA15248983
71. Zhang, H., Zhu, Y., Li, L.: Fabrication of PVDF/graphene composites with enhanced β phase via conventional melt processing assisted by solid state shear milling technology. RSC Adv. **10**(6), 3391–3401 (2020). https://doi.org/10.1039/C9RA09459H
72. Sahoo, A., et al.: Preferential perovskite surface-termination induced high piezoresponse in lead-free in situ fabricated $Cs_3Bi_2Br_9$-PVDF nanocomposites promotes biomechanical energy harvesting. Nanoscale. **15**(27), 11603–11615 (2023). https://doi.org/10.1039/D3NR01517C
73. Al-Harthi, M.A., Hussain, M.: Effect of fabrication method on the thermo mechanical and electrical properties of graphene doped PVDF nanocomposites. Nanomaterials. **12**(13), 2315 (2022). https://doi.org/10.3390/NANO12132315
74. Kaspar, P., et al.: Case study of polyvinylidene fluoride doping by carbon nanotubes. Materials (Basel). **14**(6), 1428 (2021). https://doi.org/10.3390/MA14061428
75. Adaval, A., et al.: Piezoelectric properties of noncovalently functionalized 2D nanomaterials incorporated poly(vinylidene fluoride) nanocomposites. J. Appl. Polym. Sci. **141**(3), e54814 (2024). https://doi.org/10.1002/APP.54814;SUBPAGE:STRING:FULL

Future Prospects and Challenges of PVDF-Based Materials

Ravi Ranjan Kumar and Deepak Punetha

Abstract Polyvinylidene fluoride (PVDF) and its copolymers have emerged as multifunctional polymers with exceptional piezoelectric, pyroelectric, and ferroelectric properties, coupled with mechanical flexibility, processability, and chemical stability, making them highly attractive for next-generation electronics. Their applications span flexible and wearable sensors, Internet of Things (IoT) devices, biomedical implants, energy-harvesting systems, and microelectronic memory devices, with recent advances in hybrid nanogenerators, ferroelectric transistors, structural health monitoring, and scalable thin-film processing underscoring their industrial potential. Beyond functional performance, PVDF's role in enabling low-power and self-sustaining electronic systems places it at the center of future technologies such as the Internet of Flexible Things (IoFT), smart healthcare, and sustainable energy solutions. However, challenges such as sustainability concerns, recycling difficulties due to strong C–F bonds, environmental impacts of fluoropolymer synthesis, and limitations in large-scale fabrication remain critical barriers to commercialization. Addressing these issues through green synthesis routes, eco-friendly processing technologies, recyclable PVDF-based composites, and scalable roll-to-roll manufacturing will be crucial for industrial adoption. This chapter provides a comprehensive outlook on the emerging applications of PVDF-based materials, evaluates sustainability and recyclability challenges, and proposes a roadmap for industrial integration, emphasizing the need to align advanced material engineering with circular economy principles, policy frameworks, and standardized testing to establish PVDF as a cornerstone material for sustainable, next-generation electronic technologies.

Keywords PVDF · Piezoelectric polymers · Flexible and wearable electronics · Energy harvesting · Sustainability · Recycling

R. R. Kumar · D. Punetha (✉)
Department of Electronics and Communication Engineering, Motilal Nehru National Institute of Technology, Allahabad, Prayagraj, Uttar Pradesh, India
e-mail: deepakpunetha@mnnit.ac.in

D. Punetha (ed.), *Polyvinylidene Fluoride (PVDF)*, Engineering Materials,
https://doi.org/10.1007/978-981-95-6718-8_10

1 Introduction

The polyvinylidene fluoride (PVDF) and its copolymers have gained significant attention in recent decades as multifunctional polymers with a rare combination of electrical, mechanical, and chemical properties. Unlike conventional polymers that primarily serve as structural or insulating materials, PVDF demonstrates intrinsic piezoelectric, pyroelectric, and ferroelectric responses, positioning it as one of the most versatile smart materials available today [1, 2]. These functional properties, coupled with advantages such as mechanical flexibility, lightweight nature, chemical resistance, thermal stability, and ease of processing, make PVDF a promising candidate for integration into diverse electronic, biomedical, and energy applications. PVDF is a semi-crystalline polymer that exhibits multiple electroactive properties, including piezoelectricity [1–3], pyroelectricity, and ferroelectricity. These properties originate from the high polarity of the $–CH_2/–CF_2$ dipoles in its chain structure, particularly when stabilized in the β-phase crystal configuration. PVDF and its copolymers, such as PVDF-TrFE, combine functional electrical behavior with desirable mechanical characteristics such as flexibility, lightweight nature, and resistance to chemical degradation [4]. These features make them highly suitable for modern applications where ceramic-based ferroelectrics fall short, particularly in flexible, wearable, and sustainable electronics [3–6].

One of the unique features of PVDF is its ability to crystallize into multiple polymorphs (α, β, γ, and δ phases), among which the β-phase exhibits the strongest piezoelectric and ferroelectric behavior. Engineering the crystalline phase structure through processing techniques such as electrospinning [7–11], stretching, annealing, and incorporation of nanofillers has enabled the fabrication of PVDF-based devices with tailored performance. Such advances have opened pathways for next-generation flexible electronics, wearable sensors, implantable medical devices, and energy-harvesting systems, where traditional ceramic-based materials often face limitations due to brittleness and lack of flexibility [12]. The relevance of PVDF is expanding rapidly in the context of emerging technologies such as the Internet of Things (IoT), Internet of Flexible Things (IoFT), soft robotics, and bio-integrated electronics. For instance, PVDF-based nanogenerators are being explored for self-powered IoT nodes, while PVDF-TrFE copolymers are utilized in non-volatile memory devices and ferroelectric transistors. In biomedical applications, PVDF scaffolds and membranes are being developed for tissue engineering, nerve regeneration, and implantable biosensors [13–15]. These developments highlight PVDF's pivotal role as a functional polymer bridging the gap between conventional electronics and flexible, sustainable device platforms. The transformation of PVDF into its electroactive β-phase requires specific processing techniques such as mechanical stretching, electrospinning, high-voltage poling, or nanofiller incorporation [4, 16].

At the device level, PVDF films and composites are being investigated for use in self-powered sensors, energy harvesters, biomedical implants, and next-generation memory devices as shown in Fig. 1. While these developments have significantly expanded the technological scope of PVDF, large-scale industrial implementation

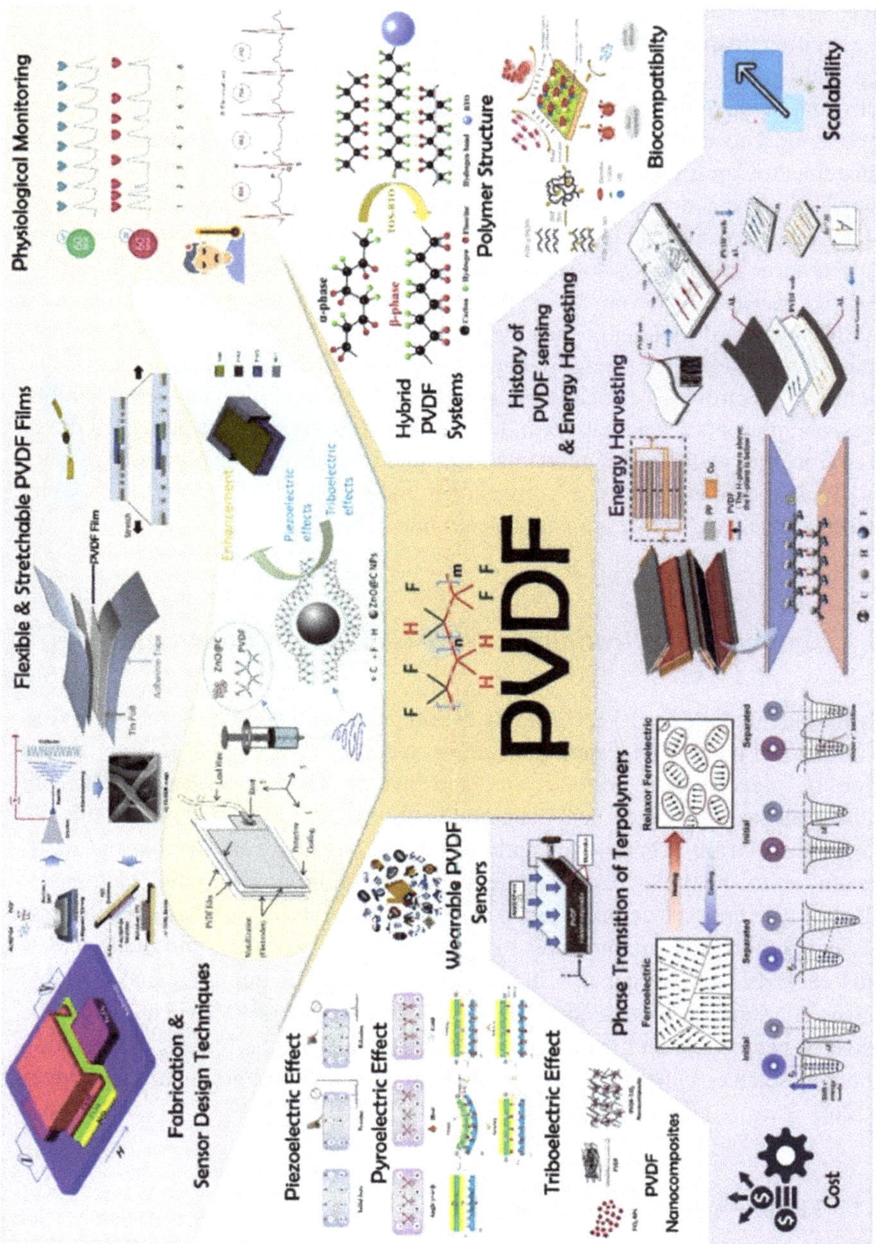

Fig. 1 Multifunctional roles and applications of PVDF in sensing, energy harvesting, and flexible electronics [1]

remains limited due to processing constraints, environmental concerns, and the absence of robust recycling frameworks. This chapter highlights technical advances in PVDF-based applications, the sustainability issues linked to fluoropolymer usage, and the roadmap required for commercial and industrial adoption. Despite these exciting prospects, the transition of PVDF from laboratory-scale innovation to large-scale industrial adoption faces significant challenges [1, 17–19]. The sustainability and recyclability of PVDF are pressing concerns, as its synthesis relies on petrochemical feedstocks and its high chemical inertness complicates end-of-life management. Conventional recycling methods often degrade its functional properties, while advanced chemical recovery techniques remain energy-intensive and costly [20, 21]. Additionally, scalable processing routes, material standardization, and long-term reliability assessments are still under development, slowing down industrial translation. Given these opportunities and challenges, it is essential to establish a forward-looking perspective on PVDF. This chapter provides a comprehensive overview of the emerging applications of PVDF-based materials in next-generation electronics, critically examines the sustainability and recyclability issues, and presents a roadmap for industrial adoption. By aligning advanced material engineering with circular economy principles and sustainable processing strategies, PVDF can evolve as a cornerstone material for the future of flexible, multifunctional, and eco-responsible electronics [16, 22, 23].

1.1 Emerging Applications in Next-Generation Electronics

Polyvinylidene fluoride (PVDF) and its copolymers are increasingly recognized as key functional materials for next-generation electronics due to their unique combination of piezoelectric, ferroelectric, and pyroelectric properties, along with mechanical flexibility and chemical stability. These attributes make them suitable for a wide spectrum of emerging applications [24], ranging from flexible and wearable sensors to energy harvesting devices and neuromorphic computing elements.

Unlike conventional ceramic-based materials, PVDF offers lightweight, processable, and conformable architectures that can seamlessly integrate into modern platforms such as the Internet of Things (IoT), biomedical implants, and portable electronic systems as illustrated in Fig. 2. The adaptability of PVDF-based systems to diverse geometries and their compatibility with large-area, low-cost fabrication techniques further enhance their potential for next-generation electronic devices [25, 26].

1.1.1 Flexible and Wearable Sensors

Flexible and wearable sensors represent one of the most mature application domains of PVDF-based materials. Their piezoelectric and pyroelectric responses enable the detection of mechanical strain, vibration, temperature changes, and bio-signals with

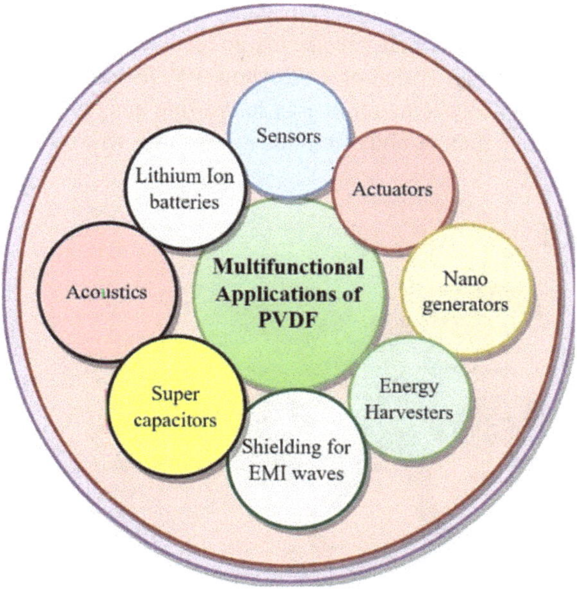

Fig. 2 Multifunctional applications of PVDF in sensors, energy devices, and EMI shielding [5]

high sensitivity. Electrospinning and thin-film casting are widely used fabrication techniques to create PVDF nanofibers and films with aligned β-phase crystallinity, essential for maximizing piezoelectric performance [3, 27, 28].

PVDF-based materials have emerged as a cornerstone in the development of flexible and wearable sensors owing to their inherent piezoelectric, pyroelectric, and triboelectric properties combined with excellent mechanical flexibility and biocompatibility. When fabricated as thin films, fibres, or nanocomposites, PVDF can conform to complex body surfaces, enabling real-time monitoring of physiological parameters such as heartbeat, respiration, body motion, and muscle activity without causing discomfort to the user [29–31]. The high sensitivity of PVDF to mechanical and thermal stimuli allows it to transduce subtle biomechanical signals into electrical outputs, making it ideal for applications in health monitoring, fitness tracking, soft robotics, and human–machine interfaces. Furthermore, its durability under repeated mechanical stress and compatibility with low-cost fabrication techniques support large-scale integration into smart textiles and next-generation wearable electronics. In the domain of human–machine interfaces, PVDF-based electronic skins (e-skins) have gained significant attention due to their ability to mimic the tactile sensing and feedback mechanisms of natural skin. By integrating conductive fillers such as carbon nanotubes (CNTs), graphene, MXenes, or metallic nanoparticles, the dielectric constant, signal-to-noise ratio, and stretchability of PVDF-based sensors can be remarkably enhanced. These hybrid nanocomposite architectures not only improve sensitivity to subtle pressure or strain variations but also ensure durability under mechanical deformation, making them highly adaptable for real-world wearable platforms. As a result, PVDF e-skins are emerging as a promising solution for advanced healthcare monitoring, soft robotics, and interactive systems where precise tactile feedback is essential [30–35].

This Fig. 3 demonstrates how PVDF-based materials can be engineered into wearable systems for energy harvesting and smart sensing applications. In part (I), various human body motions—including wrist, elbow, finger, ankle, toe, and heel

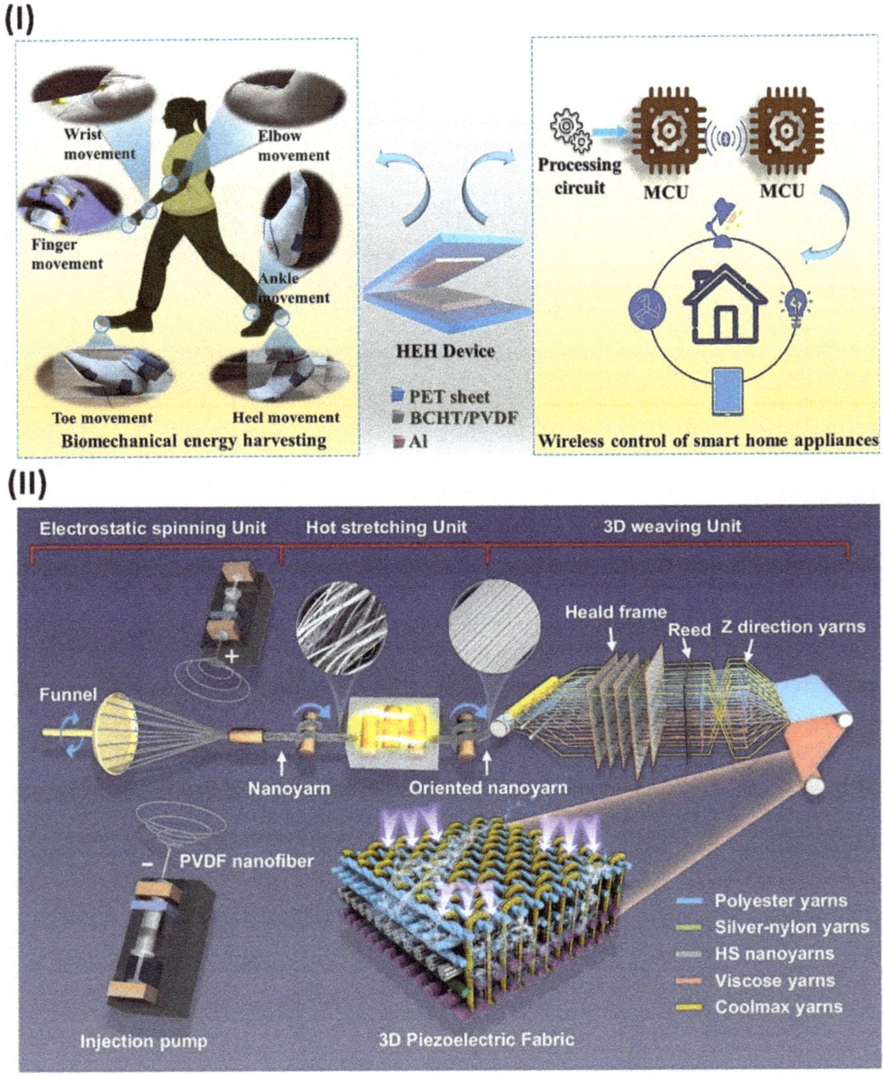

Fig. 3 (**I**) A schematic illustration of a multiphase coexisting perovskite/PVDF composite-based wearable piezo-tribo hybrid energy harvester for wireless smart applications. (**II**) Schematic depiction of the manufacturing and development of three-dimensional piezoelectric fabric (3DPF) Motion tracking: monitoring joint movements and human gestures for rehabilitation and sports training [1]

movements—are utilized to generate power through a hybrid energy harvesting (HEH) device composed of PET sheets, BCHT/PVDF films, and aluminum electrodes. The harvested biomechanical energy is then processed by microcontroller units (MCUs) to enable wireless operation of smart home appliances, showcasing its role in future IoT-enabled platforms [3, 4]. In part (II), the schematic highlights the stepwise fabrication of a 3D piezoelectric fabric: PVDF nanofibers are first produced by electrospinning, aligned into nanoyarns through hot stretching, and subsequently woven into a multifunctional textile using different yarns such as polyester, viscose, silver-nylon, and Coolmax. The resulting fabric integrates flexibility, durability, and enhanced piezoelectric response, making it highly effective for wearable electronics, energy harvesting, and self-powered sensing applications [3, 36–38].

1.1.2 Energy Harvesting Devices

PVDF's ability to convert ambient mechanical energy, acoustic waves, and thermal fluctuations into electricity has made it central to energy harvesting technologies. PVDF-based nanogenerators and piezoelectric harvesters are increasingly used for powering self-sustained devices, especially in the Internet of Things (IoT) ecosystem.

Key developments in PVDF-based energy harvesting focus on enhancing its power output and durability for practical applications [39]. The creation of bimorph cantilevers, which consist of two layers of PVDF, significantly amplifies the bending response and power density, making them particularly suitable for powering low-energy wireless sensor nodes. Advancements in electrospinning techniques have enabled the production of highly aligned nanofiber mats with a dominant piezoelectric β-phase; this optimized structure dramatically increases the piezoelectric coefficients, allowing these flexible mats to efficiently scavenge energy from subtle, low-frequency biomechanical sources like walking, heartbeat, or respiration. Furthermore, the development of hybrid harvesters, which integrate PVDF with high-performance materials such as ZnO nanorods, perovskite films, or two-dimensional nanomaterials, creates synergistic effects that not only substantially boost the generated output voltage through improved charge separation and stress concentration but also enhance the overall mechanical robustness and longevity of the devices [40–42].

PVDF energy harvesters are lightweight, biocompatible, and suitable for self-powered biomedical implants, structural health monitoring, and environmental sensing.

Polyvinylidene Fluoride (PVDF)-Based Piezoelectric Nanogenerators

PVDF based piezoelectric nanogenerators (PENGs) represent a breakthrough in portable and flexible energy harvesting technologies, leveraging PVDF's remarkable piezoelectric, pyroelectric, and dielectric properties. Known for its ease of processing, chemical resistance, and structural flexibility, PVDF serves as an ideal material for applications requiring adaptability and durability, such as electronic skins (E-skins), health monitors, and self-powered devices [6, 43]. Advanced PFM was used to analyze the piezoelectric response of the rGO/WO_3/PVDF tertiary nanocomposite film, leveraging its high spatial resolution and sensitivity to local polarization. Using a conducting tip for AC voltage application, the PFM amplitude was recorded, and the DART mode was employed due to PVDF's low piezoelectric coefficient. Out-of-plane deformation caused tip deflection, mapping z-axis amplitude and phase over a $10 \times 10\ \mu m^2$ area, as shown in Fig. 4a, b. With a 2 V AC voltage, the PFM amplitude and phase revealed high piezoelectric contrast. A DC voltage range of −7 V to +7 V was applied to measure polarization reversal and switching responsiveness, forming the butterfly curve between amplitude and bias voltage, as shown in Fig. 4c. The piezoelectric coefficient (d_{33}) reached up to 32.6 pm/V, confirming the material's piezoelectric behaviour [22, 44, 45].

Enhancements in PVDF's piezoelectric properties have been achieved through the integration of nanomaterials like reduced graphene oxide (rGO) and tungsten oxide (WO_3), forming tertiary nanocomposites with superior mechanical and electrical characteristics. As shown in Fig. 4. The synergy between rGO and WO_3 improves the dispersion and piezoelectric polar phases within PVDF, enhancing energy conversion efficiency while addressing challenges like dielectric loss and thermal management [10, 25, 26]. These advanced nanocomposites enable the fabrication of innovative devices that combine nano generation with functionalities such as ammonia gas detection, offering room-temperature operation, low power consumption, and compact design.

The interaction mechanism between rGO, SnO_2, and PVDF is illustrated in Fig. 5. The π-electrons of rGO interact with the CH_2 dipoles of PVDF, while the F/H atoms of PVDF form hydrogen bonds with the oxygen-containing OH groups in rGO as visualise through Fig. 5. Additionally, the F/H atoms of PVDF interact with the carbonyl and carboxyl groups present in rGO. SnO_2 nanoparticles, hybridized

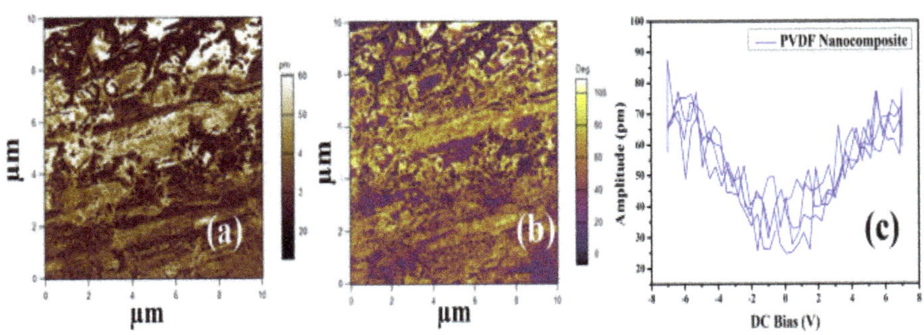

Fig. 4 Piezoresponse force microscopy (PFM) (**a**) amplitude image of rGO/WO_3/PVDF tertiary nanocomposite over a $10 \times 10\ \mu m^2$ area, (**b**) phase difference image of the same composite over a $10 \times 10\ \mu m^2$ area, and (**c**) PFM stress–strain butterfly loop under a DC bias voltage ranging from −7 V to +7 V [6]

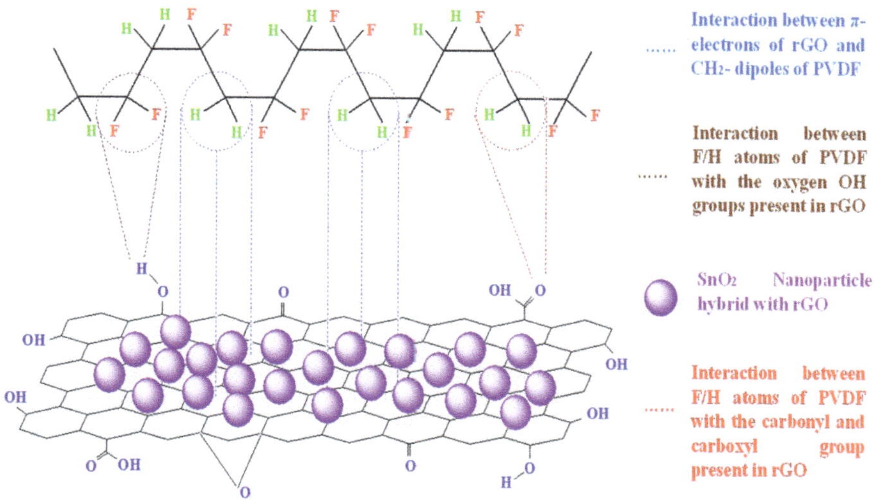

Fig. 5 The interaction mechanism between rGO, SnO_2, and PVDF involves rGO enhancing electrical conductivity, SnO_2 improving dielectric and piezoelectric properties, and PVDF acting as a matrix to synergize their functionalities [10]

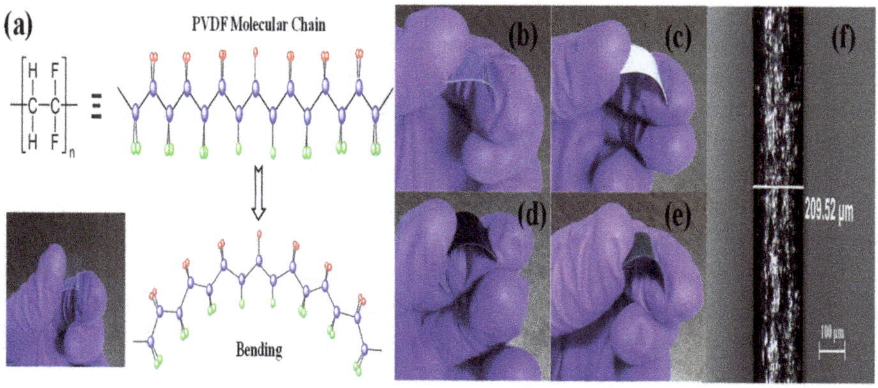

Fig. 6 (**a**) Schematic depiction of the molecular architecture of the PVDF polymer chain; high-resolution images of (**b**) pristine PVDF, (**c**) SnO_2 nanoparticles, (**d**) rGO/PVDF binary nanocomposite, (**e**) rGO/SnO_2/PVDF under mechanical flexural stress; and (**f**) transverse optical microscopy characterization of rGO/SnO_2/PVDF thick films exhibiting a measured thickness of approximately 209.52 μm [10]

with rGO, further enhance the composite's properties by providing additional active sites for interaction. This synergy strengthens the nanocomposite's overall dielectric and piezoelectric performance [6, 10]. The structural and mechanical properties of PVDF-based composites under flexural stress, along with their cross-sectional morphology, are comprehensively depicted in Fig. 6. The PVDF molecular chain is shown undergoing conformational adjustments during mechanical bending, emphasizing its inherent flexibility and adaptability. The pristine PVDF, SnO_2 nanoparticles, and rGO/PVDF binary and ternary rGO/SnO_2/PVDF composites exhibit notable resilience under applied stress. Additionally, the transverse optical micrograph details the homogenous and compact architecture of the rGO/SnO_2/PVDF composite film, with an accurately measured thickness of 209.52 μm, highlighting its suitability for advanced piezoelectric and electromechanical applications [46–48].

1.1.3 Biomedical and Implantable Electronics

The biocompatibility and chemical stability of PVDF make it a promising material for implantable biomedical systems. Its piezoelectric response can be harnessed not only for sensing but also for electrical stimulation of biological tissues, which plays a role in regenerative medicine [49].

In the biomedical field, PVDF is a cornerstone material for advanced applications due to its unique piezoelectric properties and biocompatibility. In tissue engineering, PVDF is fabricated into scaffolds that provide electromechanical stimulation to seeded cells, actively promoting critical regenerative processes such as bone growth, nerve repair, and the formation of new blood vessels (vascularization). Concurrently, its piezoelectric nature is harnessed for self-powered, implantable biosensors that can continuously monitor vital physiological parameters—including blood pressure, glucose levels, and neural activity—in vivo, eliminating the need for external power sources or frequent battery replacements. Furthermore, PVDF membranes are being innovatively incorporated into therapeutic devices, particularly smart drug delivery systems, where they act as responsive components that release therapeutic agents in a controlled manner upon receiving specific mechanical or electrical cues from the surrounding tissue [11, 50].

The use of PVDF-TrFE copolymers further enhances ferroelectric and piezoelectric effects, allowing for more sensitive implantable sensors and actuators.

Polyvinylidene fluoride (PVDF) has emerged as a highly versatile biomaterial owing to its excellent biocompatibility, mechanical stability, and unique piezoelectric and pyroelectric properties. As illustrated in the Fig. 7, PVDF plays a crucial role across a wide spectrum of biomedical applications. It is extensively used in bone and nerve regeneration, where its piezoelectric nature promotes cell proliferation and differentiation, aiding tissue healing. In wound healing and health monitoring, PVDF-based flexible sensors enable real-time tracking of physiological signals. Its adaptability also supports cardiovascular regeneration, muscle tissue regeneration, and abdominal hernia repair, ensuring functional restoration of tissues. Moreover, PVDF finds applications in surgical sutures and tissue sensors, where its stability and responsiveness enhance biomedical performance. Additionally, PVDF is being explored in drug delivery systems, enabling controlled and sustained release, further highlighting its multifunctional role in modern healthcare [51–53].

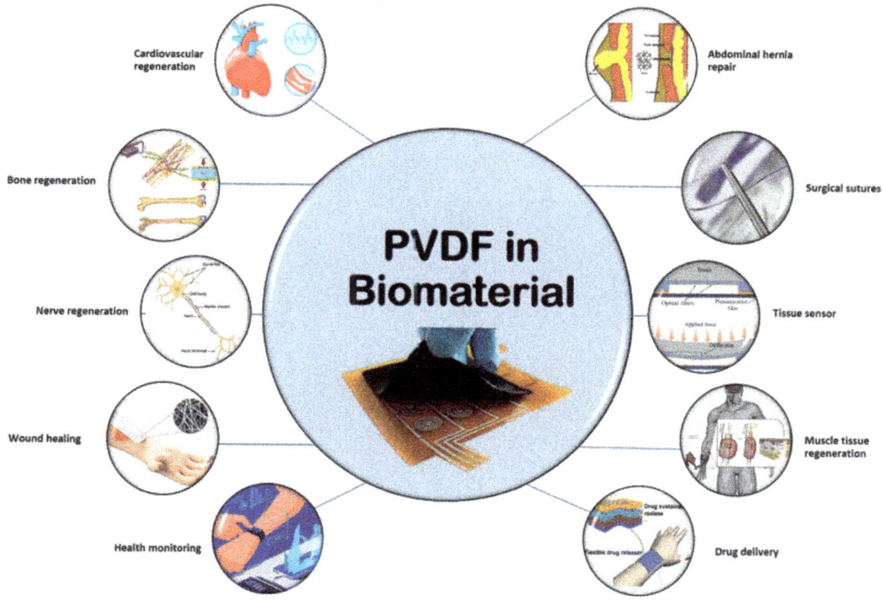

Fig. 7 Multifaceted biomedical applications of PVDF-based materials [11]

1.1.4 Memory and Neuromorphic Devices

In PVDF and particularly PVDF-TrFE copolymers exhibit ferroelectric polarization switching, enabling their use in non-volatile memory devices such as FeRAMs. Thin films of PVDF-TrFE integrated with silicon or oxide semiconductors demonstrate stable remnant polarization, low switching voltage, and scalability for next-generation memory architectures. Emerging areas include [12]:

- Ferroelectric field-effect transistors (FeFETs): Combining PVDF-TrFE with flexible substrates to achieve low-power and reconfigurable electronics.
- Neuromorphic computing: Mimicking synaptic plasticity using PVDF-TrFE films that exhibit multilevel polarization states, thus providing artificial learning and memory functionalities.
- Flexible data storage: Integration into stretchable devices where traditional ceramic ferroelectrics cannot be employed.

Figure 8 showcase PVDF-based ferroelectric materials play a crucial role in neuromorphic computing by imitating the behavior of biological synapses. A PVDF-TrFE gated synaptic transistor with a MoS_2 channel demonstrates stable conductance modulation under repeated electrical pulses, enabling the emulation of synaptic plasticity such as potentiation and depression. By closely resembling the communication between pre- and post-neurons, these devices can replicate learning and memory functions in artificial systems. The incorporation of supervised learning circuits allows fine-tuning of pulse characteristics, ensuring precise signal

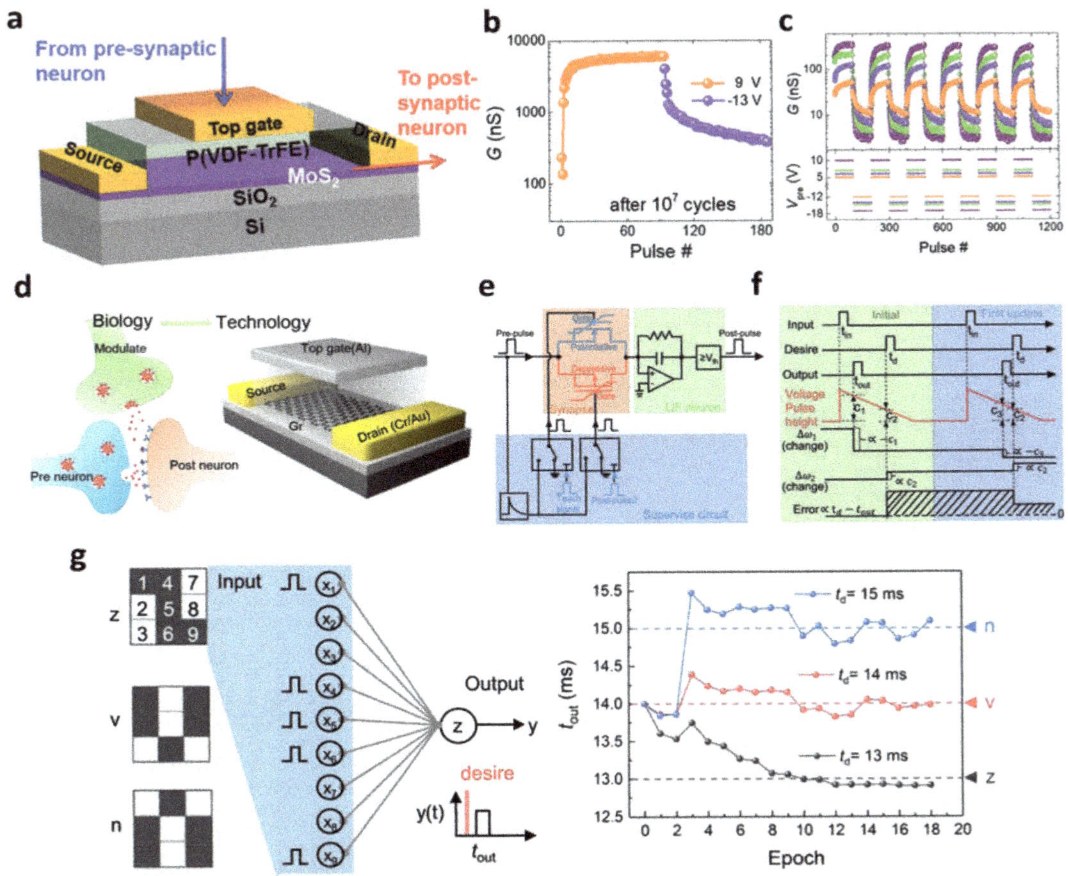

Fig. 8 P(VDF-TrFE) and 2D materials-based three-terminal devices. (**a**) Schematic illustration of the P(VDF-TrFE) and MoS_2-based synaptic transistor. (**b**, **c**) The corresponding potentiation and depression behaviors, which can still be obtained after 10^7 cycles. (**e–h**) The corresponding simulation results of recognition accuracy rate with different annealing temperature and different metal contacts. (**d**) Schematic illustration of the P(VDF-TrFE) and graphene-based electronic synapse. (**e**, **f**) The design of ReSuMe consisting of complementary P(VDF-TrFE) and graphene-based FeFET synapses and the time chart of signals, respectively [12]

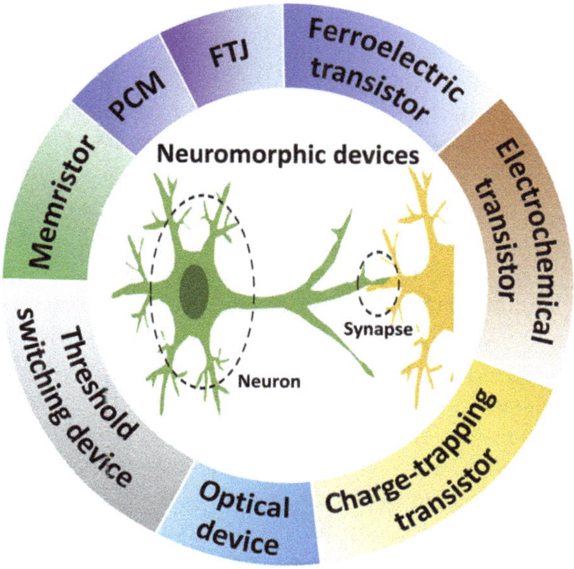

Fig. 9 Neuromorphic devices for brain-inspired computing [13]

adaptation. Neural network models built on such devices are capable of processing inputs effectively and refining outputs over multiple training epochs, highlighting the potential of PVDF-based systems in developing energy-efficient brain-inspired electronics [54].

This schematic in Fig. 9 highlights the diverse range of neuromorphic devices that mimic the biological functionalities of neurons and synapses to enable brain-like information processing. At the center, the neuron and synapse represent the fundamental biological elements responsible for signal transmission and learning. Surrounding them are various device technologies developed to emulate these processes: memristors that store and adapt resistance states, phase-change memory (PCM) for reversible structural switching, ferroelectric tunnel junctions (FTJ) and ferroelectric transistors that exploit polarization for memory and computation, and electrochemical transistors which modulate ionic transport for synaptic-like behaviour [10–13]. In addition, charge-trapping transistors store and release charges for signal modulation, optical devices mimic fast neural signaling using photons, and threshold switching devices simulate neuron firing by exhibiting abrupt resistance changes. Collectively, these device classes form the foundation of neuromorphic hardware, offering energy-efficient, parallel, and adaptive computation similar to the human brain [55–57].

1.1.5 Integration with IoT and Internet of Flexible Things (IoFT)

Piezoelectric The growing paradigm of IoT and IoFT demands scalable, self-powered, and conformable electronic components. PVDF offers unique advantages as a platform material for embedding into smart textiles, skin-mounted electronics, and large-area sensing networks. PVDF-based materials are increasingly being utilized in next-generation connected electronics due to their multifunctional properties and scalable processing capabilities. In the case of self-powered IoT nodes, PVDF nanogenerators can effectively harvest ambient mechanical energy from vibrations, human motion, or environmental sources, thereby reducing the reliance on conventional batteries and extending device lifetime in remote or inaccessible locations [58]. For smart wearable technologies, thin PVDF films can be seamlessly integrated into fabrics, adhesive skin patches, or flexible substrates to provide continuous, non-invasive health monitoring by tracking vital signals such as heart rate, respiration, and body motion. In industrial applications, PVDF sensors configured in flexible and conformable formats play a vital role in structural health monitoring, machinery vibration analysis, and predictive maintenance, ensuring safety and efficiency in real-time. Importantly, the compatibility of PVDF with roll-to-roll processing and inkjet printing techniques allows large-area, lightweight, and low-cost device fabrication, making it highly suitable for scalable production of smart, connected, and sustainable electronic platforms [59, 60].

2 Sustainability and Recyclability Concerns

The increasing integration of PVDF-based materials in next-generation electronics, biomedical systems, and energy harvesting platforms raises critical questions about sustainability and environmental responsibility. Although PVDF exhibits outstanding stability, piezoelectricity, and mechanical resilience, these very properties make it difficult to recycle and environmentally persistent. Addressing sustainability and recyclability concerns is essential to ensure that PVDF materials are not only high-performing but also aligned with global goals of green technology, circular economy, and reduced carbon emissions.

2.1 Environmental Challenges of PVDF

Polyvinylidene fluoride (PVDF) is widely recognized for its superior piezoelectricity, flexibility, and long-term durability, which make it a material of choice in electronics, biomedical devices, and energy harvesters. However, the same strong carbon–fluorine (C–F) bonds that impart high chemical and thermal stability also pose major environmental concerns. PVDF is highly resistant to natural degradation, meaning that once discarded, it can persist in ecosystems for decades without breakdown. Disposal through incineration is problematic, as the thermal decomposition of PVDF produces hazardous gases such as hydrogen fluoride (HF), carbonyl fluoride (COF_2), and other perfluorinated byproducts, which are toxic, corrosive, and environmentally persistent. Furthermore, the production of PVDF is tied to petrochemical feedstocks and fluorination processes that may release per- and polyfluoroalkyl substances (PFAS), commonly called "forever chemicals," which have been linked to bioaccumulation and long-term health risks [61].

PVDF exhibits excellent piezoelectric and mechanical properties for wearable health monitoring, but several technical challenges restrict its deployment. The absence of standardized benchmarking protocols limits comparative evaluation against commercial sensors, with variability in poling fields, film thickness, and processing routes affecting signal fidelity, stability, and reproducibility which is represented through Fig. 10. Environmental susceptibility, particularly to humidity and sweat, induces depolarization and ionic migration, degrading β-phase crystallite alignment and piezoelectric response. Scalability remains problematic, as large-area films often suffer from non-uniform charge distribution, compromised domain orientation, and thickness–porosity trade-offs, impacting electromechanical coupling [1]. Advanced fabrication strategies such as roll-to-roll processing, 3D printing, and electrospinning are needed to achieve structural control while preserving flexibility and breathability. Hybrid architectures incorporating PVDF nanofibers in elastomeric matrices show potential to enhance robustness and signal transduction, but require rigorous long-term fatigue, stability, and biocompatibility studies under in vivo conditions to validate practical integration into next-generation wearable platforms [62, 63].

Another significant challenge arises at the end-of-life stage of PVDF-based devices. Modern applications often integrate PVDF into complex composites with metallic electrodes, conductive nanofillers (such as CNTs, graphene, or MXenes),

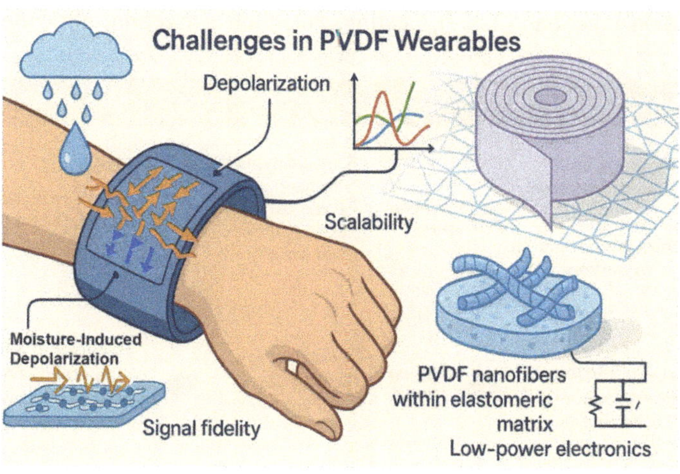

Fig. 10 Schematic representation of PVDF challenges in wearables [1]

and encapsulating polymers. This makes recovery and recycling extremely difficult, pushing most PVDF waste streams toward landfilling or unsafe burning [1]. The increasing demand for PVDF in wearables, implantable medical devices, and IoT nodes is expected to generate growing amounts of non-biodegradable waste that current recycling infrastructure cannot effectively manage. As a result, the environmental footprint of PVDF is characterized by non-biodegradability, toxic decomposition products, fossil-based production routes, and the absence of scalable end-of-life recovery strategies, highlighting the urgent need for sustainable manufacturing and greener alternatives [63–65].

2.2 *Recycling Limitations*

Recycling PVDF poses substantial technical and economic challenges due to its intrinsic material properties. Unlike commodity plastics such as polyethylene or polypropylene, which can be readily reprocessed through mechanical recycling, PVDF resists thermal re-melting because of its high crystallinity and strong C–F bonds [33]. These structural features ensure excellent stability during operation but hinder polymer chain breakdown, making conventional recycling routes inefficient. Mechanical recycling, when attempted, often leads to deterioration of piezoelectric and dielectric performance, as repeated processing alters the molecular alignment and reduces the proportion of electroactive β-phase in PVDF. Additionally, the presence of additives such as plasticizers, stabilizers, and conductive fillers complicates re-extrusion, often resulting in heterogeneous blends that no longer meet functional requirements [17, 36].

Figure 11 shows the latest methods for handling PVDF (polyvinylidene fluoride) from used lithium-ion batteries. With the rise in discarded batteries causing pollution, PVDF is separated through techniques such as physical processing, solvent use, oxidation reactions [3, 27], heating, and fluid extraction, allowing valuable

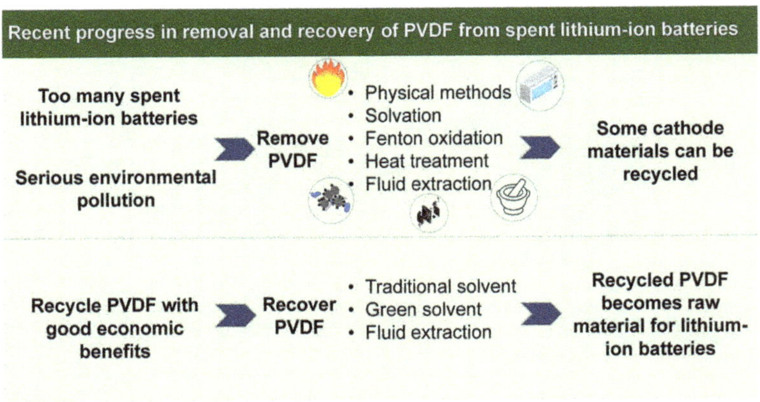

Fig. 11 Recent advances in removal and recovery of PVDF from spent lithium-ion batteries [15]

cathode materials to be reused. After separation, PVDF can be recovered using conventional solvents, eco-friendly solvents, or fluid-based extraction. The reclaimed PVDF is then repurposed as a feedstock for new batteries, which both reduces environmental damage and creates economic value [11].

In recent years, green solvents have gained attention for removing PVDF binders from spent lithium-ion batteries, offering safer and more sustainable alternatives to traditional solvents [15, 44]. Among them, methanol–citric acid (MeOH–CA) shows particular promise due to its mild reaction conditions, requiring only 45 °C and 15 min to separate PVDF from aluminum foil and other active materials. Citric acid provides protons that attack the aluminum surface and break hydrogen bonds with PVDF, while methanol forms esters that accelerate PVDF decomposition and detachment. This method is cost-effective, with a treatment cost of $4.21 per kilogram of spent batteries and a profit of $6.29 per kilogram, while also enabling recovery and reuse of the solvent. Although AlF_3 formation during the process may reduce the conductivity of recycled aluminum foil, the overall efficiency, simplicity, and economic viability make MeOH–CA a strong candidate for industrial-scale PVDF removal and cathode recycling. Apart from solvent-based approaches [45, 56], PVDF binders can also be removed through thermal decomposition methods such as non-thermal plasma, molten salt treatment, microwave roasting, and the widely applied calcination process. For example, alkali-enhanced PVDF cracking has been demonstrated by roasting waste LFP powder at 275 °C for 1.5 h, where liquid-phase PVDF at 174 °C reacts with OH^- to facilitate binder removal and decomposition efficiency increases with temperature as illustrated in Fig. 12. Although this process reduces HF emission due to absorption by alkaline solutions, fluoride corrosion can still damage high-temperature equipment and raise costs [13].

As we can see through Fig. 13 CaO or $CaCO_3$ can be added during calcination to absorb fluorine-containing gases, lower the reaction temperature to around 300 °C, and enhance environmental safety. EDAX analysis confirmed nearly 100% PVDF

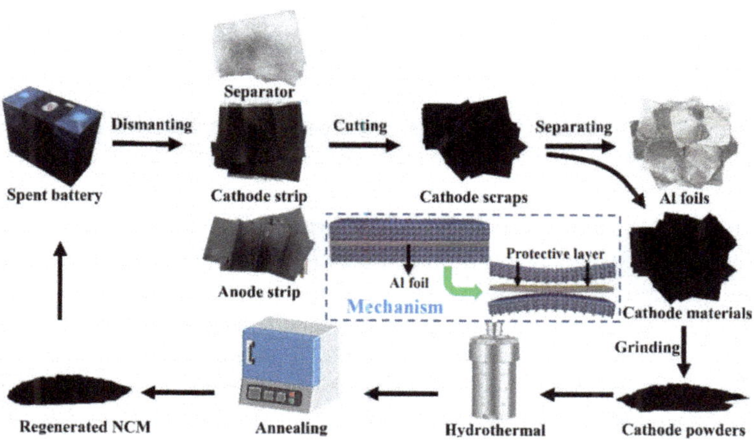

Fig. 12 Schematic illustration of the reaction pathway and underlying mechanism involved in the removal of PVDF using the MeOH–CA technique [15]

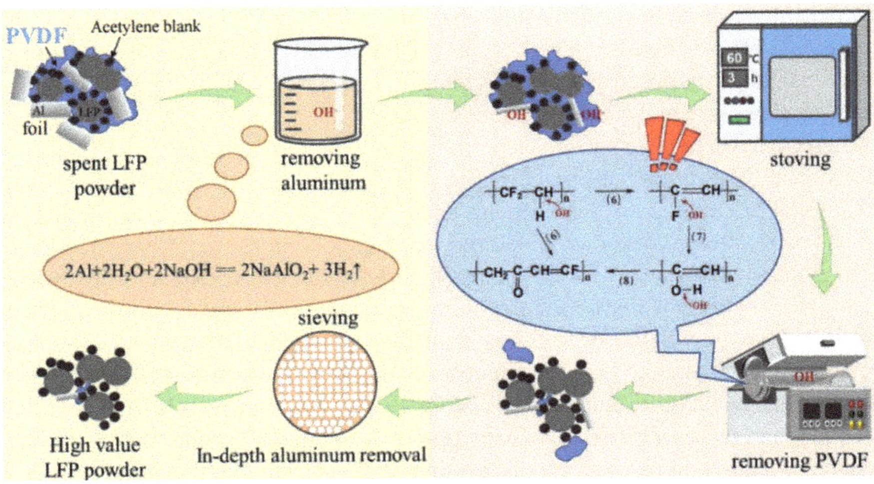

Fig. 13 Schematic diagram of the reaction process and mechanism for removing PVDF by alkali-enhanced PVDF cracking technology [15]

removal, while cost analysis showed savings of \$0.5 per kilogram of LIBs treated, with minimal additional expense for CaO. Moreover, $CaCO_3$ addition effectively fixed fluorine as stable CaF_2, reducing HF release from 72 to 0.8%, making the process more energy-efficient, cost-effective, and environmentally friendly [15, 65].

2.3 *Sustainability in Production*

The production of PVDF is energy-intensive and relies heavily on fluorinated precursors derived from non-renewable petrochemical feedstocks, which contributes significantly to its environmental footprint. The synthesis process typically involves polymerization of vinylidene fluoride monomers under high-pressure conditions, often using hazardous initiators or solvents that require careful handling and disposal. Moreover, the strong C–F bonds, while essential for PVDF's chemical stability, also demand high energy input during processing, thereby increasing greenhouse gas emissions. These factors raise concerns about the long-term sustainability of PVDF manufacturing, especially as demand rises in next-generation electronics, biomedical devices, and energy storage systems. Additionally, large-scale production often generates fluorinated byproducts and residues that can persist in the environment, raising toxicity and waste management challenges [7, 16].

Recent efforts have focused on making PVDF production more sustainable by adopting greener synthesis routes, such as solvent-free polymerization techniques, aqueous dispersion systems, and the use of safer initiators. Furthermore, research into bio-based alternatives to conventional monomers is gaining momentum, aiming to reduce dependence on fossil-derived feedstocks. Advances in processing

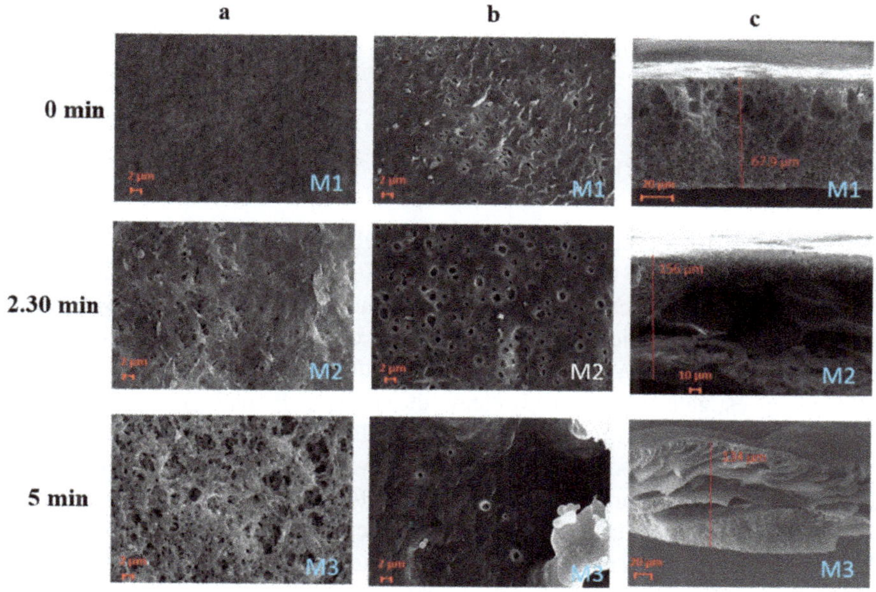

Fig. 14 SEM images illustrating the effect of the exposure time to humidity on PVDF-HFP/NMP membranes M1–M3: (**a**) top surfaces; (**b**) bottom surfaces; (**c**) cross-sections [16]

technologies, including additive manufacturing and roll-to-roll fabrication, also enable reduced material wastage and energy consumption. Incorporating renewable energy sources into PVDF manufacturing facilities further enhances sustainability by lowering the carbon footprint of production. While these strategies are still emerging, their integration into industrial-scale production could significantly improve the eco-efficiency of PVDF, paving the way for its broader adoption in environmentally conscious applications [62, 63].

The SEM analysis and characterization of PVDF-HFP/NMP membranes prepared at different humidity exposure times (0, 2.5, and 5 min) revealed clear variations in morphology and surface properties as shown in Fig. 14. All membranes showed an asymmetric structure with porous surfaces and sub-layers, while those produced by VIPS-NIPS developed macrovoids across the cross-section due to rapid phase inversion at low polymer concentration [16–18]. The membrane without humidity exposure (M1) exhibited small, uniformly distributed pores with a thin structure (68 μm), moderate porosity (69%), and the lowest contact angle (96°), reflecting relatively smoother morphology and weaker hydrophobicity. With 2.5 min exposure (M2), the membrane developed larger pores, a much greater thickness (169 μm), unchanged porosity (69%), and the highest contact angle (121°), indicating enhanced roughness and stronger hydrophobic character. At 5 min exposure (M3), the membrane showed slightly smaller pores (0.03 μm), intermediate thickness (128 μm), slightly reduced porosity (66%), and a moderate contact angle (106°), suggesting partial reduction in surface roughness compared to M2. Overall, increasing humidity exposure time promoted pore growth, modified thickness, and altered hydrophobicity, with M2 displaying the most significant structural and surface changes due to optimal water uptake during membrane formation [27, 36].

2.4 *Emerging Recycling Strategies*

In response to the recycling limitations of PVDF, several emerging strategies are being explored to enable more sustainable end-of-life management [42]. One promising approach is chemical depolymerization, where PVDF is broken down into its monomeric or oligomeric building blocks under controlled conditions. Techniques such as catalytic hydrolysis, reductive defluorination, and plasma-assisted processes are being investigated to selectively cleave the strong C–F bonds. While these methods are still at an experimental stage, they hold the potential to recover valuable fluorinated intermediates for reuse in polymer synthesis, thereby creating a partial closed-loop system. Another avenue is solvent-based dissolution and recovery, where PVDF is separated from composite structures without significant thermal degradation, enabling its reuse in secondary applications. Such methods, when optimized, can help preserve functional properties like piezoelectricity and dielectric performance, which are often lost in mechanical recycling [17, 20].

2.4.1 Sustainable Recycling of Waste Poly(vinylidene fluoride) and Rational Design of Janus Membrane

The Janus membrane was fabricated through electrospinning of hydrophobic PVDF onto the surface of a superhydrophilic PDA/CM substrate as shown in Fig. 15. The preparation process was as follows: PVDF filter membranes were first cut into small sections, thoroughly washed, and dried. Subsequently, 3.6 g of PVDF was dissolved in 44.2 g of a DMF–acetone mixed solvent (4:1, w/w%) to obtain a 7 wt% PVDF spinning solution, which was continuously stirred at 25 °C. During electrospinning [18], the distance between the needle tip and the drum collector was maintained at 10 cm, with an applied voltage of 15 kV, while the solution feed rate was controlled at 0.035 mL/min. The resulting membrane was dried at room temperature for 12 h to yield the final Janus structure. One side consisted of the PDA/CM layer with superhydrophilic properties (denoted as Janus-PDA/CM), while the opposite side was composed of hydrophobic PVDF (denoted as Janus-PVDF) [20].

The CM surface displays a hierarchical porous structure with an average pore size of about 40 μm. To improve water affinity, a PDA layer was deposited on the cellulose surface for hydrophilic modification. Compared with the original CM, the Janus-PDA/CM surface shows no significant structural alteration and maintains its honeycomb-like porous morphology, which is advantageous for allowing water droplets to penetrate during emulsion separation. In contrast, the Janus-PVDF membrane surface consists of interwoven fibers forming a network-like porous structure [9, 13]. At higher magnification, the fibers appear rough, with an average diameter of around 0.5 μm. This interconnected fibrous network enhances surface roughness, lowers surface energy, and consequently imparts hydrophobicity to the membrane which is visualised through Fig. 16.

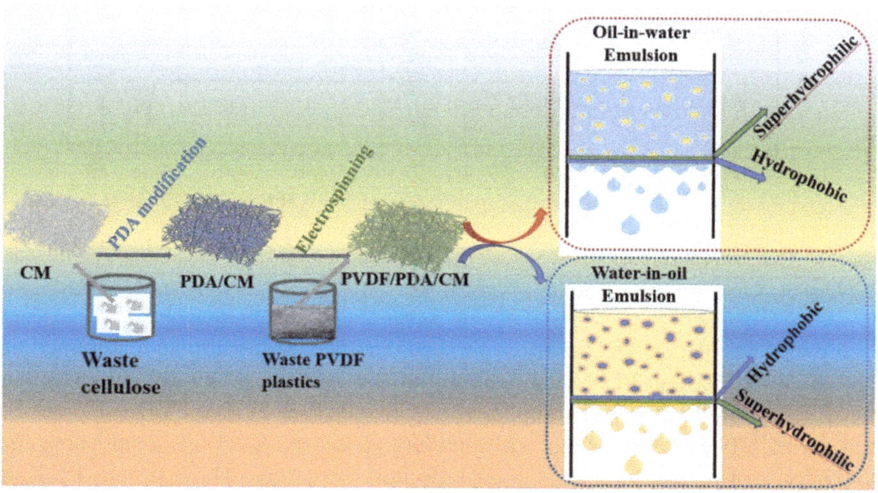

Fig. 15 The preparation process of the Janus membrane [20]

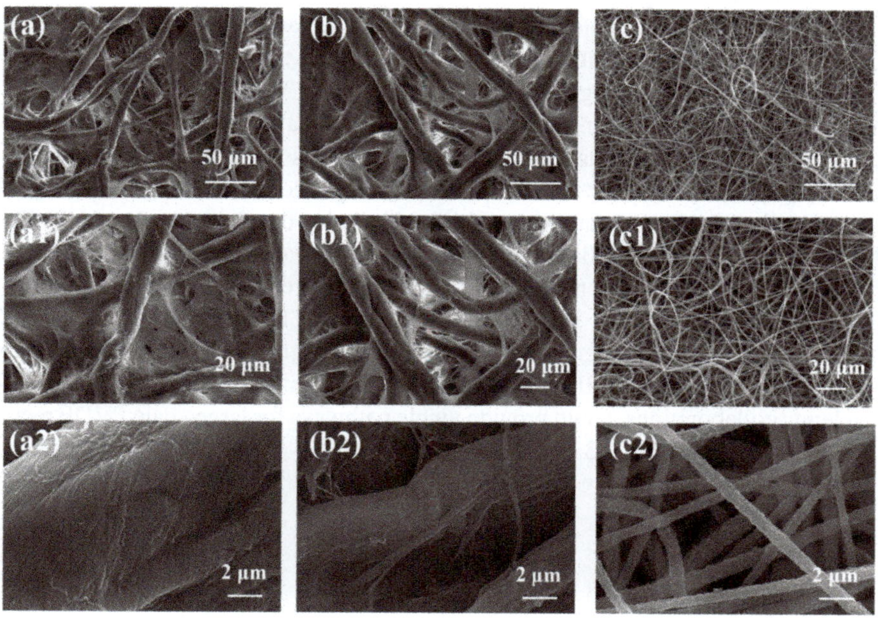

Fig. 16 SEM images of CM membrane (**a**–**a2**), Janus-PDA/CM surface (**b**–**b2**) and Janus-PVDF surface (**c**–**c2**) at different magnifications [20]

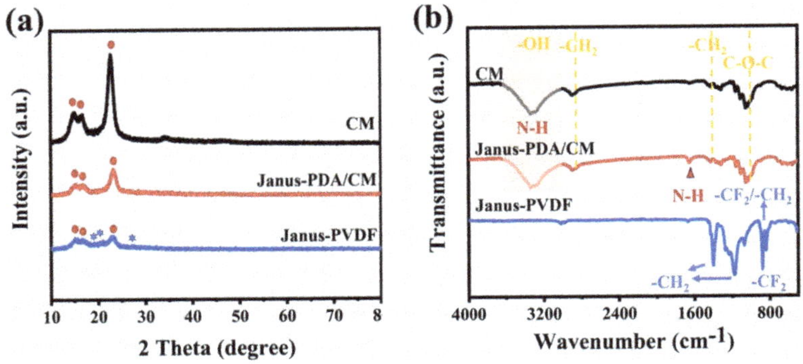

Fig. 17 XRD pattern (**a**) and FT-IR spectra (**b**) of CM, Janus-PDA/CM and Janus-PVDF membranes [20]

Figure 17 provides XRD analysis showed that the CM membrane maintained its crystalline peaks, and PDA modification did not alter the structure, ensuring that recyclability of the cellulose component remains unaffected. The PVDF side exhibited strong β-phase peaks while still retaining cellulose reflections, confirming the coexistence of both components in a single Janus structure, which is beneficial for reuse and recycling. FT-IR spectra further verified the stable incorporation of PDA on CM and the presence of characteristic CH_2 and CF_2 bands in PVDF, indicating that the structural integrity of each layer is preserved. This stability and clear phase distinction highlight the potential of the Janus membrane for sustainable applications with enhanced recyclability [17].

In parallel, hybrid strategies that combine recycling with upcycling approaches are gaining attention. For instance, pyrolysis or plasma treatment of PVDF waste can generate fluorinated gases that serve as precursors for high-value fluoropolymers or specialty chemicals, thus reducing dependence on virgin resources. Additionally, advanced separation techniques are being developed to recover PVDF from composites with metallic or ceramic fillers, a critical challenge in the recycling of flexible electronics and energy devices. Emerging green solvents and supercritical fluid systems are also being tested to minimize hazardous byproducts and improve process eco-efficiency. Although these strategies are currently limited to laboratory or pilot scales, their continued development, supported by industrial–academic collaborations, could provide viable pathways for large-scale PVDF recycling in the future [23, 25, 33].

3 Roadmap for Industrial Adoption

3.1 *Scalable Manufacturing and Process Optimization*

The transition of PVDF-based materials from laboratory-scale demonstrations to industrial adoption requires scalable and efficient manufacturing processes that preserve their unique electroactive properties. Conventional fabrication techniques such as solution casting, melt extrusion, and hot pressing provide relatively simple processing routes, but they are limited in terms of scalability, precision, and material uniformity. For instance, solution casting often produces films with variable crystallinity, while extrusion methods can restrict the induction of the β-phase, which is essential for piezoelectric activity [20–22]. To overcome these limitations, advanced processing techniques have been explored that are compatible with high-throughput production. Additive manufacturing, particularly roll-to-roll processing and inkjet printing, has emerged as a promising strategy for fabricating PVDF thin films and composites at industrial volumes. These techniques offer precise control over thickness, surface morphology, and phase composition, which directly influence the dielectric constant, ferroelectric switching, and piezoelectric coefficients. Electrospinning has also gained attention for producing PVDF nanofibers with inherently high β-phase content due to the combined effects of mechanical stretching and high-voltage polarization. Such nanofibrous mats are highly flexible, lightweight, and mechanically robust, making them suitable for wearable electronics, self-powered sensors, and biomedical implants [26, 27].

Despite these advances, challenges remain in reducing the energy and cost associated with β-phase induction, which typically requires mechanical stretching, poling under high electric fields, or incorporation of nucleating fillers. Additionally, batch-to-batch variability in film quality and device reproducibility must be minimized to meet industrial standards. Future optimization strategies should focus on integrating green chemistry approaches, such as solvent-free processing and low-temperature crystallization methods, to reduce environmental impact while ensuring consistent material performance. Ultimately, scalable process optimization will be the cornerstone for bridging laboratory research with industrial-scale manufacturing of PVDF-based electronic devices [18].

3.2 *Integration into Emerging Technologies and Application Ecosystems*

The successful industrial adoption of PVDF-based materials is closely linked to their seamless integration into emerging technologies and broader application ecosystems. PVDF's unique combination of flexibility, biocompatibility, piezoelectricity, and ferroelectric properties positions it as a cornerstone material for next-generation electronics [41]. In wearable and biomedical devices, PVDF-based sensors and energy harvesters are already being tested for continuous health monitoring, drug delivery platforms, and implantable systems. Their lightweight and conformable nature allows integration with textiles, skin patches, and soft robotic systems, thereby creating multifunctional devices that operate in dynamic environments. However, translating these laboratory demonstrations into commercially viable systems requires the establishment of standardized testing protocols, robust interfacing with microelectronics, and long-term reliability studies under real-world conditions. Beyond the biomedical sector, PVDF is playing a transformative role in the Internet of Things (IoT), smart infrastructure, and renewable energy ecosystems [63]. For instance, PVDF nanogenerators embedded in IoT nodes enable autonomous, battery-free operation by harvesting ambient vibrations, which reduces maintenance costs in large-scale deployments. In civil and industrial applications, PVDF-based structural health monitoring systems can detect strain, cracks, and vibrations in real-time, feeding into predictive maintenance frameworks supported by artificial intelligence. Furthermore, PVDF films are being investigated in flexible photovoltaics, hybrid perovskite solar cells, and electrocaloric cooling devices, where their dielectric and ferroelectric properties enhance device stability and efficiency. The challenge lies in integrating PVDF into these systems without compromising performance due to environmental degradation, interfacial incompatibility, or lack of standardized fabrication protocols [44].

As visualise through Fig. 18, accurately quantifying the environmental impact of polyvinylidene fluoride (PVDF) production is crucial due to its growing use in emerging technologies, but previous life cycle assessments have been unreliable as they used surrogate data instead of modeling the actual process [21]. This study addresses that gap by establishing a new, scientifically rigorous cradle-to-gate inventory based on PVDF's commercial synthesis, resulting in calculated impacts

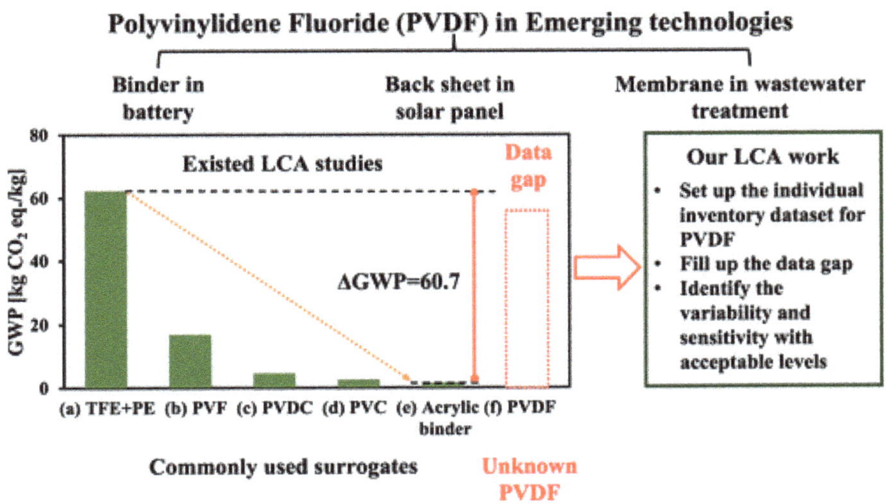

Fig. 18 Bridging the data gap: a life cycle assessment of polyvinylidene fluoride (PVDF) for emerging technologies [21]

of 55.8 kg CO_2 equivalent for global warming potential and a high cumulative energy demand of 756 MJ equivalent per kg of PVDF, primarily due to the chlorine-intensive production of its monomer. This work presents the first detailed life cycle inventory dataset for PVDF, providing a more accurate and less uncertain basis for evaluating its environmental footprint [17–19].

To fully realize its industrial potential, PVDF must be embedded within a broader application ecosystem that includes electronics, data analytics, wireless communication, and cloud-based platforms. This requires interdisciplinary collaboration between materials scientists, device engineers, and system integrators. By developing hybrid architectures where PVDF components work synergistically with silicon electronics, wireless modules, and machine learning algorithms, PVDF-based materials can move beyond niche applications toward mainstream adoption. In this context, the creation of open innovation platforms and industrial consortia will be critical to accelerate the translation of PVDF research into commercially competitive, multifunctional technologies [52].

3.3 Standardization, Sustainability, and Industrial Collaboration

One of the key challenges hindering the widespread industrial adoption of PVDF-based materials is the lack of well-defined standards and regulatory frameworks. At present, most PVDF applications remain confined to academic demonstrations or pilot-scale prototypes, with limited pathways toward commercialization due to the absence of standardized testing protocols. Mechanical durability, piezoelectric coefficient stability, and long-term environmental resistance are often evaluated under laboratory-specific conditions, making cross-comparison between studies difficult. Establishing internationally recognized standards for characterization, performance benchmarking, and reliability testing is essential to enable industry-wide confidence and promote large-scale adoption [30–32]. Furthermore, safety regulations concerning fluoropolymer degradation products must be addressed to comply with environmental and health policies, especially in regions with stringent chemical usage regulations.

Sustainability considerations further complicate the industrial deployment of PVDF. As a fluoropolymer, PVDF poses challenges in terms of recyclability and end-of-life management, particularly when used in large-scale consumer electronics, wearable devices, or energy harvesting systems [46]. A circular economy framework for PVDF is still underdeveloped, and without effective recycling strategies, the environmental footprint of PVDF-based devices may limit their long-term industrial viability. Green synthesis methods, solvent-free processing, and integration with biodegradable or recyclable substrates represent promising approaches to reduce the ecological impact of PVDF production and disposal. Collaboration between academia, industries, and policymakers will be crucial to establish eco-friendly guidelines and implement scalable recycling methods for PVDF waste streams [56, 57].

Industrial collaboration and open innovation platforms are vital to accelerate the transition of PVDF-based technologies from research labs to market-ready products. Partnerships among materials scientists, electronics manufacturers, healthcare companies, and IoT developers can help align PVDF's functional advantages with specific application needs. Such collaborations also facilitate cost reduction through shared infrastructure, optimize supply chains for large-scale deployment, and promote knowledge transfer across disciplines. International consortia and regulatory bodies can play a central role in bridging the gap between research innovations and industrial requirements [7], ensuring that PVDF-based technologies are developed in compliance with safety, sustainability, and performance standards. By combining technical standardization, sustainable production practices, and collaborative industrial frameworks, PVDF can emerge as a cornerstone material for the next generation of smart, sustainable, and connected technologies [11, 61] of PVDF that permits the piezopolymer to endure substantial temperature fluctuations without experiencing significant degradation of functionality [5, 16].

4 Summary and Outlook

The characterized by their strong piezoelectric and ferroelectric responses, remarkable chemical stability, high dielectric constant, and excellent mechanical flexibility, polyvinylidene fluoride (PVDF) and its copolymers have emerged as some of the most versatile functional polymers for advanced electronic, biomedical, and energy-related technologies. Their unique combination of lightweight and durable characteristics has enabled integration into wearable sensors, self-powered devices, structural health monitoring systems, biomedical implants, artificial muscles, smart membranes, and IoT-enabled electronics, with additional applications in soft robotics, drug delivery, and underwater transducers. Performance is closely tied to their crystalline polymorphs, with the β-phase offering the highest dipole alignment and piezoelectric response; hence, significant research has focused on enhancing β-phase content through mechanical stretching, high-voltage poling, additive blending, surface functionalization, multilayer architectures, and electrospinning. In addition to conventional processing, PVDF-based composites incorporating inorganic fillers such as $BaTiO_3$, ZnO, and graphene, or noble metals like Au and Ag, have been shown to enhance dielectric properties [17, 33], boost electromechanical coupling, and improve multifunctionality by combining piezoelectric, pyroelectric, optical, and thermal effects within a single system. Bio-inspired and hybrid architectures further extend adaptability, allowing PVDF to function as both a sensing and energy-harvesting medium in autonomous, real-time physiological monitoring and environmental surveillance applications. Thin films and copolymers are increasingly exploited in nanogenerators for powering next-generation low-power electronics, while Janus membranes demonstrate tunable wettability and directional transport for use in smart filtration and biomedical systems. Despite this broad potential, challenges remain in scalability, energy-efficient processing, and sustainability. The fabrication of pure, highly crystalline PVDF films demands energy-intensive post-processing steps such as stretching and poling, while its non-biodegradable nature and resistance to conventional recycling present environmental risks. To address these, strategies including solvent-assisted recovery, incorporation with biodegradable substrates, lifecycle-conscious design, and green processing routes have been explored. Moreover, ensuring long-term operational reliability under mechanical, thermal, and environmental stress is crucial for practical deployment [46], particularly in biomedical and outdoor applications. Industrial translation depends on the development of roll-to-roll manufacturing, additive manufacturing, and standardized composite processing techniques to ensure consistent β-phase content, mechanical robustness, and interfacial compatibility. Equally important are regulatory and policy frameworks that encourage sustainable practices and support commercialization pathways. The continued advancement of PVDF-based materials, therefore, lies in uniting material innovations with scalable, eco-friendly processing and strong academia–industry collaboration to deliver multifunctional, cost-effective, and environmentally responsible technologies for the next generation of electronics, energy systems, and biomedical devices. Emerging Applications: PVDF's versatility stems from its ability to demonstrate piezoelectric, pyroelectric, and ferroelectric properties, making it suitable for multifunctional sensors, implantable medical devices, memory architectures, and neuromorphic (brain-like) computing elements [48–50]. The polymer's adaptiveness allows integration into smart wearables for real-time physiological monitoring, flexible electronics, IoT nodes, and energy storage systems. Specialized processes such as electrospinning, mechanical stretching, and incorporation of nanofillers—enable the creation of high-performance PVDF devices with tunable features [17].

- Energy Harvesting and Biomedical Use: PVDF nanogenerators and flexible films efficiently harvest mechanical, vibrational, or ambient energy, supporting self-powered devices and enabling autonomous IoT nodes and environmental or health sensors. In biomedical engineering, PVDF scaffolds are essential for tissue regeneration, nerve growth, and smart biosensors, with added biocompatibility supporting implantable applications. Continuous health monitoring, controlled drug delivery, and smart wound detection are examples of PVDF's transformative roles in healthcare [11, 20].
- Sustainability and Recyclability Challenges: A fundamental barrier to industrial adoption lies in PVDF's environmental profile. The robust carbon–fluorine bonds that give the polymer stability also make it virtually non-biodegradable, resulting in significant waste management issues after device lifetimes end. Standard mechanical recycling is ineffective, and chemical breakdown requires energy-intensive processes or produces hazardous byproducts such as hydrogen fluoride and perfluoroalkyl substances (PFAS). Even advances in green solvent extraction and plasma treatment remain mostly at laboratory scale, calling for more scalable, cost-effective, and eco-safe recycling solutions [27].

- Scalable Manufacturing Obstacles: Current processing techniques (e.g., solution casting, extrusion, hot pressing) either cannot be scaled efficiently or fail to guarantee the consistent induction of the highly desirable electroactive phase—critical for functional properties like piezoelectricity. Innovations in additive manufacturing (roll-to-roll processes, 3D printing, inkjet printing) show promise for high-throughput, large-area, and multi-material composites, but energy and cost barriers, as well as reproducibility issues, remain unresolved for many intended applications [28–30].

4.1 Outlook

- Future Opportunities: PVDF-based materials are poised for wider adoption across flexible and wearable electronics, sustainable IoT networks, health technology, structural health monitoring, and autonomous energy platforms. Focused research into scalable, greener fabrication, functional interface design (especially for hybrid and composite architectures), and lifecycle-conscious device engineering will underpin future commercialization trends. Tailoring device durability under various environmental and operational stresses is another future research imperative [11, 17].
- Sustainability Roadmap: Integrating green chemistry approaches—such as solvent-free polymerization, renewable energy-driven manufacturing, and use of bio-based monomers—will be central to reducing PVDF's environmental footprint. Standardization of benchmarks for mechanical and piezoelectric stability, biocompatibility testing, and recyclability will be critical to industrial trust and regulatory compliance [51, 55]. Strong collaboration among academia, industry, and policymakers is needed to develop recovery, reuse, and recycling infrastructure that addresses PVDF's end-of-life concerns.

5 Conclusion

While PVDF-based materials are already a cornerstone of flexible, high-performance electronics, their long-term future hinges on developing sustainable production, robust recycling methods, scalable process optimization, standardization, and interdisciplinary innovation across the electronics and materials science landscape. Effective solutions to environmental and scalability challenges will enable PVDF to fulfill its promise as the foundational material for smart, self-powered, and eco-responsible device platforms that define the next technological era Looking forward, PVDF-based materials are poised to drive the convergence of flexible, self-powered electronics, smart sensing platforms, and eco-conscious devices. Advances in structure–property understanding, phase transformation mechanisms, and composite engineering will enable sensors, actuators, and energy harvesters with high sensitivity, efficiency, and durability. Integration of piezoelectric and pyroelectric functionalities with acoustic, optical, and thermal responses will open new avenues in smart, self-sufficient devices. Continued interdisciplinary research, sustainable fabrication practices, and industrial standardization will ensure that PVDF remains a cornerstone of multifunctional, high-performance, and environmentally responsible electronic systems in the future.

References

1. Khan, B., Khalid, R.T., Masrur, M.H., Awais, M., Imdad, N., Hilal, M.E., Khan, W.U., Riaz, Z., Ahmed, S.B., Khan, B., Amara, U., Khoo, B.L.: Advances in polyvinylidene fluoride (PVDF) for self-powered wearable physiological monitoring and energy harvesting applications. Nano Energy. **143**, 111296 (2025)
2. Gao, X., Yang, J., Wu, J., Xin, X., Li, Z., Yuan, X., Dong, S.: Piezoelectric actuators and motors: materials, designs, and applications. Adv. Mater. Technol. **5**(1), 1900716 (2020)
3. Das, A., Nandan, B., Jana, S.: Tin oxide (SnO_2) loaded PVDF composites for enhanced dielectric applications. J. Mater. Chem. C. **9**, 10125–10136 (2021)
4. Mohammadi, B., Yousefi, A.A., Bellah, S.M.: Effect of tensile strain rate and elongation on crystalline structure and piezoelectric properties of PVDF thin films. Polym. Test. **26**, 42–50 (2007)
5. Nivedhitha, D.M., Jeyanthi, S., Prabhuraj Venkatraman, A.S., Viswapriyan, S.G., Nishaanth, S.M.: Polyvinylidene fluoride as an advanced polymer for multifunctional applications—a review. Mater. Today Proc. (2023)
6. Punetha, D., Kumar, A., Pandey, S.K., et al.: Tertiary nanocomposite-based self-powered E-skin as energy harvester and electronic nose. J. Mater. Sci. Mater. Electron. **35**, 160 (2024)
7. Barrau, S., Ferri, A., Da Costa, A., Defebvin, J., Leroy, S., Desfeux, R., Lefebvre, J.M.: Nanoscale investigations of α-and γ-crystal phases in PVDF-based nanocomposites. ACS Appl. Mater. Interfaces. **10**(15), 13092–13099 (2018)
8. Gebrekrstos, A., Muzata, T.S., Ray, S.S.: Nanoparticle-enhanced β-phase formation in electroactive PVDF composites: a review of systems for applications in energy harvesting, EMI shielding, and membrane technology. ACS Appl. Nano Mater. **5**(6), 7632–7651 (2022)
9. Zhang, L., Li, S., Zhu, Z., Rui, G., Du, B., Chen, D., Zhu, L.: Recent progress on structure manipulation of poly(vinylidene fluoride)-based ferroelectric polymers for enhanced piezoelectricity and applications. Adv. Funct. Mater. **33**(38), 2301302 (2023)
10. Punetha, D., Kar, M., Pandey, S.K.: A new type low-cost, flexible and wearable tertiary nanocomposite sensor for room temperature hydrogen gas sensing. Sci. Rep. **10**, 2151 (2020)
11. Mohammadpourfazeli, S., et al.: Future prospects and recent developments of polyvinylidene fluoride (PVDF) piezoelectric polymer; fabrication methods, structure, and electro-mechanical properties. RSC Adv. **13**(1), 370–387 (2023)
12. Boyer, C., Ameduri, B., Boutevin, B., Dolbier, W.R., Winter, R., Gard, G.: Radical terpolymerization of 1,1,2-trifluoro-2-pentafluorosulfanylethylene and pentafluorosulfanylethylene in the presence of vinylidene fluoride and hexafluoropropylene by iodine transfer polymerization. Macromolecules. **41**(4), 1254–1263 (2008)
13. Pradhan, L.K., Pandey, R., Kumar, S., Kar, M.: Lead-free $Bi_{0.5}Na_{0.5}TiO_3$ (BNT) and polyvinylidene fluoride (PVDF) based nanocomposite for energy storage applications. AIP Conf. Proc. **1953**, 090036 (2018)
14. An, H., Park, H., Kim, Y., Lee, S.: Cellulose nanofiber reinforced PVDF composites for enhanced dielectric performance. Carbohydr. Polym. **264**, 118043 (2021)

15. Li, A., Sun, Z., Lv, C., Zhongtian, F.: Recent progress in removal and recovery of PVDF from spent lithium-ion batteries. Sep. Purif. Technol. **380**, 135193 (2026)
16. Kumar, R.R., Punetha, D.: Computational prognosis and efficiency augmentation of lead-free and legacy piezoelectric bimorph cantilevers for low-power energy scavenging in IoT and wearable systems. Energy Technol. **13**, 2402193 (2025)
17. Ahbab, N., Xu, T.B.: Multifunctional polyvinylidene fluoride (PVDF)-based polymer, copolymer, and composites for aerospace applications: a comprehensive technical review. In: ASME Aerospace Structures, Structural Dynamics, and Materials Conference, vol. 87745, p. V001T02A001. American Society of Mechanical Engineers (2024)
18. Xu, T.B., Naz, S.: Review of cryogenic piezoelectric ultrasonic motors for aerospace applications. In: AIAA SCITECH 2025 Forum, p. 1090 (2025)
19. Xu, J., Xiong, Q., Liu, Q., Jiang, Y., Yue, X., Yang, D., Zhang, T., Qiu, F.: Sustainable recycling of waste poly(vinylidene fluoride) and rational design of Janus membrane with superhydrophilic/hydrophobic asymmetric wettability for efficient separation of surfactant-stabilized water-in-oil and oil-in-water emulsions. Colloids Surf. A Physicochem. Eng. Asp. **684**, 133237 (2024)
20. Hu, X., An, A.K.J., Chopra, S.S.: Life cycle assessment of the polyvinylidene fluoride polymer with applications in various emerging technologies. ACS Sustain. Chem. Eng. **10**(18), 5708–5718 (2022)
21. Zhang, X., Jiang, J., Liu, H.: Effects of rGO on PVDF nanocomposites. Mater. Res. Express. **4**, 095033 (2017)
22. Kumar, R.R., Punetha, D.: Fermi level modulation for enhanced graphene-based ultra-sensitive gas detection. Diam. Relat. Mater. **154**, 112214 (2025)
23. He, D., et al.: Significantly enhanced electromechanical performance of PDMS crosslinked PVDF hybrids. Polymers. **10**(7), 714 (2018)
24. Zhu, Y., et al.: Wireless and opto-stimulated flexible implants: artificial retina constructed by ferroelectric $BiFeO_3$-$BaTiO_3$/P(VDF-TrFE) composites. ACS Appl. Mater. Interfaces. **16**(36), 48395–48405 (2024)
25. Sharma, T., et al.: Aligned PVDF-TrFE nanofibers with high-density PVDF nanofibers and PVDF core–shell structures for endovascular pressure sensing. IEEE Trans. Biomed. Eng. **62**(1), 188–195 (2014)
26. Gusarov, B.: PVDF piezoelectric polymers: characterization and application to thermal energy harvesting. (Doctoral dissertation, Université Grenoble Alpes) (2015)
27. Ünsal, Ö.F., Altın, Y., Çelik Bedeloğlu, A.: Flexible electrospun PVDF piezoelectric nanogenerators with electrospray-deposited graphene electrodes. J. Electron. Mater. **52**(3), 2053–2061 (2023)
28. Xin, Y., Qi, X., Qian, C., Tian, H., Ling, Z., Jiang, Z.: A wearable respiration and pulse monitoring system based on PVDF piezoelectric film. Integr. Ferroelectr. **158**(1), 43–51 (2014)
29. Zhu, Y., et al.: Polar phase formation and competition in the melt crystallization of poly(vinylidene fluoride) containing an ionic liquid. Mater. Chem. Phys. **144**(1–2), 194–198 (2014)
30. Aubry, G., Lee, H.J., Lu, H.: Advances in microfluidics: technical innovations and applications in diagnostics and therapeutics. Anal. Chem. **95**(1), 444–467 (2023)
31. Hossain, N., Al Mahmud, M.Z., Hossain, A., Rahman, M.K., Islama, M.S., Tasnim, R., Mobarak, M.H.: Advances of materials science in mems applications: a review. Results Eng. **22**, 102115 (2024)
32. Singh, M., Apata, I.E., Samant, S., Wu, W., Tawade, B.V., Pradhan, N., Raghavan, D., Karim, A.: Nanoscale strategies to enhance the energy storage capacity of polymeric dielectric capacitors: review of recent advances. Polym. Rev. **62**(2), 211–260 (2022)
33. Wu, Y., Ma, Y., Zheng, H., Ramakrishna, S.: Piezoelectric materials for flexible and wearable electronics: a review. Mater. Des. **211**, 110164 (2021)

34. Sengupta, A., Das, S., Dasgupta, S., Sengupta, P., Datta, P.: Flexible nanogenerator from electrospun PVDF–polycarbazole nanofiber membranes for human motion energy-harvesting device applications. ACS Biomater Sci. Eng. **7**(4), 1673–1685 (2021)
35. Concha, V.O., Timóteo, L., Duarte, L.A., Bahú, J.O., Munoz, F.L., Silva, A.P., et al.: Properties, characterization and biomedical applications of polyvinylidene fluoride (PVDF): a review. J. Mater. Sci. **59**(31), 14185–14204 (2024)
36. Ma, Z., Lu, X., Wu, C., Gao, Q., Zhao, L., Zhang, H., Liu, Z.: Functional surface modification of PVDF membrane for chemical pulse cleaning. J. Membr. Sci. **524**, 389–399 (2017)
37. Ma, Y.-F., Wang, Y.-M., Wen, J., et al.: Review of roll-to-roll fabrication techniques for colloidal quantum dot solar cells. J. Electron. Sci. Technol. **21**, 100189 (2023)
38. Hu, Z., Zhang, J., Xiong, S., Zhao, Y.: Performance of polymer solar cells fabricated by dip coating process. Sol. Energy Mater. Sol. Cells. **99**, 221–225 (2012)
39. Yin, Z., Tian, B., Zhu, Q., Duan, C.: Characterization and application of PVDF and its copolymer films prepared by spin-coating and Langmuir–Blodgett method. Polymers (Basel). **11**, 2033 (2019)
40. Parlato, S., Esposito, D., Centracchio, J., Andreozzi, E., Gragnaniello, M., Riccio, M., Bifulco, P.: A new, simple wrist-mounted PVDF sensor for continuous heart rate monitoring. In: In 2024 IEEE Sensors Applications Symposium (SAS), pp. 1–6. IEEE (2024)
41. Pei, H., Jing, J., Chen, Y., Guo, J., Chen, N.: 3D printing of PVDF-based piezoelectric nanogenerator from programmable metamaterial design: promising strategy for flexible electronic skin. Nano Energy. **109**, 108303 (2023)
42. Spanggaard, H., Krebs, F.C.: A brief history of the development of organic and polymeric photovoltaics. Sol. Energy Mater. Sol. Cells. **83**, 125–146 (2004)
43. Won, D., et al.: Digital selective reversible phase control of monolithically integrated heterogeneous piezoelectric polymer for frequency dependent unimorph. Adv. Opt. Mater. **10**(24), 2201206 (2022)
44. Tan, K., et al.: Pyroelectricity enhancement of PVDF nanocomposite thin films doped with ZnO nanoparticles. Smart Mater. Struct. **23**(12), 125006 (2014)
45. Wang, G., et al.: Fabrication of radial ZnO nanowire clusters and radial ZnO/PVDF composites with enhanced dielectric properties. Adv. Funct. Mater. **18**(17), 2584–2592 (2008)
46. Toinet, S., et al.: Design optimization of printed multi-layered electroactive actuators used for steerable guidewire in micro-invasive surgery. Materials. **17**(9), 2135 (2024)
47. Li, N., Xie, K., Huang, H.: Ferroelectric materials for high energy density batteries: progress and outlook. ACS Energy Lett. **8**, 4357–4370 (2023)
48. Liu, H., Luo, S., Yu, S., Ding, S., Shen, Y., Sun, R., Wong, C.P.: Flexible $BaTiO_3$nf-Ag/PVDF nanocomposite films with high dielectric constant and energy density. IEEE Trans. Dielectr. Electr. Insul. **24**(2), 757–763 (2017)
49. Guo, S., Duan, X., Xie, M., Aw, K.C., Xue, Q.: Composites, fabrication and application of polyvinylidene fluoride for flexible electromechanical devices: a review. Micromachines. **11**, 1076 (2020)
50. Tressler, J.F., Alkoy, S., Newnham, R.E.: Piezoelectric sensors and sensor materials. J. Electroceram. **2**, 257–272 (1998)
51. Panda, S., Pasha, S.K.K.: Amplified dielectric properties of PVDF–HFP/$SrTiO_3$ nanocomposites for a flexible film capacitor. Langmuir. **39**(37), 13345–13358 (2023)
52. Abu-Zurayk, R., Alnairat, N., Bozeya, A., Khalaf, A., Abu-Dalo, D.: Enhanced properties of PVDF membranes using green Ag-nanoclay composite nanoarchitectonics. Mater. Res. Express. **11**(4), 045007 (2024)
53. Karan, S.K., Mandal, D., Khatua, B.B.: Self-powered flexible Fe-doped RGO/PVDF nanocomposite: an excellent material for a piezoelectric energy harvester. Nanoscale. **7**(24), 10655–10666 (2015)
54. Yang, X., Wang, Y., Qing, X.: A flexible capacitive sensor based on the electrospun PVDF nanofiber membrane with carbon nanotubes. Sens. Actuators A Phys. **299**, 111579 (2019)

55. Hoshyarmanesh, H., Maddahi, Y.: Poling process of composite piezoelectric sensors for structural health monitoring: a pilot comparative study. IEEE Sens. Lett. **2**(1), 1–4 (2018)
56. Xu, R., Kim, S.G.: Figures of merits of piezoelectric materials in energy harvesters. Proc. PowerMEMS. **12**, 1–467 (2012)
57. Kumar, V., Jeyanthi, S.: A study to determine electromagnetic interference-shielding effectiveness on bio-based polyurethane foam reinforced with PVDF/MgO/Ni for emerging applications. J. Appl. Polym. Sci. **140**(5), e53386 (2022)
58. Veved, A., Ejuh, G.W., Djongyang, N.: Review of emerging materials for PVDF-based energy harvesting. Energy Rep. **8**, 12853–12870 (2022)
59. Saxena, P., Shukla, P.: A comprehensive review on fundamental properties and applications of poly(vinylidene fluoride) (PVDF). Adv. Compos. Hybrid Mater. **4**, 8–26 (2021)
60. Haddadi, S.A., Ghaderi, S., Amini, M., Ramazani, S.A.A.: Mechanical and piezoelectric characterizations of electrospun PVDF-nanosilica fibrous scaffolds for biomedical applications. Mater. Today Proc. **5**, 15710–15716 (2018)
61. Martins, P., Moya, X., Ghidini, M., Sencadas, V., Botelho, G., et al.: Magnetoelectric $CoFe_2O_4$/polyvinylidene fluoride electrospun nanofibres. Nanoscale. **7**(17), 8058–8061 (2015)
62. Palneedi, H., Annapureddy, V., Priya, S., Ryu, J.: Status and perspectives of multiferroic magnetoelectric composite materials and applications. In: Actuators, vol. 5, p. 9. MDPI (2016)
63. Kang, S.J., Park, Y.J., Bae, I., Kim, K.J., Kim, H.C., Bauer, S., et al.: Printable ferroelectric PVDF/PMMA blend films with ultralow roughness for low voltage non-volatile polymer memory. Adv. Funct. Mater. **19**(17), 2812–2818 (2009)
64. Ruan, L., Yao, X., Chang, Y., Zhou, L., Qin, G., Zhang, X.: Properties and applications of the β phase poly(vinylidene fluoride). Polymers. **10**(3), 228 (2018)
65. Cao, X., Ma, J., Shi, X., Ren, Z.: Effect of TiO_2 nanoparticle size on the performance of PVDF membrane. Appl. Surf. Sci. **253**(4), 2003–2010 (2006)

GPSR Compliance
The European Union's (EU) General Product Safety Regulation (GPSR) is a set of rules that requires consumer products to be safe and our obligations to ensure this.

If you have any concerns about our products, you can contact us on

ProductSafety@springernature.com

In case Publisher is established outside the EU, the EU authorized representative is:

Springer Nature Customer Service Center GmbH
Europaplatz 3
69115 Heidelberg, Germany

www.ingramcontent.com/pod-product-compliance
Ingram Content Group UK Ltd.
Pitfield, Milton Keynes, MK11 3LW, UK
UKHW020631260926
475829UK00003BA/28
* 9 7 8 9 8 1 9 5 6 7 1 7 1 *